Lecture Notes in Control and Information Sciences

Edited by M. Thoma and A. Wyner

Vol. 62: Analysis and Optimization
of Systems
Proceedings of the Sixth International
Conference on Analysis and Optimization
of Systems
Nice, June 19–22, 1984
Edited by A. Bensoussan, J. L. Lions
XIX, 591 pages. 1984.

Vol. 63: Analysis and Optimization
of Systems
Proceedings of the Sixth International
Conference on Analysis and Optimization
of Systems
Nice, June 19–22, 1984
Edited by A. Bensoussan, J. L. Lions
XIX, 700 pages. 1984.

Vol. 64: Arunabha Bagchi
Stackelberg Differential Games
in Economic Models
VIII, 203 pages, 1984

Vol. 65: Yaakov Yavin
Numerical Studies
in Nonlinear Filtering
VIII, 273 pages, 1985.

Vol. 66: Systems and Optimization
Proceedings of the Twente Workshop
Enschede, The Netherlands, April 16–18, 1984
Edited by A. Bagchi, H. Th. Jongen
X, 206 pages, 1985.

Vol. 67: Real Time Control of Large Scale Systems
Proceedings of the First European Workshop
University of Patras, Greece, Juli 9–12, 1984
Edited by G. Schmidt, M. Singh, A. Titli,
S. Tzafestas
XI, 650 pages, 1985.

Vol. 68: T. Kaczorek
Two-Dimensional Linear Systems
IX, 397 pages, 1985.

Vol. 69: Stochastic Differential Systems –
Filtering and Control
Proceedings of the IFIP-WG 7/1 Working Conference
Marseille-Luminy, France, March 12-17, 1984
Edited by M. Metivier, E. Pardoux
X, 310 pages, 1985.

Vol. 70: Uncertainty and Control
Proceedings of a DFVLR International Colloquium
Bonn, Germany, March, 1985
Edited by J. Ackermann
IV, 236 pages, 1985.

Vol. 71: N. Baba
New Topics in Learning Automata
Theory and Applications
VII, 231 pages, 1985.

Vol. 72: A. Isidori
Nonlinear Control Systems:
An Introduction
VI, 297 pages, 1985.

Vol. 73: J. Zarzycki
Nonlinear Prediction
Ladder-Filters for Higher-Order
Stochastic Sequences
V, 132 pages, 1985.

Vol. 74: K. Ichikawa
Control System Design based on
Exact Model Matching Techniques
VII, 129 pages, 1985.

Vol. 75: Distributed Parameter
Systems
Proceedings of the 2nd International
Conference, Vorau, Austria 1984
Edited by F. Kappel, K. Kunisch,
W. Schappacher
VIII, 460 pages, 1985.

Vol. 76: Stochastic Programming
Edited by F. Archetti, G. Di Pillo,
M. Lucertini
V, 285 pages, 1986.

Vol. 77: Detection of
Abrupt Changes in Signals
and Dynamical Systems
Edited by M. Basseville,
A. Benveniste
X, 373 pages, 1986.

Vol. 78: Stochastic
Differential Systems
Proceedings of the 3rd Bad Honnef
Conference, June 3–7, 1985
Edited by N. Christopeit, K. Helmes,
M. Kohlmann
V, 372 pages, 1986.

Vol. 79: Signal
Processing for Control
Edited by K. Godfrey, P. Jones
XVIII, 413 pages, 1986.

Vol. 80: Artificial Intelligence
and Man-Machine Systems
Edited by H. Winter
IV, 211 pages, 1986.

For information about Vols. 1–61 please contact your bookseller or Springer-Verlag.

Lecture Notes in Control and Information Sciences

Edited by M. Thoma and A. Wyner

118

L. Dai

Singular Control Systems

Springer-Verlag
Berlin Heidelberg GmbH

Author
Liyi Dai
Division of Applied Sciences
Harvard University
Cambridge, MA 02138
USA

ISBN 978-3-540-50724-6

Library of Congress Cataloging in Publication Data

Dai, L. (Liyi)
Singular control systems / L. Dai.
(Lecture notes in control and information sciences ; 118)
Bibliography: p.
Includes index.
ISBN 978-3-540-50724-6 ISBN 978-3-540-46074-9 (eBook)
DOI 10.1007/978-3-540-46074-9

1. Automatic control. 2. Control theory. 3. Linear systems.
I. Title. II. Series.
TJ213.D25 1989
629.8–dc19 89-4091

Originally published by Springer-Verlag Berlin Heidelberg New York in 1989

2161/3020-543210

To Yun Wang

PREFACE

Singular systems are found in engineering systems (such as electrical circuit network, power systems, aerospace engineering, and chemical processing), social systems, economic systems, biological systems, network analysis, time-series analysis, singular singularly perturbed systems with which the singular system has a great deal to do, etc. Their form also makes them useful in system modeling. In many articles singular systems are called the descriptor variable systems, the differential/algebraic systems, the generalized state space systems, etc. Singular systems are governed by the so-called singular differential equations, which endow the systems with many special features that are not found in classical systems. Among these are impulse terms and input derivatives in the state response, nonproperness of transfer matrix, noncausality between input and state (or output), consistent initial conditions, etc., making the study of singular systems more sophisticated than of classical linear systems. In my opinion, this is the reason why singular systems have attracted interest in recent years. Currently, although much effort has been made in the exploration of special properties for singular systems, the studies are mostly confined to the generalization of classical system theory; of course, this is also an important task in control theory.

There are mainly two approaches in the studies: geometric and algebraic. The latter approach is adopted in this book. The primary purpose of writing this book is summing up the development of singular control system theory and providing the control circle with a systematic theory of the systems. Some acquaintance with linear algebra and linear system theory is assumed. Those not familiar with these prerequisites can read the book admitting the results in appendices. The systems are all linear and time invariant. Technically, the book focuses on the mathematical treatment of singular systems from the point of view of the system, or, in other words, I make an exclusive effort to discuss the analysis and design of singular control systems. This makes possible the systematic design of singular control systems in practical applications. Some examples are used only to illustrate the results and design procedures in this book.

The book is organized as follows. Chapter 1 discusses the state response structure for singular systems and the state space equivalent forms needed for later discussion. Chapter 2 provides some fundamental concepts in the system analysis, such as reachability, controllability, observability, realization theory, transfer matrix, and so on. These concepts are essential for system analysis and design. Chapter 3 deals with the control problem. The closed-loop behavior is given under state feedback controls and detailed information on finite or infinite pole placement may be obtained. Chapters 4 and 5 investigate some aspects of control system design for singular systems; the state observation problem is discussed in Chapter 4 and the dynamic compensation problem is treated in Chapter 5. Results in these two chapters show that under fairly general conditions, we may design normal observers and dynamic compensators which may be realized as in linear system theory. Chapter 6 is devoted to a special problem --- the structural stability, or, more precisely, the robustness of stability, which is not as crucial in classical linear system theory as in the singular system acse. It is shown that to guarantee structural stability, great care must be excercised in singular systems. Chapter 7 presents results on the system analysis and synthesis through the transfer matrix method. Three problems are solved: the transfer matrix structure under state feedback control, decoupling control, and input function observers. Since the digital technique holds an important position in practical system design. Chapter 8 explores discrete-time singular systems. Only basic design concepts are studied. Chapter 9 contains basic results on optimal control theory and Chapter 10 contains a preliminary discussion on topics for further research.

Liyi Dai

Beijing, PRC
August, 1988

CONTENTS

CHAPTER 1

SOLUTIONS OF LINEAR SINGULAR SYSTEMS

1-1. Introduction

Based on state space models, system analysis and synthesis are the core features in modern control theory, which was developed at the end of the 1950s and the beginning of the 1960s. State space models of systems are obtained mainly using the so-called state space variable method, whose advantage is that it not only provides us with a completely new method for system analysis and synthesis, but it also offers us more understanding of systems.

To get a state space model, we need to choose some variables such as speed, weight, temperature, and acceleration, which must be sufficient to characterize the system of interest. Then several equations are established according to relationships among the variables. Naturally, it is usually differential or algebraic equations that form the mathematical model of the system, or the description equation. Its general form is as follows:

$$f(\dot{x}(t),\ x(t),\ u(t),\ t) = 0 \qquad (1\text{-}1.1a)$$

$$g(x(t),\ u(t),\ y(t),\ t) = 0 \qquad (1\text{-}1.1b)$$

where $x(t)$ is the state of the system composed of state variables; $u(t)$ is the control input; $y(t)$ is the measure output; and f and g are vector functions of $\dot{x}(t)$, $x(t)$, $u(t)$, $y(t)$, and t, of appropriate dimensions. Generally (1-1.1a) and (1-1.1b) are called state and output equations, respectively.

A special case of (1-1.1) used to describe the system of interest is

$$\begin{aligned} E(t)\dot{x}(t) &= H(x(t),\ u(t),\ t) \\ y(t) &= J(x(t),\ u(t),\ t) \end{aligned} \qquad (1\text{-}1.2)$$

where H, J are appropriate dimensional vector functions in $x(t)$, $u(t)$, and t. The matrix $E(t)$ may be singular. Systems described by (1-1.2) are the general form of the so-called singular systems proposed recently.

If H, J are linear functions of $x(t)$ and $u(t)$, another special form of (1-1.2) is the linear singular system:

$$\begin{aligned} E\dot{x}(t) &= Ax(t) + Bu(t) \\ y(t) &= Cx(t) \end{aligned} \qquad (1\text{-}1.3)$$

which is the main system studied in this book. Here, $x(t) \in \mathbb{R}^n$, $u(t) \in \mathbb{R}^m$, $y(t) \in \mathbb{R}^r$. E, $A \in \mathbb{R}^{n \times n}$, $B \in \mathbb{R}^{n \times m}$, and $C \in \mathbb{R}^{n \times r}$ are constant matrices.

When E is nonsingular, system (1-1.3) becomes

$$\begin{aligned} \dot{x}(t) &= E^{-1}Ax(t) + E^{-1}Bu(t) \\ y(t) &= Cx(t). \end{aligned} \tag{1-1.4}$$

It is the well-known classical linear system that will be termed normal system here. We shall not attempt to introduce the abundant results for normal systems here, they may be found in any book on control theory. For this reason, when we mention singular systems we always mean the singular E, i.e., $\text{rank}E \triangleq q < n$.

In practical system analysis and control system design, many system models may be established in the form of (1-1.2), while they could not be described by (1-1.4). Thus, the studies of singular systems began at the end of 1970s, although they were first mentioned in 1973 (Singh and Liu, 1973). In many articles, singular systems are called descriptor variable systems, generalized state space systems, semistate systems, differential-algebraic systems, singular singularly perturbed systems, degenerate systems, constrained systems, etc.

Singular systems appear in many systems, such as engineering systems (for example, power system, electrical networks, aerospace engineering, chemical processes), social economic systems, network analysis, biological systems, time-series analysis, system modeling, and so on.

Example 1-1.1. Consider a class of interconnected large-scale systems with subsystems of

$$\begin{aligned} \dot{x}_i(t) &= A_i x_i(t) + B_i a_i(t) \\ b_i(t) &= C_i x_i(t) + D_i a_i(t), \quad i = 1, 2, \ldots, N \end{aligned} \tag{1-1.5}$$

where $x_i(t)$, $a_i(t)$, and $b_i(t)$ are the substate, control input, and output, respectively, of the ith subsystem. By denoting

$$x(t) = \begin{bmatrix} x_1(t) \\ x_2(t) \\ \vdots \\ x_N(t) \end{bmatrix}, \qquad a(t) = \begin{bmatrix} a_1(t) \\ a_2(t) \\ \vdots \\ a_N(t) \end{bmatrix}, \qquad b(t) = \begin{bmatrix} b_1(t) \\ b_2(t) \\ \vdots \\ b_N(t) \end{bmatrix},$$

$$A = \text{diag}(A_1, A_2, \ldots, A_N), \qquad B = \text{diag}(B_1, B_2, \ldots, B_N)$$

$$C = \text{diag}(C_1, C_2, \ldots, C_N), \qquad D = \text{diag}(D_1, D_2, \ldots, D_N)$$

(1-1.5) may be rewritten as

$$\begin{aligned} \dot{x}(t) &= Ax(t) + Ba(t) \\ b(t) &= Cx(t) + Da(t) \end{aligned} \tag{1-1.6}$$

Assume that the subsystem interconnection is the linear interconnection:

$$\begin{aligned} a(t) &= L_{11}b(t) + L_{12}u(t) + R_{11}a(t) + R_{12}y(t) \\ y(t) &= L_{21}b(t) + L_{22}u(t) + R_{21}a(t) + R_{22}y(t) \end{aligned} \qquad (1\text{-}1.7)$$

where u(t) is the overall input of the large-scale system; y(t) is its overall measure output; L_{ij}, R_{ij}, i, j = 1, 2, are constant matrices of appropriate dimensions. Equations (1-1.6) and (1-1.7) form a large-scale system which is not in the form of (1-1.4). In fact, Singh and Liu (1973) and Petzold (1982) have proven that the system composed by (1-1.6) and (1-1.7) could not be equivalent to a normal system. On the other hand, if we choose the state variable $[x^\tau(t) \ \ a^\tau(t) \ \ b^\tau(t) \ \ y^\tau(t)]^\tau$, (1-1.6) and (1-1.7) form the system

$$\begin{bmatrix} I & 0 & 0 & 0 \\ 0 & 0 & 0 & 0 \\ 0 & 0 & 0 & 0 \\ 0 & 0 & 0 & 0 \end{bmatrix} \begin{bmatrix} \dot{x}(t) \\ \dot{a}(t) \\ \dot{b}(t) \\ \dot{y}(t) \end{bmatrix} = \begin{bmatrix} A & B & 0 & 0 \\ C & D & -I & 0 \\ 0 & R_{11}-I & L_{11} & R_{12} \\ 0 & R_{21} & L_{21} & R_{22}-I \end{bmatrix} \begin{bmatrix} x(t) \\ a(t) \\ b(t) \\ y(t) \end{bmatrix} + \begin{bmatrix} 0 \\ 0 \\ L_{12} \\ L_{22} \end{bmatrix} u(t)$$

$$y(t) = [0 \ 0 \ 0 \ I][x^\tau(t) \ \ a^\tau(t) \ \ b^\tau(t) \ \ y^\tau(t)]^\tau$$

which is in the form of (1-1.3).

Example 1-1.2. The fundamental dynamic Leontief model of economic systems is a singular system. Its description model is (Luenberger, 1977):

$$x(k) = Ax(k) + B[x(k+1) - x(k)] + d(k) \qquad (1\text{-}1.8)$$

where x(k) is the n dimensional production vector of n sectors; $A \in \mathbb{R}^{n \times n}$ is an input-output (or production) matrix; Ax(k) stands for the fraction of production required as input for the current production, $B \in \mathbb{R}^{n \times n}$ is the capital coefficient matrix, and B[x(k+1)-x(k)] is the amount for capacity expansion (Stengel et al, 1979), which often appears in the form of capital. d(k) is the vector that includes demand or consumption. Equation (1-1.8) may be rewritten as

$$Bx(k+1) = (I-A+B)x(k) - d(k).$$

In multisector economic systems, production augmention in one sector often doesn't need the investment from all other sectors, and moreover, in practical cases only a few sectors can offer investment in capital to other sectors. Thus, most of the elements in B are zero except for a few. B is often singular. In this sense the system (1-1.8) is a typical discrete-time singular system.

Example 1-1.3. When administration is included, the oil catalytic craking is an extremely complicated process. Some companies reportedly found a description model for this process, whose simplication is

$$\begin{aligned} \dot{x}_1 &= A_{11}x_1 + A_{12}x_2 + B_1u + F_1f \\ 0 &= A_{21}x_1 + A_{22}x_2 + B_2u + F_2f \end{aligned} \qquad (1\text{-}1.9)$$

where x_1 is a vector to be regulated, such as regenerate temperature, valve position, blower capacity, etc; x_2 is the vector reflecting business benefits, administration, policy, etc; u is the regulation value; and f represents extra disturbances. Equation (1-1.9) is a standard singular system.

1-2. Regularity of General Singular Equations

Let us consider the following general singular equation:

$$E\dot{x}(t) = Ax(t) + f(t) \tag{1-2.1}$$

where $E, A \in \mathbb{R}^{n\times n}$, and $f(t) \in \mathbb{R}^n$, which is a special case of (1-1.3) when $B = I_n$. The function f(t) is assumed to be sufficiently differential; it has as many derivatives as needed. For (1-2.1), we are mainly concerned with the existence, uniqueness, and structure of its solution.

Lemma 1-2.1 (Gantmacher, 1974). For any two matrices $E, A \in \mathbb{R}^{m\times n}$, there always exist two nonsingular matrices Q, P such that

$$\begin{aligned} \tilde{E} &\triangleq QEP = \mathrm{diag}(0, L_1, L_2, \ldots, L_p, L_1', L_2', \ldots, L_q', I, N) \\ \tilde{A} &\triangleq QAP = \mathrm{diag}(0, J_1, J_2, \ldots, J_p, J_1', J_2', \ldots, J_q', A_1, I) \end{aligned} \tag{1-2.2}$$

where

$$0 \in \mathbb{R}^{m_o\times n_o}, \qquad A_1 \in \mathbb{R}^{h\times h}$$

$$L_i = \begin{bmatrix} 1 & 0 & & & \\ & 1 & 0 & & \\ & & \ddots & \ddots & \\ & & & 1 & 0 \end{bmatrix}, \qquad J_i = \begin{bmatrix} 0 & 1 & & & \\ & 0 & 1 & & \\ & & \ddots & \ddots & \\ & & & 0 & 1 \end{bmatrix} \in \mathbb{R}^{m_i\times(m_i+1)}$$

$$i = 1, 2, \ldots, p,$$

$$L_j' = \begin{bmatrix} 1 & & & \\ 0 & 1 & & \\ & \ddots & \ddots & \\ & & & 1 \\ & & & 0 \end{bmatrix}, \qquad J_j' = \begin{bmatrix} 0 & & & \\ 1 & 0 & & \\ & \ddots & \ddots & \\ & & & 0 \\ & & & 1 \end{bmatrix} \in \mathbb{R}^{(n_j+1)\times n_j}$$

$$j = 1, 2, \ldots, q,$$

$$N = \mathrm{diag}(N_{k_1}, N_{k_r}, \ldots, N_{k_r}) \in \mathbb{R}^{g\times g},$$

$$N_{k_i} = \begin{bmatrix} 0 & 1 & & & \\ & 0 & 1 & & \\ & & \ddots & \ddots & \\ & & & & 1 \\ & & & & 0 \end{bmatrix} \in \mathbb{R}^{k_i\times k_i}, \qquad i = 1, 2, \ldots, r,$$

$$m_o + \sum_i m_i + \sum_j (n_j+1) + \sum_i k_i + h = m$$

$$n_o + \sum_j n_j + \sum_i (m_i+1) + \sum_i k_i + h = n$$

$$\sum_i k_i = g.$$

Taking the coordinate transformation

$$x(t) = P\tilde{x}(t)$$

and left multiplying (1-2.1) by the nonsingular matrix Q, we have

$$\tilde{E}\dot{\tilde{x}}(t) = \tilde{A}\tilde{x}(t) + \tilde{f}(t), \qquad \tilde{f}(t) = Qf(t). \tag{1-2.3}$$

By expanding (1-2.3) according to (1-2.2), it may be rewritten as

$$0\dot{x}_{n_o}(t) = f_{m_o}(t) \tag{1-2.4a}$$

$$L_i\dot{x}_{m_i}(t) = J_i x_{m_i}(t) + f_{m_i}(t), \quad i = 1, 2, \dots, p \tag{1-2.4b}$$

$$L_j'\dot{x}_{n_j}(t) = J_j' x_{n_j}(t) + f_{n_j}(t), \quad j = 1, 2, \dots, q \tag{1-2.4c}$$

$$N_{k_i}\dot{x}_{k_i}(t) = x_{k_i}(t) + f_{k_i}(t), \quad i = 1, 2, \dots, r \tag{1-2.4d}$$

$$\dot{x}_h(t) = A_1 x_h(t) + f_h(t) \tag{1-2.4e}$$

where $x_k(t) \in \mathbb{R}^k$, $f_k(t) \in \mathbb{R}^k$, and

$$\tilde{x}(t) = [x_{n_o}^\tau\ x_{m_1}^\tau\ \dots\ x_{m_p}^\tau\ x_{n_1}^\tau\ \dots\ x_{n_q}^\tau\ x_{k_1}^\tau\ \dots\ x_{k_r}^\tau\ x_h^\tau]^\tau$$

$$\tilde{f}(t) = [f_{n_o}^\tau\ f_{m_1}^\tau\ \dots\ f_{m_p}^\tau\ f_{n_1}^\tau\ \dots\ f_{n_q}^\tau\ f_{k_1}^\tau\ \dots\ f_{k_r}^\tau\ f_h^\tau]^\tau.$$

Thus, solving equation (1-2.2) is equivalent to solving the equations in (1-2.4a$\sim$e). Next we will examine the solvability of each equation.

1. Equation (1-2.4a) can be solved only when $f_{m_o}(t) = 0$. In this case, (1-2.4a) is an identical equation for any differentiable function $x_{n_o}(t)$. Therefore, (1-2.4a) has either no solution or an infinite number of solutions.

2. Equation (1-2.4b) is composed of a set of equations of the following form:

$$\begin{bmatrix} 1 & 0 & & & \\ & 1 & 0 & & \\ & & \ddots & \ddots & \\ & & & 1 & 0 \end{bmatrix} \dot{z}(t) = \begin{bmatrix} 0 & 1 & & & \\ & 0 & 1 & & \\ & & \ddots & \ddots & \\ & & & 0 & 1 \end{bmatrix} z(t) + f_z(t).$$

Expanding it, we obtain

$$\dot{z}_1(t) = z_2(t)+f_1(t), \quad \dot{z}_2(t) = z_3(t)+f_2(t), \quad \dots, \quad \dot{z}_{k-1}(t) = z_k(t)+f_{k-1}(t).$$

Clearly, for any sufficiently differentiable function, $z_1(t), z_2(t), \dots, z_k(t)$ may be determined successively. Therefore, such equations have an infinite number of solutions.

3. Every equation in (1-2.4c) is of the following form:

$$\begin{bmatrix} 1 & & & & \\ 0 & 1 & & & \\ & 0 & \ddots & & \\ & & \ddots & 1 & \\ & & & 0 & \end{bmatrix} \dot{z}(t) = \begin{bmatrix} 0 & & & & \\ 1 & 0 & & & \\ & 1 & \ddots & & \\ & & \ddots & 0 & \\ & & & 1 & \end{bmatrix} z(t) + f_z(t) \ .$$

By expanding, we know that this equation is equivalent to

$$\dot{z}_1(t) = f_1(t), \quad \dot{z}_2(t) = z_1(t) + f_2(t), \quad \dots$$
$$\dot{z}_k(t) = z_{k-1}(t) + f_k(t), \quad \dot{z}_k(t) + f_{k+1}(t) = 0.$$

Except the last one, the equations may determine unique functions $z_1(t)$, $z_2(t)$, ... , $z_k(t)$. But accounting for the last equation, $z_k(t)$ must satisfy $\dot{z}_k(t) + f_{k+1}(t) = 0$. This shows that the equation has no solution unless $f_{k+1}(t)$ satisfies the consistent condition $\dot{z}_k(t) + f_{k+1}(t) = 0$.

4. Equation (1-2.4d) consists of the following equations:

$$\begin{bmatrix} 0 & 1 & & & \\ & 0 & 1 & & \\ & & \ddots & \ddots & \\ & & & & 1 \\ & & & & 0 \end{bmatrix} \dot{z}(t) = z(t) + f_z(t)$$

which may be rewritten as

$$\dot{z}_2(t) = z_1(t) + f_1(t), \quad \dot{z}_3(t) = z_2(t) + f_2(t), \quad \dots$$
$$\dot{z}_k(t) = z_{k-1}(t) + f_{k-1}(t), \quad \dot{z}_k(t) + f_k(t) = 0.$$

Beginning with the last one, $z_1(t)$, ... , $z_k(t)$ may be successively determined for sufficiently differentiable function $f_z(t)$. Hence (1-2.4d) has a unique solution for any such $f_z(t)$.

5. Equation (1-2.4e) is an ordinary differential equation, which has a unique solution for any piece-wise continuous function f(t).

To sum up, the necessary and sufficient condition for the existence and uniqueness of the solution to equation (1-2.1) is the disappearance of (1-2.4a)-(1-2.4c), or, in other words,

$$QEP = \mathrm{diag}(I, N), \qquad QAP = \mathrm{diag}(A_1, I). \tag{1-2.5}$$

This condition is somewhat not so obvious and difficult to verify. To obtain one that is simpler to verify, we define the following.

<u>Definition 1-2.1</u>. For any given two matrices E, $A \in \mathbb{R}^{n\times n}$, the pencil (E,A) is call-

ed regular if there exists a constant scalar $\alpha \in \mathbb{C}$ such that $|\alpha E+A| \neq 0$, or the polynomial $|sE-A| \not\equiv 0$.

Lemma 1-2.2. (E,A) is regular if and only if two nonsingular matrices Q, P may be chosen such that

$$QEP = \mathrm{diag}(I_{n_1}, N), \qquad QAP = \mathrm{diag}(A_1, I_{n_2}) \tag{1-2.6}$$

where $n_1 + n_2 = n$, $A_1 \in \mathbb{R}^{n_1 \times n_1}$, $N \in \mathbb{R}^{n_2 \times n_2}$ is nilpotent.

Proof. Sufficiency: Let (1-2.6) be true. We choose $\alpha \notin \sigma(A_1)$ (such αs are numerous). Then

$$|\alpha E+A| = |Q^{-1}||P^{-1}||\alpha QEP+QAP| = |Q^{-1}||P^{-1}||\alpha I+A_1| \neq 0.$$

Thus (E,A) is regular.

Necessity: If (E,A) is regular, the definition shows the existence of α such that $|\alpha E+A| \neq 0$. Consider the pencil

$$\hat{E} = (\alpha E+A)^{-1}E, \qquad \hat{A} = (\alpha E+A)^{-1}A.$$

It is easy to obtain

$$\hat{A} = I - \alpha \hat{E}. \tag{1-2.7}$$

On the other hand, it follows immediately from the Jordan canonical form decomposition in matrix theory that there exists a nonsingular matrix T such that

$$T\hat{E}T^{-1} = \mathrm{diag}(\hat{E}_1, \hat{E}_2)$$

where $\hat{E}_1 \in \mathbb{R}^{n_1 \times n_1}$ is nonsingular, $\hat{E}_2 \in \mathbb{R}^{n_2 \times n_2}$ is nilpotent. This means that $I-\alpha\hat{E}_2$ is nonsingular. Let $Q = \mathrm{diag}(\hat{E}_1^{-1}, (I-\alpha\hat{E}_2)^{-1})T(\alpha E+A)^{-1}$, and $P = T^{-1}$. From (1-2.7) we have $QEP = \mathrm{diag}(I_{n_1}, N)$, $QAP = \mathrm{diag}(A_1, I_{n_2})$, where $A_1 = \hat{E}_1^{-1}(I-\alpha\hat{E}_2)$, $N = (I-\alpha\hat{E}_2)^{-1}\hat{E}_2$ is nilpotent. Q.E.D.

Equation (1-2.5) shows that if the differential equation (1-2.1) has a unique solution for any sufficiently differentiable function f(t), the matrices must be square and (1-2.5) holds. This is equivalent to the regularity of pencil (E,A) by Lemma 1-2.2. Hence, to guarantee the existence and uniqueness of solution for system (1-1.3), we shall hereafter assume that E and A are square and regular matrices, unless otherwise specified. In this case, system (1-1.3) will be termed "regular".

We often use (E,A,B,C) to denote the system (1-1.3) for convenience.

The following theorem gives an easy criterion for regularity.

Theorem 1-2.1.(Yip and Sincovec, 1981). The following statements are equivalent

(1). System (1-1.3) is regular;

(2). If $X_0 = \mathrm{Ker}A$, $X_i = \{x \mid Ax \in EX_{i-1}\}$, it must be $\mathrm{Ker}E \cap X_i = 0$, i= 0, 1, 2, ...;

(3). If $Y_0 = \mathrm{Ker}A^\tau$, $Y_i = \{x \mid A^\tau x \in E^\tau Y_{i-1}\}$, $\mathrm{Ker}E^\tau \cap Y_i = 0$, i = 0, 1, 2, ...;

(4). Let

$$G(k) = \begin{bmatrix} E & & & \\ A & E & & \\ & A & \ddots & \\ & & \ddots & E \\ & & & A \end{bmatrix} \in \mathbb{R}^{(k+1)n \times nk}.$$

Then rank$G(k) = nk$, $k = 1, 2, \ldots$;

(5). Let

$$F(k) = \begin{bmatrix} E & A & & & \\ & E & A & & \\ & & \ddots & \ddots & \\ & & & \ddots & A \\ & & & E & A \end{bmatrix} \in \mathbb{R}^{nk \times n(k+1)}.$$

Then rank$F(k) = nk$, $k = 1, 2, \ldots$.

Example 1-2.1. Consider the following matrix pencil:

$$E = \begin{bmatrix} 0 & 1 & 1 \\ 1 & 1 & 0 \\ -1 & 0 & 1 \end{bmatrix}, \qquad A = \begin{bmatrix} 1 & 0 & 1 \\ 0 & 2 & 0 \\ -1 & 0 & 1 \end{bmatrix}.$$

Direct computation shows that $|sE-A| = -4(s-1)^2 \not\equiv 0$. Thus (E,A) is regular.

This method verifying regularity is inconvenient when E and A are of high orders. From the viewpoint of computation, Luenberger (1978) proposed another criterion, which is called the shuffle algorithm.

Consider the nx2n matrix

$$[E \quad A] \tag{1-2.8}$$

If E is nonsingular, (E,A) is regular and the algorithm stops. Otherwise, E is singular and by taking row transformation we may change (1-2.8) into the block form of

$$\begin{bmatrix} E_1 & A_1 \\ 0 & A_2 \end{bmatrix} \tag{1-2.9}$$

where $E_1 \in \mathbb{R}^{q \times n}$ is a full row rank with $q = \text{rank}E$. Then we "shuffle" the second block row in (1-2.9) to obtain

$$\begin{bmatrix} E_1 & A_1 \\ A_2 & 0 \end{bmatrix} \tag{1-2.10}$$

If $[E_1/A_2]$ is nonsingular, (E,A) is regular and the algorithm stops. Otherwise, we repeat the algorithm. It has been proven that the algorithm would terminate in either of the following ways: The matrix formed by the left n columns is nonsingular, indicating (E,A) is regular; or at least one zero row appears in the matrix (1-2.10), which would indicate that (E,A) is not regular (Stengel et al, 1979).

Example 1-2.2. Consider the matrix pencil (E,A) in Example 1-2.1. Then

$$[E \quad A] = \left[\begin{array}{ccc|ccc} 0 & 1 & 1 & 1 & 0 & 1 \\ 1 & 1 & 0 & 0 & 2 & 0 \\ -1 & 0 & 1 & -1 & 0 & 1 \end{array}\right], \qquad (1\text{-}2.11)$$

where E is singular. A row transformation in (1-2.11) yields

$$\left[\begin{array}{ccc|ccc} 0 & 1 & 1 & 1 & 0 & 1 \\ 1 & 1 & 0 & 0 & 2 & 0 \\ 0 & 0 & 0 & -2 & 2 & 0 \end{array}\right], \qquad (1\text{-}2.12)$$

and shuffle of (1-2.12) gives:

$$\left[\begin{array}{ccc|ccc} 0 & 1 & 1 & 1 & 0 & 1 \\ 1 & 1 & 0 & 0 & 2 & 0 \\ -2 & 2 & 0 & 0 & 0 & 0 \end{array}\right]$$

The left 3x3 matrix is nonsingular, indicating the regularity of pencil (E,A).

<u>Comments on shuffle algorithm</u>. In the shuffle algorithm, we see that the row transformation on [E A] is equivalent to the row transformation on sE-A with a nonzero constant multiplying $|sE-A|$, i.e.,

$$\left| s\begin{bmatrix} E_1 \\ 0 \end{bmatrix} - \begin{bmatrix} A_1 \\ A_2 \end{bmatrix} \right| = \begin{vmatrix} sE_1 - A_1 \\ -A_2 \end{vmatrix} . \qquad (1\text{-}2.13)$$

Note the equation

$$\left| \begin{bmatrix} I & 0 \\ 0 & -sI \end{bmatrix} \begin{bmatrix} sE_1 - A_1 \\ -A_2 \end{bmatrix} \right| = \left| \begin{bmatrix} E_1 \\ A_2 \end{bmatrix} s - \begin{bmatrix} A_1 \\ 0 \end{bmatrix} \right| . \qquad (1\text{-}2.14)$$

From the second part in (1-2.14) we understand that the shuffle algorithm at one step is just multiplying the lower rows in sE-A , obtained from the row transformation, by (-s). This results in multiplying the polynomial $|sE-A|$ by the same factor, when (E,A) is regular. The final step constitutes the following equation

$$P(s)(sE-A) = sI - \tilde{E} , \qquad |P(s)| = \beta s^m , \quad \beta = \text{constant} \neq 0.$$

Thus we may conclude that $|sE-A| \not\equiv 0$ from above equation, or in other words, (E,A) is regular.

The advantage of shuffle algorithm lies in its ease of computation.

1-3. Equivalence of Singular Systems

As pointed out in the introduction, the description equation for a real system or mathematical model may be obtained by selecting appropriate state variables. The selection is certainly not unique; this in turn causes the nonuniqueness of the model. We will now study the relationships between the state variables and models. If this problem is clear, an appropriate model may be established that is easy and convenient for system analysis and/or design.

Definition 1-3.1. Let $\mathcal{F}$ denote a set, in which a certain relationship π is established between any two elements $\Sigma_2, \Sigma_2 \in \mathcal{F}$. The sign "$\Sigma_1 \overset{\pi}{\sim} \Sigma_2$" will be used herein to represent a π relationship between Σ_1 and Σ_2. If π satisfies

(1). (Reflexivity): $\forall\ \Sigma_1 \in \mathcal{F}, \Sigma_1 \overset{\pi}{\sim} \Sigma_1$,

(2). (Transitivity): $\forall\ \Sigma_1, \Sigma_2, \Sigma_3 \in \mathcal{F}$, if $\Sigma_1 \overset{\pi}{\sim} \Sigma_2, \Sigma_2 \overset{\pi}{\sim} \Sigma_3, \Sigma_1 \overset{\pi}{\sim} \Sigma_3$ is true,

(3). (Invertibility): $\forall\ \Sigma_1, \Sigma_2 \in \mathcal{F}$, if $\Sigma_1 \overset{\pi}{\sim} \Sigma_2$, it must be $\Sigma_2 \overset{\pi}{\sim} \Sigma_1$.

The relationship π will be considered equivalence.

For simplicity, the time variable will be omitted in our discussion later unless it is necessary to include it. The initial time is taken as $t_0 = 0$.

Consider the two singular systems:

$$\begin{aligned} E\dot{x} &= Ax + Bu \\ y &= Cx \end{aligned} \tag{1-3.1}$$

and

$$\begin{aligned} \tilde{E}\dot{\tilde{x}} &= \tilde{A}\tilde{x} + \tilde{B}u \\ y &= \tilde{C}\tilde{x}\ . \end{aligned} \tag{1-3.2}$$

Assume that there exist two nonsingular matrices Q and P such that

$$x = P\tilde{x} \tag{1-3.3}$$

and $QEP = \tilde{E}$, $QAP = \tilde{A}$, $QB = \tilde{B}$, $CP = \tilde{C}$. Systems (1-3.1) and (1-3.2) will be called restricted system equivalent (r.s.e.). (1-3.3) is the coordinate transformation.

From the definition, restricted system equivalence is an equivalent relationship that possesses reflexivity, transitivity, and invertibility.

Example 1-3.1. Consider a simple circuit network as shown in Figure 1-3.1, where voltage source $V_S(t)$ is the driver (control input), R, L, and C stand for the resistor, inductor, and capacity, respectively, as well as their quantities, and their voltages are denoted by $V_R(t)$, $V_L(t)$, $V_C(t)$, respectively. Then from Kirchoff's laws, we have the following circuit equation (description equations):

$$\begin{bmatrix} L & 0 & 0 & 0 \\ 0 & 0 & 1 & 0 \\ 0 & 0 & 0 & 0 \\ 0 & 0 & 0 & 0 \end{bmatrix}\begin{bmatrix} \dot{I}(t) \\ \dot{V}_L(t) \\ \dot{V}_C(t) \\ \dot{V}_R(t) \end{bmatrix} = \begin{bmatrix} 0 & 1 & 0 & 0 \\ 1/C & 0 & 0 & 0 \\ -R & 0 & 0 & 1 \\ 0 & 1 & 1 & 1 \end{bmatrix}\begin{bmatrix} I(t) \\ V_L(t) \\ V_C(t) \\ V_R(t) \end{bmatrix} + \begin{bmatrix} 0 \\ 0 \\ 0 \\ -1 \end{bmatrix} V_s(t). \qquad (1\text{-}3.4)$$

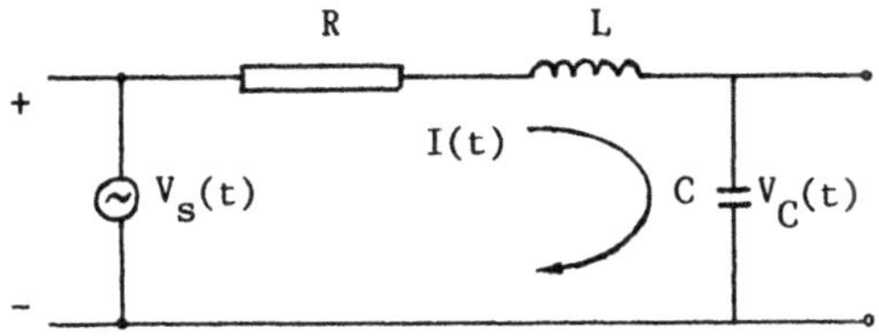

Figure 1-3.1.

On the other hand, if we choose

$$\tilde{x} = \begin{bmatrix} I(t) \\ V_R(t)+V_L(t) \\ V_R(t)+V_L(t)+V_C(t) \\ V_R(t) \end{bmatrix}$$

as state variable, it will be

$$\begin{bmatrix} L & 0 & 0 & 0 \\ L & 1 & -1 & 0 \\ 0 & -1 & 1 & 0 \\ L & 0 & 0 & 0 \end{bmatrix}\dot{\tilde{x}} = \begin{bmatrix} 0 & 1 & 1 & -1 \\ -1/C & 1 & 1 & -1 \\ 1/C-2R & 0 & 0 & 2 \\ -R & 1 & 0 & 0 \end{bmatrix}\tilde{x} + \begin{bmatrix} -1 \\ -1 \\ 0 \\ 0 \end{bmatrix} V_s(t). \qquad (1\text{-}3.5)$$

Both (1-3.4) and (1-3.5) are mathematical models that describe the circuit network in Figure 1-3.1. Although (1-3.4) and (1-3.5) are different, we can easily verify that they are r.s.e., with the transformation matrices

$$Q = \begin{bmatrix} 1 & 0 & 0 & 1 \\ 1 & -1 & 0 & 1 \\ 0 & 1 & 2 & 0 \\ 1 & 0 & 1 & 0 \end{bmatrix} \qquad P = \begin{bmatrix} 1 & 0 & 0 & 0 \\ 0 & 1 & 0 & -1 \\ 0 & -1 & 1 & 0 \\ 0 & 0 & 0 & 1 \end{bmatrix}.$$

Now we will introduce some common r.s.e. forms in singular systems. Consider the singular system

$$E\dot{x} = Ax + Bu \qquad (1\text{-}3.6)$$
$$y = Cx$$

where $x \in \mathbb{R}^n$, $u \in \mathbb{R}^m$, $y \in \mathbb{R}^r$; E, $A \in \mathbb{R}^{nxn}$, $B \in \mathbb{R}^{nxm}$, and $C \in \mathbb{R}^{rxn}$ are constant matrices.

1-3.1. First Equivalent Form (EF1)

Lemma 1-2.2 shows that: For any singular system (1-3.6), there exist two nonsingular matrices Q and P such that (1-3.6) is r.s.e. to

$$\begin{aligned} \dot{x}_1 &= A_1 x_1 + B_1 u \\ y_1 &= C_1 x_1 \end{aligned} \qquad (1\text{-}3.7a)$$

$$\begin{aligned} N\dot{x}_2 &= x_2 + B_2 u \\ y_2 &= C_2 x_2 \end{aligned} \qquad (1\text{-}3.7b)$$

$$y = C_1 x_1 + C_2 x_2 = y_1 + y_2 \qquad (1\text{-}3.7c)$$

with the coordinate transformation

$$\begin{bmatrix} x_1 \\ x_2 \end{bmatrix} = P^{-1}x, \qquad x_1 \in \mathbb{R}^{n_1 x n_2}, \qquad x_2 \in \mathbb{R}^{n_2 x n_2} \qquad (1\text{-}3.8)$$

and

$$QEP = \mathrm{diag}(I_{n_1}, N), \quad QAP = \mathrm{diag}(A_1, I_{n_2}), \quad QB = [B_1/B_2], \quad CP = [C_1 \quad C_2], \qquad (1\text{-}3.9)$$

where $n_1+n_2 = n$, $N \in \mathbb{R}^{n_2 x n_2}$ is nilpotent.

Equation (1-3.7) is the EF1, usually called the standard decomposition. In this form, subsystems (1-3.7a) and (1-3.7b) are called slow and fast subsystems respectively; x_1, x_2 are the slow and fast substates, respectively.

Generally, the matrices Q and P, which transfer a singular system into its standard EF1, are not unique, resulting in the nonuniqueness of EF1, i.e., A_1, B_1, B_2, C_1, C_2, N. Assume that $\bar{Q}$ and $\bar{P}$ are nonsingular and system (1-3.6) is transferred into its EF1, in other words, system (1-3.5) is r.s.e. to:

$$\begin{aligned} \dot{\bar{x}}_1 &= \bar{A}_1 \bar{x}_1 + \bar{B}_1 u \\ \bar{y} &= \bar{C}\,\bar{x} \end{aligned} \qquad (1\text{-}3.10a)$$

$$\begin{aligned} \bar{N}\dot{\bar{x}}_2 &= \bar{x}_2 + \bar{B}_2 u \\ \bar{y}_2 &= \bar{C}_2 \bar{x}_2 \end{aligned} \qquad (1\text{-}3.10b)$$

$$y = \bar{C}_1 \bar{x}_1 + \bar{C}_2 \bar{x}_2 = \bar{y}_1 + \bar{y}_2 \qquad (1\text{-}3.10c)$$

with the coordinate transformation $\bar{P}^{-1}x = [\bar{x}_1/\bar{x}_2]$.

The following theorem shows the relationship between the matrices Q, P and $\bar{Q}$, $\bar{P}$, and the system coefficient matrices.

Theorem 1-3.1. Suppose that (1-3.7)-(1-3.9) and (1-3.10) are the EF1s for system

(1-3.6). Then $n_1 = \bar{n}_1$, $n_2 = \bar{n}_2$, and there exist nonsingular matrices $T_1 \in \mathbb{R}^{n_1 x n_1}$, $T_2 \in \mathbb{R}^{n_2 x n_2}$ such that

$$\begin{aligned} &Q = \mathrm{diag}(T_1, T_2)\bar{Q}, && P = \bar{P}\mathrm{diag}(T_1, T_2) \\ &A_1 = T_1\bar{A}_1T_1^{-1}, && N = T_2\bar{N}T_2^{-1} \\ &B_i = T_i\bar{B}_i, && C_i = \bar{C}_iT_i^{-1}, \quad i = 1, 2. \end{aligned} \tag{1-3.11}$$

Proof. Since

$$|sE-A| = |Q^{-1}||P^{-1}||sI-A_1| = |\bar{Q}^{-1}||\bar{P}^{-1}||sI-\bar{A}_1| \tag{1-3.12}$$

and $n_1 + n_2 = n$, and $\bar{n}_1 + \bar{n}_2 = n$, we know that $n_1 = \bar{n}_1$, $n_2 = \bar{n}_2$.

Furthermore, if we denote

$$Q\bar{Q}^{-1} = \begin{bmatrix} Q_{11} & Q_{12} \\ Q_{21} & Q_{22} \end{bmatrix}, \qquad P\bar{P}^{-1} = \begin{bmatrix} P_{11} & P_{12} \\ P_{21} & P_{22} \end{bmatrix} \tag{1-3.13}$$

the equations (1-3.8) and (1-3.10) show that

$$\begin{aligned} &Q^{-1}\mathrm{diag}(I, N)P^{-1} = \bar{Q}^{-1}\mathrm{diag}(I, \bar{N})\bar{P}^{-1}, \\ &Q^{-1}\mathrm{diag}(A_1, I)P^{-1} = \bar{Q}^{-1}\mathrm{diag}(\bar{A}_1, I)\bar{P}^{-1}. \end{aligned} \tag{1-3.14}$$

Substituting (1-3.13) into (1-3.14), we know

$$Q_{11} = P_{11}, \qquad Q_{22} = P_{22} \tag{1-3.15a}$$

$$Q_{12} = A_1P_{12}, \qquad Q_{21}\bar{A}_1 = P_{21}, \tag{1-3.15b}$$

$$Q_{21} = NP_{21}, \qquad Q_{12}\bar{N} = P_{12}, \tag{1-3.15c}$$

$$Q_{11}\bar{A}_1 = A_1P_{11}, \qquad Q_{22}\bar{N} = NP_{22}. \tag{1-3.15d}$$

The substitution of (1-3.15c) into (1-3.15b) yields

$$P_{21} = NP_{21}\bar{A}_1, \qquad P_{12} = A_1P_{12}\bar{N}. \tag{1-3.16}$$

Note that matrices N and $\bar{N}$ are nilpotent. Thus

$$P_{21} = NP_{21}\bar{A}_1 = \dots = N^nP_{21}\bar{A}_1^n = 0, \qquad P_{12} = 0 \tag{1-3.17}$$

which, in conjunction with (1-3.15b) and (1-3.15c), yields

$$Q_{21} = 0, \qquad Q_{12} = 0. \tag{1-3.18}$$

Denoting $T_1 = Q_{11}$ and $T_2 = Q_{22}$, from above equation (1-3.18) and the nonsingularity of Q, P, $\bar{Q}$ and $\bar{P}$, we may conclude that T_1 and T_2 are invertible. Equations (1-3.13), (1-3.15a), and (1-3.16)-(1-3.18) further show that

$$Q = \mathrm{diag}(T_1, T_2)\bar{Q}, \quad P = \bar{P}\mathrm{diag}(T_1^{-1}, T_2^{-1}).$$

Hence, equation (1-3.11) holds. Q.E.D.

Although different EF1s may be obtained under different coordinate transformation matrices, this theorem assures the similarity property among these EF1s. This means that EF1 is unique in the sense of similar equivalence.

Example 1-3.2. Consider the circuit network in Example 1-3.1, with L = C = R = 1, and a measure equation:

$$y(t) = V_C(t) = [0\ 0\ 1\ 0]x(t), \quad x(t) = [I(t)\ \ V_L(t)\ \ V_C(t)\ \ V_R(t)]^{\tau}.$$

Then the system becomes

$$\begin{bmatrix} 1 & 0 & 0 & 0 \\ 0 & 0 & 1 & 0 \\ 0 & 0 & 0 & 0 \\ 0 & 0 & 0 & 0 \end{bmatrix} \dot{x} = \begin{bmatrix} 0 & 1 & 0 & 0 \\ 1 & 0 & 0 & 0 \\ -1 & 0 & 0 & 1 \\ 0 & 1 & 1 & 1 \end{bmatrix} x + \begin{bmatrix} 0 \\ 0 \\ 0 \\ -1 \end{bmatrix} V_s(t) \tag{1-3.19}$$

$$y = [0\ \ 0\ \ 1\ \ 0]x\ .$$

If we choose the following transformation

$$P^{-1}x = [x_1/x_2], \quad x_1 \in \mathbb{R}^2, \quad x_2 \in \mathbb{R}^2,$$

$$Q = \begin{bmatrix} 1 & 0 & 1 & -1 \\ 0 & 1 & 0 & 0 \\ 0 & 0 & -1 & 1 \\ 0 & 0 & 1 & 0 \end{bmatrix} \qquad P = \begin{bmatrix} 1 & 0 & 0 & 0 \\ -1 & -1 & 1 & 0 \\ 0 & 1 & 0 & 0 \\ 1 & 0 & 0 & 1 \end{bmatrix}$$

it will become

$$\dot{x}_1 = \begin{bmatrix} -1 & -1 \\ 1 & 0 \end{bmatrix} x_1 + \begin{bmatrix} 1 \\ 0 \end{bmatrix} V_s(t)$$

$$0 = x_2 + \begin{bmatrix} -1 \\ 0 \end{bmatrix} V_s(t) \tag{1-3.20}$$

$$y = [0\ \ 1]x_1\ .$$

This is the EF1 for system (1-3.19).

1-3.2. Second Equivalent Form (EF2)

Let q = rankE. From matrix theory, we know that there exist nonsingular matrices Q_1 and P_1 such that $Q_1EP_1 = \mathrm{diag}(I_q, 0)$. By taking the coordinate transformation $P_1^{-1}x = [x_1\ /x_2]$, $x_1 \in \mathbb{R}^q$, $x_2 \in \mathbb{R}^{n-q}$, system (1-3.6) is r.s.e. to

$$\dot{x}_1 = A_{11}x_1 + A_{12}x_2 + B_1u$$

$$0 = A_{21}x_1 + A_{22}x_2 + B_2u \tag{1-3.21}$$

$$y = C_1x_1 + C_2x_2$$

where

$$Q_1AP_1 = \begin{bmatrix} A_{11} & A_{12} \\ A_{21} & A_{22} \end{bmatrix}, \quad Q_1B = \begin{bmatrix} B_1 \\ B_2 \end{bmatrix}, \quad CP_1 = [C_1 \quad C_2] .$$

Equation (1-3.21) is the second equivalent form (EF2) for system (1-3.6). In this transformation, matrices Q_1 and P_1 are not unique, which results in the nonuniqueness of EF2. Two EF2s may have a relationship too complicated to merit further study.

The EF2 clearly reflects the physical meaning of singular systems. In (1-3.21), the first equation is a differential one composed of dynamic subsystems, and the second is an algebraic equation that represents the connection between subsystems. Thus, singular systems may be viewed as a composite system formed by several interconnected subsystems. Furthermore, substates x_1 and x_2 reflect a layer property in some singular systems: one layer has a dynamic property (described by the differential equation); the other has interconnection, constraint, and administration properties (described by the algebraic equation).

Singular value decomposition may be used in the decomposition of E instead of rank decomposition here.

Example 1-3.3. Consider the system in Example 1-3.2. Under the coordinate transformation $P_1^{-1}x = [x_1/x_2]$, $x_1 \in \mathbb{R}^2$, $x_2 \in \mathbb{R}^2$, with

$$Q_1 = I_4, \qquad P_1 = \begin{bmatrix} 1 & 0 & 0 & 0 \\ 0 & 0 & 1 & 0 \\ 0 & 1 & 0 & 0 \\ 0 & 0 & 0 & 1 \end{bmatrix}$$

system (1-3.19) is r.s.e. to

$$\dot{x}_1 = \begin{bmatrix} 0 & 0 \\ 1 & 0 \end{bmatrix} x_1 + \begin{bmatrix} 1 & 0 \\ 0 & 0 \end{bmatrix} x_2 + \begin{bmatrix} 0 \\ 0 \end{bmatrix} V_s$$

$$0 = \begin{bmatrix} -1 & 0 \\ 0 & 1 \end{bmatrix} x_1 + \begin{bmatrix} 0 & 1 \\ 1 & 1 \end{bmatrix} x_2 + \begin{bmatrix} 0 \\ -1 \end{bmatrix} V_s$$

$$y = [0 \quad 1] x_1 + [0 \quad 0] x_2 .$$

1-3.3. Third Equivalent Form (EF3)

Under the regularity assumption, there always exists a scalar α (such αs are numerous) such that $|\alpha E+A| \neq 0$. Let the transformation matrices be

$$Q_2 = (\alpha E+A)^{-1}, \qquad P_2 = I_n .$$

Then $Q_2A = I_n - \alpha(\alpha E+A)^{-1}E$. Thus, system (1-3.6) is r.s.e. to

$$\hat{E}\dot{x} = (I - \alpha\hat{E})x + \hat{B}u \qquad (1\text{-}3.22)$$
$$y = Cx$$

where $\hat{E} = Q_2E$ and $\hat{B} = Q_2B$, which is the third equivalent form (EF3) for (1-3.6). Obviously, for a fixed α, the EF3 is unique in the sense of algebraic equivalence (similarity).

Example 1-3.4. Consider system (1-3.19). Note that A^{-1} exists. Then α may be chosen as zero, and

$$Q_2 = A^{-1} = \begin{bmatrix} 0 & 1 & 0 & 0 \\ 1 & 0 & 0 & 0 \\ -1 & -1 & -1 & 1 \\ 0 & 1 & 1 & 0 \end{bmatrix} \qquad P_2 = I_4 .$$

Thus, system (1-3.6) is r.s.e. to

$$\hat{E}\dot{x} = x + \hat{B}V_s(t) \qquad (1\text{-}3.23)$$
$$y = Cx$$

where

$$\hat{E} = \begin{bmatrix} 0 & 0 & 1 & 0 \\ 1 & 0 & 0 & 0 \\ -1 & 0 & -1 & 0 \\ 0 & 0 & 1 & 0 \end{bmatrix}, \qquad \hat{B} = \begin{bmatrix} 0 \\ 0 \\ -1 \\ 0 \end{bmatrix}, \qquad C = [0 \quad 0 \quad 1 \quad 0].$$

1-4. State Response

Consider the system

$$E\dot{x} = Ax + Bu \qquad (1\text{-}4.1)$$
$$y = Cx$$

where $x \in \mathbb{R}^n$, $u \in \mathbb{R}^m$, $y \in \mathbb{R}^r$. As pointed out in Section 1-3.1, under the regularity assumption on (1-4.1), there exist two nonsingular matrices Q and P, such that system (1-4.1) is r.s.e. to

$$\dot{x}_1 = A_1x_1 + B_1u, \qquad y_1 = C_1x_1 \qquad (1\text{-}4.2a)$$

$$N\dot{x}_2 = x_2 + B_2u, \qquad y_2 = C_2x_2 \qquad (1\text{-}4.2b)$$

$$y = C_1x_1 + C_2x_2 = y_1 + y_2 \qquad (1\text{-}4.2c)$$

where $x_1 \in \mathbb{R}^{n_1}$, $x_2 \in \mathbb{R}^{n_2}$, $n_1 + n_2 = n$, and N is nilpotent whose nilpotent index is denoted by h, and

$$QEP = \text{diag}(I, N), \quad QAP = \text{diag}(A_1, I), \quad CP = [C_1 \quad C_2], \qquad (1\text{-}4.3a)$$

$$P^{-1}x = [x_1/x_2], \qquad QB = [B_1/B_2] \ . \tag{1-4.3b}$$

Note that the slow subsystem (1-4.2) is an ordinary differential equation. It has a unique solution with any initial condition $x_1(0)$ (the initial time is supposed to be zero) for any piecewise continuous function $u(t)$:

$$x_1(t) = e^{A_1 t}x_1(0) + \int_0^t e^{A_1(t-\tau)}B_1 u(\tau)d\tau \ . \tag{1-4.4}$$

Thus, $x_1(t)$ is completely determined by $x_1(0)$, $u(\tau)(0 \leqslant \tau \leqslant t)$.

Suppose that the function $u(t)$ is h times piecewise continuously differentiable. By continuously taking derivatives with respect to t on both sides of (1-4.2b), and left multiplying both sides by matrix N, we obtain the following equations:

$$\begin{aligned} N\dot{x}_2 &= x_2 + B_2 u \\ N^2 x_2^{(2)} &= N\dot{x}_2 + NB_2\dot{u} \\ &\cdots \\ N^h x_2^{(h)} &= N^{h-1}x_2^{(h-1)} + N^{h-1}B_2 u^{(h-1)} \end{aligned} \tag{1-4.5}$$

where $x_2^{(i)}$ stands for the ith derivative of $x_2(t)$. From the addition of these equations and the fact $N^h = 0$, we know readily that the solution $x_2(t)$ is given by

$$x_2(t) = -\sum_{i=o}^{h-1} N^i B_2 u^{(i)}. \tag{1-4.6}$$

$x_2(t)$ is a linear combination of derivatives of $u(t)$ at time t, which is certainly determined by the derivatives at time t. For any scalar $\varepsilon > 0$, the properties of $u(\tau)$, $0 \leqslant \tau \leqslant t-\varepsilon$ have no contribution to $x_2(t)$. This shows an interesting phenomenon between the substates $x_1(t)$ and $x_2(t)$: $x_1(t)$ represents a cummulative effect of $u(\tau)$, $0 \leqslant \tau \leqslant t$, with no relation to $u(t)$; while on the contrary, $x_2(t)$ response so rapidly that it insistently reflects the properties of $u(t)$ at time t. This is why we call (1-4.2a) and (1-4.2b) the slow and fast subsystems, respectively.

To sum up, the state and output responses of singular system (1-4.1) is given by

$$\begin{aligned} x(t) &= P[I/0]x_1(t) + P[0/I]x_2(t) \\ &= P[I/0](e^{A_1 t}x_1(0) + \int_0^t e^{A_1(t-\tau)}B_1 u(\tau)d\tau) - P[0/I]\sum_{i=o}^{h-1} N^i B_2 u^{(i)}(t) \end{aligned} \tag{1-4.7a}$$

$$y(t) = Cx(t) = CP\begin{bmatrix} I \\ 0 \end{bmatrix}(e^{A_1 t}x_1(0) + \int_0^t e^{A_1(t-\tau)}B_1 u(\tau)d\tau) - CP\begin{bmatrix} 0 \\ I \end{bmatrix}\sum_{i=o}^{h-1} N^i B_2 u^{(i)}(t). \tag{1-4.7b}$$

Particularly by setting $t > 0$, $t \to 0^+$, we obtain

$$x(0^+) = P\begin{bmatrix} I \\ 0 \end{bmatrix} x_1(0) - P\begin{bmatrix} 0 \\ I \end{bmatrix} \sum_{i=o}^{h-1} N^i B_2 u^{(i)}(0^+) \tag{1-4.8}$$

which is the so-called consistent initial condition imposed on initial value x(0).

Compared with results in linear system theory, the normal system always has one unique solution if the input function is piecewise continuous, or weaker. However, equation (1-4.8) shows that the singular system (1-4.1) has a unique solution only for the consistent initial vector x(0), and for the h times piecewise continuously differentiable input function u(t). The latter is stronger than that in the normal system case. Such characters stand for the special feature of singular systems.

On the other hand, we treat the singular system from another point of view. Note the EF2 for system (1-4.1), whose first differential equation is of the order q = rankE. Some argue that system (1-4.1) should have at least q freedom to choose in its solution, which implies that system (1-4.1) should have a unique solution for some initial conditions that don't satisfy the consistent condition in equation (1-4.8). Moreover, systems of electric circuits formed at the starting time would allow any initial value to be present. Thus, it has been proposed that arbitrary initial conditions have physical meaning for systems that are, in one sense, created or structurally changed at t = 0. Hence, it has been suggested that the singular systems should adopt a generalized solution ---- the distribution as solution. Now we simply state the main idea of distribution as a solution for singular systems. This was discussed in Cobb (1983b, 1984), Verghese et al (1981), Kecs and Teodorescu (1974).

Suppose that u(t) is h times piecewise continuously differentiable. The notations $\dot{x}_2'(t)$ and $\dot{x}_2(t)$ are used to refer to the derivatives of $x_2(t)$ in the distributional and normal senses, respectively. From (A.4) in Appendix A, the distributional equivalence of (1-4.1) is

$$N\dot{x}_2'(t) - Nx_2(0)\delta(t) = x_2(t) + B_2u(t)$$

where $\delta(t)$ is the Dirac function at time point t. It has its infinite value at 0 and zero value at other points. Rewritting the above equation as

$$N\dot{x}_2'(t) = x_2(t) + B_2u(t) + Nx_2(0)\delta(t)$$

from (1-4.6) we have

$$x_2(t) = -\sum_{i=1}^{h-1} \delta^{(i-1)}(t)N^i x_2(0) - \sum_{i=o}^{h-1} N^i B_2 u^{(i)}(t) \ . \tag{1-4.9}$$

This is the general form of the solution of fast subsystem (1-4.2b) in the sense of distribution. We will call it the distribution solution, in which impulse terms appear.

It follows from (1-4.4) and (1-4.9) that the general state response is

$$x(t) = P\begin{bmatrix} I \\ 0 \end{bmatrix} e^{A_1 t}[I \; 0]P^{-1}x(0) + P\begin{bmatrix} I \\ 0 \end{bmatrix}\int_0^t e^{A_1(t-\tau)}B_1 u(\tau)d\tau$$

$$- P\begin{bmatrix} 0 \\ I \end{bmatrix}\sum_{i=1}^{h-1} \delta^{(i-1)}(t)N^i[0 \; I]P^{-1}x(0) - P\begin{bmatrix} 0 \\ I \end{bmatrix}\sum_{i=o}^{h-1} N^i B_2 u^{(i)}(t). \qquad (1\text{-}4.10)$$

Unlike in the normal system theory, the distribution solutions for singular systems have a complicated form. It includes not only the normal exponential response part, which is created by the slow subsystem (a normal one), but also the impulse and input derivative portions (due to the fast subsystem). The complicated state response form (1-4.10) just characterizes the singular system, and differs itself from the normal one. Obviously, (1-4.10) and (1-4.7a) become the same when $t > 0$.

The substate of fast subsystem (1-4.2b) may also be obtained in another way. Taking the Laplace transformation on both sides of (1-4.2b), we have

$$(sN-I)x_2(s) = Nx_2(0) + B_2 u(s)$$

where $x_2(s)$ and $u(s)$ stand for the Laplace transformation of $x_2(t)$ and $u(t)$, respectively. From this equation we have

$$x_2(s) = (sN-I)^{-1}(Nx_2(0)+B_2u(s)) = -\sum_{i=o}^{h-1} N^i s^i (Nx_2(0)+B_2u(s)).$$

Note that the Dirac function $\delta(t)$ has the Laplace transformation of

$$L[\delta^{(i)}(t)] = s^i.$$

Hence the inverse Laplace transformation of $x_2(s)$ yields:

$$x_2(t) = -\sum_{i=o}^{h-1} \delta^{(i)}(t)N^{i+1}x_2(0) - \sum_{i=o}^{h-1} N^i B_2 u^{(i)}(t)$$

which is just the same as (1-4.9).

Example 1-4.1. Consider the circuit network system in Example 1-3.2:

$$\begin{bmatrix} 1 & 0 & 0 & 0 \\ 0 & 0 & 1 & 0 \\ 0 & 0 & 0 & 0 \\ 0 & 0 & 0 & 0 \end{bmatrix}\dot{x} = \begin{bmatrix} 0 & 1 & 0 & 0 \\ 1 & 0 & 0 & 0 \\ -1 & 0 & 0 & 1 \\ 0 & 1 & 1 & 1 \end{bmatrix}x + \begin{bmatrix} 0 \\ 0 \\ 0 \\ -1 \end{bmatrix}V_s(t)$$

$$y = [0 \;\; 0 \;\; 1 \;\; 0]x \qquad (1\text{-}4.11)$$

with the standard decomposition shown in (1-3.20). From (1-4.6) and (1-4.4), the slow and fast subsystems have the following state responses:

$$x_1(t) = e^{A_1 t}x_1(0) + \int_0^t e^{A_1(t-\tau)}\begin{bmatrix} 1 \\ 0 \end{bmatrix} V_s(\tau)d\tau$$

$$x_2(t) = \begin{bmatrix} 1 \\ 0 \end{bmatrix} V_s(t) \qquad (1\text{-}4.12)$$

respectively.

Furthermore, direct computation shows the state response of system (1-4.11) to be:

$$x(t) = \begin{bmatrix} x_1(t) \\ x_2(t) + \begin{bmatrix} -1 & -1 \\ 1 & 0 \end{bmatrix} x_1(t) \end{bmatrix}$$

where

$$x_1(t) = e^{-\frac{1}{2}t} \begin{bmatrix} -\sin\frac{\sqrt{3}}{2}t - \sqrt{3}\cos\frac{\sqrt{3}}{2}t & -2\sin\frac{\sqrt{3}}{2}t \\ 2\sin\frac{\sqrt{3}}{2}t & \sin\frac{\sqrt{3}}{2}t + \sqrt{3}\cos\frac{\sqrt{3}}{2}t \end{bmatrix} x_1(0)$$

$$+ \int_0^t e^{-\frac{1}{2}(t-\tau)} \begin{bmatrix} -\sin\frac{\sqrt{3}}{2}(t-\tau) - \sqrt{3}\cos\frac{\sqrt{3}}{2}(t-\tau) \\ 2\sin\frac{\sqrt{3}}{2}(t-\tau) \end{bmatrix} V_s(\tau)d\tau$$

$$x_2(t) = \begin{bmatrix} 1 \\ 0 \end{bmatrix} V_s(t) .$$

In this example, $N = 0$; thus no impulse terms appear apparently in fast state response, nor do the derivative portions.

Example 1-4.2. The following 2-order system

$$\begin{bmatrix} 0 & 1 \\ 0 & 0 \end{bmatrix} \begin{bmatrix} \dot{x}_1(t) \\ \dot{x}_2(t) \end{bmatrix} = \begin{bmatrix} x_1(t) \\ x_2(t) \end{bmatrix} + \begin{bmatrix} -1 \\ -1 \end{bmatrix} u(t) \qquad (1\text{-}4.13)$$

has only fast subsystem, with coefficients

$$N = \begin{bmatrix} 0 & 1 \\ 0 & 0 \end{bmatrix}, \qquad B = \begin{bmatrix} -1 \\ -1 \end{bmatrix} .$$

Therefore, its solution is

$$\begin{bmatrix} x_1 \\ x_2 \end{bmatrix} = -\sum_{i=1}^{1} N^i \delta^{(i-1)}(t) \begin{bmatrix} x_1(0) \\ x\ (0) \end{bmatrix} - \sum_{i=o}^{1} N^i B u^{(i)}(t)$$

$$= \begin{bmatrix} -x_2(0)\delta(t) + u(t) + \dot{u}(t) \\ u(t) \end{bmatrix} \qquad (1\text{-}4.14)$$

Thus

$$\begin{aligned} x_1(t) &= -x_2(0)\delta(t) + u(t) + \dot{u}(t) \\ x_2(t) &= u(t). \end{aligned} \qquad (1\text{-}4.15)$$

Two special cases are studied herein.

1. Let the input function be the unit jump function

$$u(t) = f(t-\alpha) = \begin{cases} 1 & t > \alpha > 0 \\ 0 & t \leqslant \alpha , \quad \alpha \text{ is a constant.} \end{cases}$$

Then

$$x_1(t) = - x_2(0)\delta(t) + f(t) + \delta(t-\alpha)$$

$$x_2(t) = f(t).$$

They are illustrated in Figure 1-4.1.

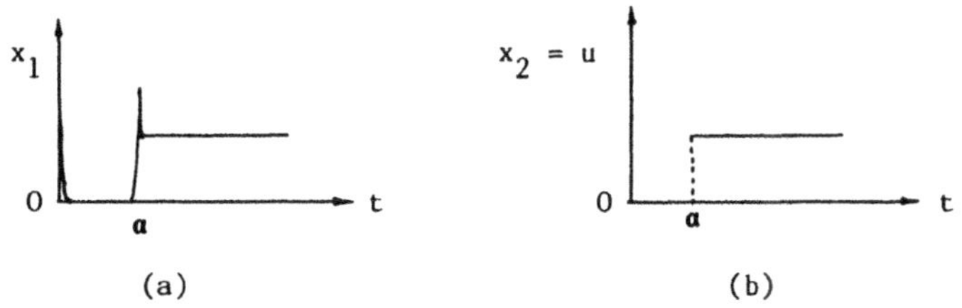

Figure 1-4.1.

2. If the input function is

$$u(t) = g(t) = \begin{cases} 0 & t \leqslant \beta \\ t-\beta & t > \beta \end{cases} \qquad \beta > 0 \text{ is a constant}$$

then u(t) is continuous. But

$$x_1(t) = - x_2(0)\delta(t) + g(t) + f(t-\beta)$$

$$x_2(t) = g(t)$$

whose locus is shown in Figure 1-4.2.

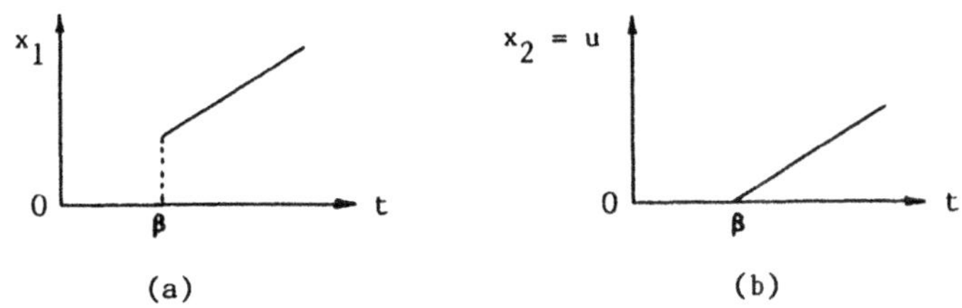

Figure 1-4.2.

Classical system theory confirms that the state response is continuous, provided that the input function is piecewise continuous (or weaker). However, this is not true in singular systems. Figures 1-4.1 and 1-4.2 show that when input derivatives are involved, either impulse terms may be created in the state response by any jump behavior in input due to the operation at the starting and closing switch actions in practice (Figure 1-4.1(a)); or jump behavior appears in the state response, although the input function is continuous, as described in Figure 1-4.2, which shows jump behavior appears at $t = \beta$ in $x_1(t)$ due to the discontinuous property in input derivatives.

It is worth pointing out that (1-4.10) is only a solution for (1-4.1) in the distribution sense, not a solution in the ordinary sense, i.e., when applied, equation (1-4.10) doesn't make the equality hold in the ordinary differential equation sense. However, the equality holds in the distribution sense.

1-5. Notes and References

Currently, there are mainly two opinions concerning the initial condition. One argues that the initial condition should be restricted to the consistent conditions; while the other says that any possible initial condition should be acceptable. Because of its convenience in the system analysis, this book adopts the latter. Knowledge of distribution may be found in Appendix A, but one may read this section accepting only distribution solutions.

Practical examples of singular systems may be found in Campbell (1982b), Campbell and Rose (1982), Haggman and Bryant (1984), Kiruthi et al. (1980), Lewis (1986), Luenberger (1977), Luenberger and Arbel (1977), Martens et al. (1984), Newcomb (1981, 1982), Petzold (1982), Singh and Liu (1973), Stengel et al. (1979), and Wang and Dai (1986a).

Typical papers concerning regularity are Yip and Sincovec (1981), and Fletcher (1986).

Among papers on restricted system equivalence, typical ones are Campbell (1982b), Cullen (1984), Dai (1987b), Fuhymann (1977), Hayton et al. (1986), Pugh and Shelton (1978), Verghese et al. (1981), Zhou et al. (1987).

On the solution of distribution one may refer to Cobb (1983b, 1984), Verghese et al. (1981), and Cullen (1984).

CHAPTER 2

TIME DOMAIN ANALYSIS

The description equation of a practical system may be established through selection of the proper state variables. Time domain analysis is the method of analyzing the system based on this description equation, through which we may gain a fair understanding of the system's structural features as well as its internal properties. Using time domain analysis, this chapter studies the fundamentals in system theory such as reachability, controllability, observability, system decomposition, transfer matrix, and minimal realization. The concepts explained in this chapter are essential for later chapters.

2-1. State Reachability

Consider the regular singular system:

$$\begin{aligned} E\dot{x} &= Ax + Bu \\ y &= Cx \end{aligned} \tag{2-1.1}$$

where $x \in \mathbb{R}^n$, $u \in \mathbb{R}^m$, $y \in \mathbb{R}^r$ are its state, control input, and measure output, respectively; and E, $A \in \mathbb{R}^{n\times n}$, $B \in \mathbb{R}^{n\times m}$, $C \in \mathbb{R}^{r\times n}$ are constant matrices. Without loss of generality, $q \triangleq \operatorname{rank} E < n$ is supposed. For system, as pointed out in the previous section, two nonsingular matrices Q and P exist such that it is r.s.e. to (EF1):

$$\begin{aligned} \dot{x}_1 &= A_1x_1 + B_1u \\ N\dot{x}_2 &= x_2 + B_2u \\ y &= C_1x_1 + C_2x_2 \end{aligned} \tag{2-1.2}$$

where $x_1 \in \mathbb{R}^{n_1}$, $x_2 \in \mathbb{R}^{n_2}$, $n_1 + n_2 = n$, $N \in \mathbb{R}^{n_2\times n_2}$ is nilpotent, the nilpotent index is denoted by h, and

$$QEP = \operatorname{diag}(I, N),\ QAP = \operatorname{diag}(A_1, I),\quad CP = [C_1\ \ C_2]$$

$$P^{-1}x = [x_1/x_2],\qquad QB = [B_1/B_2].$$

For the sake of simplicity, in this chapter, system (2-1.1) is assumed to be in its standard decomposition form (2-1.2) unless specified. However, the results are applicable to systems in the general form of (2-1.1).

We start from the concept of reachable set to study the state structure. Assume that admissible control input is confined to $u(t) \in \mathbf{C}_p^{h-1}$, which represents the (h-1)-times piecewise continuously differentiable function set. Then from Section 1-4 we know that, when $t > 0$, the state response for system (2-1.2) is

$$\begin{aligned} x_1(t) &= e^{A_1 t}x\ (0) + \int_0^t e^{A_1(t-\tau)}B_1 u(\tau)d\tau \\ x_2(t) &= -\sum_{i=0}^{h-1} N^i B_2 u^{(i)}(t). \end{aligned} \tag{2-1.3}$$

Equation (2-1.3) indicates that when $t > 0$, state response $x(t) = [x_1/x_2]$ is uniquely determined by the initial condition $x_1(0)$, control input $u(\tau)$, $0 \leqslant \tau \leqslant t$, and time point t. The reachable set may be defined as follows.

Definition 2-1.1. Any vector $w \in \mathbb{R}^n$ in n-dimensional vector space is said to be reachable, if there exists an initial condition $x_1(0)$, admissible control input $u(t) \in \mathbf{C}_p^{h-1}$, and $t_1 > 0$ such that

$$x(t_1) = [x_1(t_1)/x_2(t_1)] = w.$$

Let $R(x_1(0))$ denote the reachable state from the initial condition $x_1(0)$. Then

$$R(x_1(0)) = \{ w \in \mathbb{R}^n \mid \text{there exists an admissible control } u(t) \in \mathbf{C}_p^{h-1} \text{ and } t_1 > 0 \text{ such that } x(t_1) = w \}$$

which is a subspace in n vector space.

If we define

$$\langle E \mid B \rangle = \mathrm{Im}[B,\ EB,\ \dots\ ,\ E^{n-1}B]$$

where n is the dimension of E,

$$\mathrm{Im}B = \{ y \mid y = Bx,\ x \in \mathbb{R}^m \}, \qquad \mathrm{Ker}E = \{ x \mid Ex = 0,\ x \in \mathbb{R}^n \}$$

we have the following lemma.

Lemma 2-1.1. For any nonidentically zero polynomial f(t), we define

$$W(f,t) = \int_0^t f(s)e^{A_1 s}B_1 B_1^{\tau} e^{A_1^{\tau} s} f(s)ds.$$

Then $\mathrm{Im}W(f,t) = \langle A_1 \mid B_1 \rangle$ for any $t > 0$.

Proof. To prove the result equals to the proof of

$$\mathrm{Ker}W(f,t) = \bigcap_{i=0}^{n_1-1} \mathrm{Ker}B_1^{\tau}(A_1^{\tau})^i \tag{2-1.4}$$

First, when $x \in \mathrm{Ker}W(f,t)$ the definition indicates:

$$x^{\tau}W(f,t)x = x^{\tau}0 = 0$$

i.e.,

$$x^{\tau}W(f,t)x = \int_0^t x^{\tau}f(s)e^{A_1 s}B_1B_1^{\tau}e^{A_1^{\tau}s}f(s)x\,ds = \int_0^t \|B_1^{\tau}e^{A_1^{\tau}s}f(s)x\|_2^2 ds = 0 \qquad (2\text{-}1.5)$$

where $\|\cdot\|_2$ represents the spectral vector norm in the definition of $\|x\|_2 = (x^{\tau}x)^{\frac{1}{2}}$. Since

$$\|B_1^{\tau}e^{A_1^{\tau}s}f(s)x\|_2^2 \geqslant 0$$

is continuous, we know from (2-1.5) that

$$B_1^{\tau}e^{A_1^{\tau}s}f(s)x = 0, \qquad 0 \leqslant s \leqslant t.$$

Since polynomial $f(s)$ has a finite number of zeros on $0 \leqslant s \leqslant t$, from the preceding equation, we immediately have

$$B_1^{\tau}e^{A_1^{\tau}s}x = 0, \qquad 0 \leqslant s \leqslant t.$$

The arbitrariness of s yields

$$x \in \bigcap_0^{n_1-1} \operatorname{Ker}(B_1^{\tau}(A_1^{\tau})^i).$$

Thus

$$\operatorname{Ker}W(f,t) \subseteq \bigcap_0^{n_1-1} \operatorname{Ker}(B_1^{\tau}(A_1^{\tau})^i).$$

If $x \in \bigcap_0^{n_1-1} \operatorname{Ker}(B_1^{\tau}(A_1^{\tau})^i)$, the reverse of this process will be easy to prove $x \in \operatorname{Ker}W(f,t)$, i.e.,

$$\bigcap_0^{n_1-1} \operatorname{Ker}(B_1^{\tau}(A_1^{\tau})^i) \subseteq \operatorname{Ker}W(f,t).$$

Hence, the lemma is easy to obtain by combining these results.Q.E.D.

Obviously, this process is similar to the one in the normal case.

Lemma 2-1.2. For any h vectors $x_i \in \mathbb{R}^n$, $i = 0, 1, 2, \ldots, h-1$, and $t_1 > 0$, there always exists a vector polynomial $f(t) \in \mathbb{R}^n$ whose order is h-1, such that

$$f^{(i)}(t_1) = x_i, \qquad i = 0, 1, 2, \ldots, h-1.$$

Proof. The proof is straightforward by setting

$$f(t) = x_0 + x_1(t-t_1) + \ldots + x_{h-1}(t-t_1)^{h-1}. \quad \text{Q.E.D.}$$

Theorem 2-1.1. Let R(0) represent the state reachable set from the zero initial condition ($x_1(0) = 0$). Then

$$R(0) = \langle A_1 \mid B_1 \rangle \oplus \langle N \mid B_2 \rangle$$

where " $\oplus$ " is the direct sum in vector space.

Proof. When $x_1(0) = 0$, equation (2-1.3) shows

$$x_1(t) = \int_0^t e^{A_1(t-\tau)} B_1 u(\tau) d\tau , \qquad x_2(t) = -\sum_{i=o}^{h-1} N^i B_2 u^{(i)}(t), \quad t > 0.$$

Thus, obviously, $x_2(t) \in \langle N \mid B_2 \rangle$. Further use of the results in Appendix B shows that

$$e^{A_1 t} = \beta_o(t) I + \beta_1(t) A_1 + \dots + \beta_{n_1-1}(t) A_1^{n_1-1}.$$

Thus

$$x_1(t) = \int_0^t e^{A_1(t-\tau)} B_1 u(\tau) d\tau = \sum_{i=o}^{n_1-1} A_1^i B_1 \int_0^t \beta_i(t-\tau) d\tau \in \langle A_1 \mid B_1 \rangle$$

or in other words, $x(t) \in \langle A_1 \mid B_1 \rangle \oplus \langle N \mid B_2 \rangle$. Therefore,

$$R(0) \subseteq \langle A_1 \mid B_1 \rangle \oplus \langle N \mid B_2 \rangle. \qquad (2\text{-}1.6)$$

On the other hand, for any $x = [x_1/x_2] \in \langle A_1 \mid B_1 \rangle \oplus \langle N \mid B_2 \rangle$, with $x_1 \in \langle A_1 \mid B_1 \rangle$, $x_2 \in \langle N \mid B_2 \rangle$, from $x_2 \in \langle N \| B_2 \rangle$ there exists $x_{2i} \in \mathbb{R}^{n_2}$, $i = 0, 1, 2, \dots, h-1$ such that

$$x_2 = -\sum_{i=o}^{h-1} N^i B_2 x_{2i} .$$

Also from Lemma 2-1.2, for any fixed $t > 0$, there exists a polynomial $f_2(s)$ of order $h-1$ such that $f_2^{(i)}(t) = x_{2i}$. Thus, if we impose the input control

$$u(t) = u_1(t) + f_2(t)$$

it will be that

$$x_1(t) = \int_0^t e^{A_1(t-\tau)} B_1 u_1(\tau) d\tau + \int_0^t e^{A_1(t-\tau)} B_2 f_2(\tau) d\tau ,$$

implying

$$\tilde{x}_1 \in \langle A_1 \mid B_1 \rangle$$

where $\tilde{x}_1$ is defined as

$$\tilde{x}_1 = x_1 - \int_0^t e^{A_1(t-\tau)} B_2 f_2(\tau) d\tau .$$

For any fixed $t > 0$, let $f_1(s) = s^h (s-t)^h$. Then $f_1(s)$ is not indentically equal to zero and Lemma 2-1.1 shows that a vector $z \in \mathbb{R}^{n_1}$ may be chosen such that

$$W(f_1, t) z = \tilde{x}_1 .$$

For the $z \in \mathbb{R}^{n_1}$ chosen, we construct

$$u_1(s) = f_1^2(s)B_1^\tau e^{A_1^\tau(t-s)}z\ , \qquad 0 \leqslant s \leqslant t.$$

Direct computation would verify that

$$x_1(t) = x_1 - \tilde{x}_1 + \int_0^t e^{A_1(t-\tau)}B_1u_1(\tau)d\tau = x_1 - \tilde{x}_1 + W(f_1,t)z = x_1$$

and

$$x_2(t) = x_2 - \sum_{i=0}^{h-1} N^iB_2u^{(i)}(t).$$

Since

$$u_1^{(i)}(t) = 0, \quad i = 0, 1, \dots , h-1,$$

it must be $x_2(t) = x_2$.

Thus $R(0) \supseteq \langle A_1 \mid B_1\rangle \oplus \langle N \mid B_2\rangle$. The combination of it with (2-1.6) results in $R(0) = \langle A_1 \mid B_1\rangle \oplus \langle N \mid B_2\rangle$. Q.E.D.

Denote

$$H(x_1(0)) = \{x = [x_1/x_2] \mid x_1 = e^{A_1t}x_1(0) \in \mathbb{R}^{n_1},\ x_2 = 0 \in \mathbb{R}^{n_2}\}$$

which represents the free state reachable set from starting point $x_1(0)$.

If R is the reachable set for system (2-1.2) defined as the union of all reachable set from all possible initial condition $x_1(0) \in \mathbb{R}^{n_1}$, i.e.,

$$R = \bigcup_{x_1(0)} R(x_1(0)).$$

Earlier discussion shows that R takes the form of

$$R = \bigcup_{x_1(0)} (R(0)+H(x_1(0))) = R^{n_1} \oplus \langle N \mid B_2\rangle\ . \tag{2-1.7}$$

Obviously $0 \in R$.

Equation (2-1.7) provides us with a precise form of reachable set. While the state x(t) at any time point t lies in R, any vector in R may be reached at a certain time $t_1 > 0$ by the state response, starting from a certain $x_1(0)$ and driven by an appropriate control input u(t). Furthermore, the proof process also shows that such vectors may be reached in any short period with a suitably chosen control u(t).

Another direct phenomenon may be seen in represention (2-1.7): Instead of filling up the whole space as in the normal case, the state response for singular systems lies on a fluid manifold in the space. This is a feature that differentiates the singular system from the normal one.

Example 2-1.1. Consider the system described by (1-3.20):

$$\dot{x}_1 = \begin{bmatrix} -1 & -1 \\ 1 & 0 \end{bmatrix} x_1 + \begin{bmatrix} 1 \\ 0 \end{bmatrix} V_s(t)$$

$$0 = x_2 + \begin{bmatrix} -1 \\ 0 \end{bmatrix} V_s(t) \qquad (2\text{-}1.8)$$

$$y = [0 \quad 1] x_1$$

where $x_1, x_2 \in \mathbb{R}^2$, which is in the standard form for the circuit network in Example 1-3.2, with coefficient matrices

$$A_1 = \begin{bmatrix} -1 & -1 \\ 1 & 0 \end{bmatrix}, \quad B_1 = \begin{bmatrix} 1 \\ 0 \end{bmatrix}, \quad B_2 = \begin{bmatrix} -1 \\ 0 \end{bmatrix}, \quad N = 0, \quad C_1 = [0 \quad 1], \quad C_2 = 0.$$

According to (2-1.7), the reachable set from the zero initial condition is

$$R(0) = \langle A_1 \mid B_1 \rangle \oplus \langle N \mid B_2 \rangle$$

with

$$\langle A_1 \mid B_1 \rangle = \mathrm{Im}[B_1 \quad A_1B_1] = \mathbb{R}^2, \qquad \langle N \mid B_2 \rangle = \mathrm{Im}B_2 = \mathbb{R}^1 \oplus \{0\}.$$

Thus $R(0) = \mathbb{R}^3 \oplus \{0\}$, and $R = \mathbb{R}^2 \oplus \langle N \mid B_2 \rangle = R(0)$.

In this example, $R = R(0)$, or in other words, the reachable set for the whole system is identical to the reachable set from zero initial condition.

Example 2-1.2. Consider the second order system in Example 1-4.2:

$$\begin{bmatrix} 0 & 1 \\ 0 & 0 \end{bmatrix} \dot{x}(t) = x(t) + \begin{bmatrix} -1 \\ -1 \end{bmatrix} u(t). \qquad (2\text{-}1.9)$$

For this system, the reachable set for the system and from the zero initial condition are, respectively,

$$R = R(0) = \mathbb{R}^2, \qquad R(0) = \langle N \mid B_2 \rangle = \mathbb{R}^2.$$

Both are whole vector space. Now, we will examine the state response of this system

$$x(t) = -\sum_{i=o}^{h-1} N^i B_2 u^{(i)}(t) = \begin{bmatrix} u(t)+\dot{u}(t) \\ u(t) \end{bmatrix}.$$

Thus, for any $w = [w_1/w_2] \in \mathbb{R}^2$ and $t_1 > 0$, the equality $x(t_1) = w$ implies

$$u(t_1) = w_2, \qquad \dot{u}(t_1) = w_1 - w_2 \qquad (2\text{-}1.10)$$

whose first equation fixes the value of $u(t)$ at time point t_1 and the second assumes the potential variation of $u(t)$ at t_1. Clearly, $u(t)$ will increase or decrease greatly at t_1 if w_1 and w_2 defer remarkably. This control strategy is difficult to realize in the slow changing mechanical or chemical system, and high energy may be required, even though it may be realized.

2-2. Controllability, R-Controllability, and Impulse Controllability

Characterizing the structural properties from various views, controllability, R-controllability, and impulse controllability are three important concepts in singular systems. By using these concepts, we may have a fair understanding of controllability by control input. The controllabilities also reflect the difference of singular systems from normal ones. As before, only (2-1.2) is considered.

Definition 2-2.1. System (2-1.2) is called controllable if, for any $t_1 > 0$, $x(0) \in \mathbb{R}^n$ and $w \in \mathbb{R}^n$, there exists a control input $u(t) \in \mathbf{C}_p^{h-1}$ such that $x(t_1) = w$.

This definition states that under the controllability assumption, for any initial condition $x(0)$, we may always choose a control input such that the state response starting from $x(0)$ may arrive at any prescribed position in $\mathbb{R}^n$ in any given time period. It is easy to see that the definition is a natural generalization of controllability concepts in the normal case.

Rewriting system (2-1.2) in the slow-fast subsystem form, we get:

$$\begin{aligned} \dot{x}_1 &= A_1x_1 + B_1u \\ y_1 &= C_1x_1 \end{aligned} \tag{2-2.1a}$$

$$\begin{aligned} N\dot{x}_2 &= x_2 + B_2u \\ y_2 &= C_2x_2 \end{aligned} \tag{2-2.1b}$$

$$y = C_1x_1 + C_2x_2 = y_1 + y_2. \tag{2-2.1c}$$

Then we have the following theorem.

Theorem 2-2.1.

(1). Slow subsystem (2-2.1a) is controllable iff

$$\mathrm{rank}[sE-A,\ B] = n, \qquad \forall\ s \in \mathbf{C},\ s \text{ finite} \tag{2-2.2}$$

where $\mathbf{C}$ represents the complex plane.

(2). The following statements are equivalent.

(a). The fast subsystem (2-2.1b) is controllable.

(b). $\mathrm{rank}[B_2, NB_2, \dots, N^{h-1}B_2] = n_2$.

(c). $\mathrm{rank}[N \quad B_2] = n_2$.

(d). $\mathrm{rank}[E \quad B] = n$.

(e). For any nonsingular matrices Q_1 and P_1 satisfying $E = Q_1\mathrm{diag}(I, 0)P_1$, Let $Q_1B = [\tilde{B}_1/\tilde{B}_2]$. Then $\tilde{B}_2$ is of full row rank, $\mathrm{rank}\tilde{B}_2 = n - \mathrm{rank}E$.

(3). The following statements are equivalent.

(a). System (2-1.2) is controllable.

(b). Both its slow and fast subsystems are controllable.

(c). $\text{rank}[B_1, A_1B_1, \dots, A_1^{n_1-1}B_1] = n_1$ and $\text{rank}[B_2, NB_2, \dots, N^{h-1}B_2] = n_2$.

(d). $\text{rank}[sE-A \;\; B] = n$, $\forall\, s \in \mathbb{C}$, s finite, and $\text{rank}[E \;\; B] = n$.

(e). The matrix

$$D_1 = \begin{bmatrix} -A & & & & B & & & \\ E & -A & & & & B & & \\ & E & \ddots & -A & & & B & \ddots \\ & & \ddots & E & & & & B \end{bmatrix}_{n^2 \times (n+m-1)n}$$

or, equivalently,

$$D_1 = \begin{bmatrix} \begin{bmatrix} -A & B \\ E & 0 \end{bmatrix} & \begin{bmatrix} -A & B \\ E & 0 \end{bmatrix} & & \\ & & \ddots & \begin{bmatrix} -A & B \\ E & 0 \end{bmatrix} \end{bmatrix}_{n^2 \times (n+m-1)n}$$

is of full row rank n^2.

Proof. We will prove the statements one by one.

(1). The slow subsystem (2-2.1a) is a normal one, for which Definition 2-2.1 becomes the controllability definition in the linear system theory. Thus, (2-2.1a) is controllable iff

$$\text{rank}[sI-A_1, B_1] = n_1, \quad \forall \; s \in \mathbb{C}, \text{ s finite.} \tag{2-2.3}$$

Noticing that

$$\text{rank}[sE-A, B] = \text{rank}[sQEP-QAP, QB]$$

$$= \text{rank}\begin{bmatrix} sI-A_1 & 0 & B_1 \\ 0 & sN-I & B_2 \end{bmatrix}$$

and $sN-I$ is invertible for any finite $s \in \mathbb{C}$, we have

$$\text{rank}[sE-A, B] = n_2 + \text{rank}[sI-A_1, B_1]$$

which indicates that (2-2.3) holds if and only if $\text{rank}[sE-A, B] = n$, $\forall\, s \in \mathbb{C}$, and s is finite. This is (2-2.2).

(2). According to the definition, if the fast subsystem (2-2.1b) is controllable $\langle N \mid B_2 \rangle = \mathbb{R}^{n_2}$, or $\text{rank}[B_2, NB_2, \dots, N^{h-1}B_2] = n_2$. Thus (a) and (b) are equivalent.

Furthermore, (N, B_2) is controllable if and only if

$$\text{rank}[sI-N, B_2] = n_2, \quad \forall \; s \in \sigma(N) \tag{2-2.4}$$

where $\sigma(N) = \{ s \mid s \in \mathbb{C}, |sI-N| = 0 \}$. Since N is nilpotent, $\sigma(N) = \{0\}$. Equation (2-2.4) is true if and only if $\text{rank}[-N \;\; B_2] = \text{rank}[N, B_2] = n_2$. Thus proves the equivalence between (b) and (c).

To prove the equivalence between (c) and (d), we only need to note the fact

$$\operatorname{rank}[E \quad B] = \operatorname{rank}[QEP \quad QB] = n_1 + \operatorname{rank}[N \quad B_2]$$

which means that $\operatorname{rank}[N \quad B_2] = n_2$ is equal to $\operatorname{rank}[E \quad B] = n$.

The equivalence between (d) and (e) may be proven in a similar way.

(3). Let (a) hold and $x_1(0)=0$, The controllability definition states that for any $t_1 > 0$ and $w \in \mathbb{R}^n$, there exists an admissible control input $u(t) \in \mathbf{C}_p^{h-1}$ such that $x(t_1) = w$. This in turn assures that the reachable set from the zero initial condition R(0) has the property $R(0) = \langle A_1 \mid B_1 \rangle \oplus \langle N \mid B_2 \rangle = \mathbb{R}^n$, or more precisely, $\operatorname{rank}[B_1, A_1B_1, \ldots A_1^{n_1-1}B_1] = n_1$ and $\operatorname{rank}[B_2, NB_2, \ldots, N^{h-1}B_2] = n_2$, which is (c). On the other hand, if (c) holds, from the equation

$$R(x_1(0)) = R(0) + H(x_1(0))$$

we know immediately that $R(x_1(0)) = \mathbb{R}^n$. The system is controllable. Preceding discussion proves the equivalence of (a) and (c).

The combination of these results with (1) and (2) proves the equivalence between (b) and (c) as well between (a) and (c).

Next we will prove the equivalence of (c) and (e). Let Q and P be the transformation matrices from (2-1.1) to (2-1.2), and

$$\hat{Q} = \operatorname{diag}(Q, Q, \ldots, Q) \in \mathbb{R}^{n^2 \times n^2}$$

$$\hat{P} = \operatorname{diag}(P, P, \ldots, P, I_{nm}) \in \mathbb{R}^{(n+m-1)n \times (n+m-1)n}.$$

Then both $\hat{Q}$ and $\hat{P}$ are nonsingular. Since

$$\hat{Q}D_1\hat{P} = \begin{bmatrix}
-A_1 & 0 & & & & & & B_1 & & & \\
0 & -I & & & & & & B_2 & & & \\
I & 0 & -A_1 & 0 & & & & & B_1 & & \\
0 & N & 0 & -I & & & & & B_2 & & \\
 & & I & 0 & & & & & & \ddots & \\
 & & 0 & N & & & & & & & \\
 & & & & \ddots & -A_1 & 0 & & & B_1 & \\
 & & & & & 0 & -I & & & B_2 & \\
 & & & & & I & 0 & & & & B_1 \\
 & & & & & 0 & N & & & & B_2
\end{bmatrix}$$

taking each matrix block as an element, the first row plus the third row multiplied by A_1, and then the fourth row plus the second row multiplied by N, we have

$$\operatorname{rank}D_1 = \operatorname{rank}\hat{Q}D_1\hat{P}$$

$$= \text{rank}\begin{bmatrix} 0 & 0 & -A_1^2 & 0 & & & & B_1 & A_1B_1 & & \\ 0 & -I & I & 0 & & & & B_2 & 0 & & \\ I & 0 & -A_1 & 0 & & & & 0 & B_1 & & \\ 0 & 0 & 0 & -I & & & & NB_2 & B_2 & & \\ & & & & \ddots & & & & & \ddots & \\ & & & & & I & 0 & & & & B_1 \\ & & & & & 0 & N & & & & B_2 \end{bmatrix}.$$

Thus, D_1 is of full row rank if and only if

$$\begin{bmatrix} -A_1^2 & 0 & & & & B_1 & A_1B_1 & & \\ 0 & I & & & & NB_2 & B_2 & & \\ I & 0 & -A_1 & 0 & & & & B_1 & \\ 0 & N & 0 & I & & & & B_2 & \\ & & & & \ddots & & & & \ddots \end{bmatrix}$$

is of full row rank. Continuing the process, we know that the full row rank of matrix D_1 is equivalent to the full row rank of

$$\bar{D}_1 = \begin{bmatrix} B_1 & A_1B_1 & \cdots & A_1^{n-2}B_1 & A_1^{n-1}B_1 \\ N^{n-1}B_2 & N^{n-2}B_2 & \cdots & NB_2 & B_2 \end{bmatrix}.$$

Noticing the fact that $N^i = 0$, $i \geq h$ ($h \leq n_2$) and Hamilton theorem, which assures the existence of $\beta_0, \beta_1, \ldots, \beta_{n_1-1}$, such that

$$A_1^i = \beta_0 I + \beta_1 A_1 + \ldots + \beta_{n_1-1}A_1^{n_1-1}, \qquad i \geq n_1,$$

it must be that

$$\begin{aligned} \text{rank}\bar{D}_1 &= \text{rank}[B_1, A_1B_1, \ldots, A_1^{n_1-1}B_1] + \text{rank}[B_2, NB_2, \ldots, N^{h-1}B_2] \\ &\leq n_1 + n_2 = n. \end{aligned}$$

Therefore, $\bar{D}_1$ (thus D_1) is of full row rank if and only if both (A_1, B_1) and (N, B_2) are controllable, in other words, system (2-1.1) is controllable. This proves that (a) and (e) are equivalent.

Thus we complete our proof. Q.E.D.

Example 2-2.1. Consider system (2-1.8). Direct computation shows that

$$\text{rank}[B_1, A_1B_1] = \text{rank}\begin{bmatrix} 1 & -1 \\ 0 & 1 \end{bmatrix} = 2$$

$$\text{rank}[B_2, NB_2] = \text{rank}\begin{bmatrix} -1 & 0 \\ 0 & 0 \end{bmatrix} = 1 < 2.$$

From Theorem 2-2.1 we know system (2-1.8) is not controllable, while its slow subsystem is controllable. Noticing that (2-1.8) is the standard decomposition of the system in Example 1-3.2, we know that the circuit network in Example 1-3.2 is not controllable. In fact, in the descriptor equation (1-3.4), $V_R(t) = RI(t)$ is an identical equation. Thus $V_R(t) - RI(t) = 0$ ($V_R(t) - I(t) = 0$ in Example 1-3.2) is not controllable by control input.

Example 2-2.2. Consider the circuit systen shown in Figure 2-2.1.

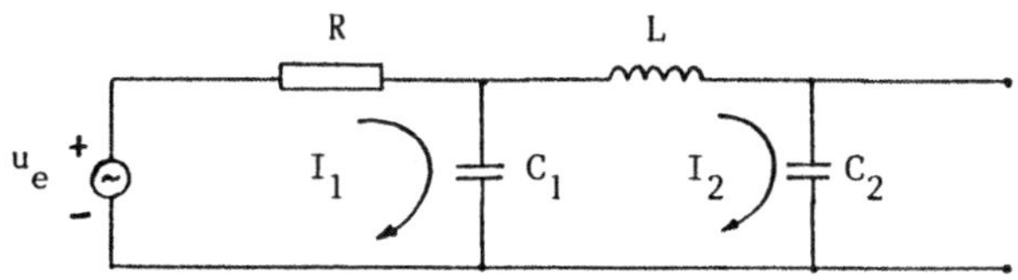

Figure 2-2.1.

in which the voltage sourse u_e is the control input. Choose the state variable

$$x = [u_{c_1} \quad u_{c2} \quad I_2 \quad I_1]$$

where u_{c1}, u_{c2}, I_1, I_2 are the voltages of C_1, C_2 and the amperage of the currents flowing over them. According to Kirchoff's second law (Smith, 1966) we may establish the following state equation

$$\begin{bmatrix} C_1 & 0 & 0 & 0 \\ 0 & C_2 & 0 & 0 \\ 0 & 0 & -L & 0 \\ 0 & 0 & 0 & 0 \end{bmatrix} \dot{x} = \begin{bmatrix} 0 & 0 & 0 & 1 \\ 0 & 0 & 1 & 0 \\ -1 & 1 & 0 & 0 \\ 1 & 0 & 0 & R \end{bmatrix} x + \begin{bmatrix} 0 \\ 0 \\ 0 \\ -1 \end{bmatrix} u \tag{2-2.5}$$

with the measure output:

$$y = u_{c2} = [0 \quad 1 \quad 0 \quad 0]x. \tag{2-2.6}$$

In this description equation, we have

$$\mathrm{rank}[sE-A,\ B] = \mathrm{rank}\begin{bmatrix} sC_1 & 0 & 0 & -1 & 0 \\ 0 & sC_2 & -1 & 0 & 0 \\ 1 & -1 & -sL & 0 & 0 \\ -1 & 0 & 0 & -R & -1 \end{bmatrix} = 4,\ \forall\ s \in \mathbb{C},\ s \text{ finite}$$

and

$$\mathrm{rank}[E,\ B] = \mathrm{rank}\begin{bmatrix} C_1 & 0 & 0 & 0 & 0 \\ 0 & C_2 & 0 & 0 & 0 \\ 0 & 0 & -L & 0 & 0 \\ 0 & 0 & 0 & 0 & -1 \end{bmatrix} = 4.$$

Therefore, by Theorem 2-2.1, the circuit network (2-2.5)-(2-2.6) is controllable.

Genrally speaking, in the case of high-dimension n for system (2-1.2), criteria in (1) and (2) in Theorem 2-2.1 are not suitable for computation since the system decomposition or its eigenvalues are needed, and are not easy to obtain. In such cases, the matrix criteria in (3) are of special importance in the verification of controllability.

Definition 2-2.2. Singular system (2-1.2) is called R-controllable, if it is controllable in the reachable set, or more precisely, for any prescribed $t_1 > 0$, $x_1(0) \in R$ and $w \in R$, there always exists an admissible control input $u(t) \in \mathbf{C}_p^{h-1}$ such that $x(t_1) = w$.

The R-controllability guarantees our controllability for the system from any admissible initial condition $x_1(0)$ to any reachable state and this process will be finished in any given time period if the control u(t) is suitably chosen.

Theorem 2-2.2. The following statements are equivalent.

(i). Singular system (2-1.2) is R-controllable.

(ii). The slow subsystem (2-2.1a) is controllable.

(iii). $\text{rank}[sE-A, B] = n$, $\forall s \in \mathbf{C}$, and s is finite.

(iv). $\text{rank}[B_1, A_1B_1, \ldots, A_1^{n_1-1}B_1] = n_1$.

(v). Denote

$$D_2 = \begin{bmatrix} -A & & & & & B & & & & \\ E & -A & & & & & B & & & \\ & E & \ddots & & & & & B & & \\ & & \ddots & & & & & & \ddots & \\ & & & -A & & & & & & B \\ & & & E & -A & & & & & & B \end{bmatrix}_{kn \times (n+m)k}$$

where $k \geq n_1$ may be any positive number; for example, k = rankE or n. The matrix D_2 is of full row rank kn.

Proof. We first prove the equivalence between (i) and (ii).

By definition, for $x(0) = 0 \in R$, the R-controllability states that $R(0) = \langle A_1 \mid B_1 \rangle \oplus \langle N \mid B_2 \rangle = \mathbb{R}^{n_1} \oplus \langle N \mid B_2 \rangle$. Thus, $\langle A_1 \mid B_1 \rangle = \mathbb{R}^{n_1}$. Slow subsystem (2-2.1a) is controllable and vice versa. Therefore, (i) and (ii) are equivalent.

Equivalence among (ii), (iii), and (iv) are direct results of Theorem 2-2.1.

Criteria (iv) and (v) may be proven in the same way as in the proof of (3) (e) in Theorem 2-2.1. Q.E.D.

This theorem, combined with Theorem 2-2.1, shows that system (2-1.2) is R-controllable if it is controllable, but its inverse is not true.

Example 2-2.3. According to Theorem 2-2.2, systems (2-1.8) and (2-2.5)-(2-2.6) are R-controllable. But system (2-1.8) is not controllable.

Example 2-2.4. From the definition of R-controllability, the following singular system with only fast subsystem:

$$\begin{aligned} N\dot{x} &= x + Bu \\ y &= Cx \end{aligned} \tag{2-2.7}$$

where N is nilpotent, is always R-controllable.

Definitions 2-2.1 and 2-2.2 become the results in linear system theory for normal systems.

Clearly, the concepts of controllability and R-controllability are concerned with only the terminal behavior for the system, which is an element in vector space. But, on the other hand, we have previously noted that in the state response, there also exist impulse terms that are set out either by the initial condition or by the possible jump behavior in control input $u \in \mathbf{C}_p^{h-1}$ and its derivatives. Therefore, it is necessary to analyze the control effect on impulse terms in the state response.

Recalling the state structure in Section 1-4, there are no impulse terms in substate $x_1(t)$ when $u \in \mathbf{C}_p^{h-1}$ and the impulse part in $x_2(t)$ is determined by

$$x_{2\tau}(t) = -\sum_{i=1}^{h-1} \delta^{(i-1)}(t)N^i x_2(0) - \sum_{i=o}^{h-1} N^i B_2 \Delta_\tau(u^{(i)}(t)) \tag{2-2.8}$$

and when $u(t) \equiv 0$, $\Delta_\tau(u^{(i)}(t)) = 0$, we have

$$x_{2\tau} = -\sum_{i=1}^{h-1} \delta^{(i-1)}(t)N^i x_2(0). \tag{2-2.9}$$

Clearly, $x_{2\tau} = 0$ when $x_2(0) = 0$. For any element w in vector space, denote

$$I_\tau(w,t) = \begin{bmatrix} 0 \\ I_{2\tau}(w,t) \end{bmatrix} \qquad I_{2\tau}(w,t) = \sum_{i=1}^{h-1} \delta^{(i-1)}(t-\tau)N^i w. \tag{2-2.10}$$

Then $I_{\tau=0}(x_2(0),t)$ represents the impulse behavior in $x(t)$ at the start point caused by only the initial condition $x_2(0)$. $I_\tau(w,t)$ includes all possible impulse terms in $x(t)$ at τ.

Since the impulse terms in state $x(t)$ only appear in the substate $x_2(t)$, $x_\tau(t) = [0/x_{2\tau}(t)]$.

Definition 2-2.3. System (2-1.2) is termed impulse controllable, if for any initial condition $x(0)$, $\tau \in \mathbb{R}$ and $w \in \mathbb{R}^{n_2}$, there exists an admissible control input $u \in \mathbf{C}_p^{h-1}$ such that

$$x_\tau(t) = I_\tau(w,t). \tag{2-2.11}$$

This definition is similar to the controllability definition. It characterizes the ability to generate impulse terms by admissible control. Under the impulse controllability, a suitable control in admissible set may be selected such that impulse terms in $x(t)$ may "arrives" at any "element" in $I_\tau(\mathbb{R}^{n_2},t)$ at any given time point τ. Here $I_\tau(\mathbb{R}^{n_2},t) = \{I_\tau(w,t) \mid w \in \mathbb{R}^{n_2}\}$. Thus x_τ may take any possible"value".

Impulse controllability is important for the necessity to eliminate the impulse portions in a system in which impulse terms are generally not expected to appear. Otherwise, strong impulse behavior may stop the system from working or even destroy it. For example, in the singular system $E\dot{x} = Ax + Bu + v$, where v is white noise, any possible impulse terms may appear in $x_\tau(t)$ (Cobb, 1984). Thus, it requires that we must eliminate these impulse terms by imposing appropriate control inputs. This point will be seen more clearly later. Now we will explore the criteria for impulse controllability.

Theorem 2-2.3. The following statements are equivalent.

(a) System (2-1.2) is impulse controllable.

(b) Its fast subsystem (2-2.1b) is impulse controllable.

(c) $\mathrm{Ker}N + \mathrm{Im}[B_2, NB_2, \ldots, N^{h-1}B_2] = \mathbb{R}^{n_2}$.

(d) $\mathrm{Im}N = \mathrm{Im}[NB_2, N^2B_2, \ldots, N^{h-1}B_2]$.

(e) $\mathrm{Im}N + \mathrm{Im}B_2 + \mathrm{Ker}N = \mathbb{R}^{n_2}$.

(f) Let the controllability decomposition of (N, B_2) be

$$\left(\begin{bmatrix} N_{11} & N_{12} \\ 0 & N_{22} \end{bmatrix}, \begin{bmatrix} B_{21} \\ 0 \end{bmatrix}\right).$$

Then either N_{22} disappears or a matrix M exists such that $[N_{12}/N_{22}] = [N_{11}/0]M$.

(g) $$\mathrm{rank}\begin{bmatrix} E & 0 & 0 \\ A & E & B \end{bmatrix} = n + \mathrm{rank}E.$$

Proof. The equivalence between (a) and (b) is obvious.

The equivalence among (b)-(e) are given in Cobb (1984).

Now we prove the equivalence between (d) and (f). Assume that $N_{11} \in \mathbb{R}^{\tilde{n}_1 x\tilde{n}_1}$, $N_{22} \in \mathbb{R}^{\tilde{n}_2 x\tilde{n}_2}$. Since (N_{11}, B_{21}) is controllable, $\mathrm{Im}[B_{21}, N_{11}B_{21}, \ldots, N_{11}^{h-1}B_{21}] = R^{\tilde{n}_1}$. In this case, (d) holds iff

$$\mathrm{Im}N = \mathrm{Im}\begin{bmatrix} N_{11} & N_{12} \\ 0 & N_{22} \end{bmatrix} = \mathrm{Im}[NB_2, N^2B_2, \ldots, N^{h-1}B_2] = \mathrm{Im}\begin{bmatrix} N_{11} \\ 0 \end{bmatrix}.$$

This equation is true iff either N_{22} disappears or there exists an M such that $[N_{12}/N_{22}] = [N_{11}/0]M$, which is the equivalence between (d) and (f).

Moreover, note that

$$\mathrm{rank}\begin{bmatrix} E & 0 & 0 \\ A & E & B \end{bmatrix} = \mathrm{rank}\begin{bmatrix} QEP & 0 & 0 \\ QAP & QEP & QB \end{bmatrix} = 2n_1 + \mathrm{rank}\begin{bmatrix} N & 0 & 0 \\ I & N & B_2 \end{bmatrix}$$

$$= 2n_1 + \operatorname{rank}\begin{bmatrix} N_{11} & N_{12} & 0 & 0 & 0 \\ 0 & N_{22} & 0 & 0 & 0 \\ I & 0 & N_{11} & N_{12} & B_{21} \\ 0 & I & 0 & N_{22} & 0 \end{bmatrix}$$

the controllability of (N_{11}, B_{21}) and the nilpotent property of N_{11}, N_{22}, we know that $\operatorname{rank}[N_{11}, B_{21}] = \tilde{n}_1$ according to Theorem 2-2.1 (2). This results in:

$$\operatorname{rank}\begin{bmatrix} E & 0 & 0 \\ A & E & B \end{bmatrix} = n + n_1 + \operatorname{rank}\begin{bmatrix} N_{11} & -N_{12}N_{22} \\ 0 & -N_{22}^2 \end{bmatrix} .$$

Hence

$$\operatorname{rank}\begin{bmatrix} E & 0 & 0 \\ A & E & B \end{bmatrix} = n + \operatorname{rank}E = n + n_1 + \operatorname{rank}\begin{bmatrix} N_{11} & N_{12} \\ 0 & N_{22} \end{bmatrix}$$

holds if and only if

$$\operatorname{rank}\begin{bmatrix} N_{11} & N_{12} \\ 0 & N_{22} \end{bmatrix} = \operatorname{rank}\begin{bmatrix} N_{11} & N_{12}N_{22} \\ 0 & N_{22}^2 \end{bmatrix} .$$

Since N_{22} is nilpotent, this equation holds if and only if N_{22} disappears or $N_{22} = 0$. In this case, there exists an M, $N_{12} = N_{11}M$. This is the equivalence between (f) and (g). Q.E.D.

Clearly from Theorems 2-2.3 and 2-2.2, a system is impulse controllable if it is controllable. Its inverse is false.

The relationships among these three controllability concepts may be described by the following diagram:

controllability ⟶ R-controllability

controllability ⟶ impulse controllability

Here A ⟶ B represents that B may be deduced from A.

Example 2-2.5. The singular system with only slow subsystem (or normal system):

$$\dot{x} = Ax + Bu$$
$$y = Cx$$

is always impulse controllable according to the definition.

Example 2-2.6. For the circuit network (2-2.5)-(2-2.6), we have

$$\operatorname{rank}\begin{bmatrix} E & 0 & 0 \\ A & E & B \end{bmatrix}$$

$$= \text{rank}\begin{bmatrix} C_1 & 0 & 0 & 0 & & & & & \\ 0 & C_2 & 0 & 0 & & & 0 & & \\ 0 & 0 & -L & 0 & & & & & \\ 0 & 0 & 0 & 0 & & & & & \\ 0 & 0 & 0 & 0 & C_1 & 0 & 0 & 0 & 0 \\ 0 & 0 & 1 & 0 & 0 & C_2 & 0 & 0 & 0 \\ -1 & 1 & 0 & 0 & 0 & 0 & -L & 0 & 0 \\ 1 & 0 & 0 & R & 0 & 0 & 0 & 0 & -1 \end{bmatrix} = 7 = n + \text{rank}E.$$

Thus, it is impulse controllable. Moreover, its controllability has been proven in Example 2-2.2.

Example 2-2.7. Example 2-2.1 has proven that system (2-1.8) is not controllable. But since

$$\text{rank}\begin{bmatrix} E & 0 & 0 \\ A & E & B \end{bmatrix} = 6 = n + \text{rank}E,$$

it is impulse controllable.

2-3. Observability, R-Observability and Impulse Observability

Having discussed various controllabilities, we now introduce the dual concepts --- observability, R-observability and impulse observability, which characterize the ability of state reconstruction from measure outputs.

Definition 2-3.1. System (2-1.2) is observable if the initial condition x(0) may be uniquely determined by u(t), y(t), $0 \leqslant t < \infty$.

The observability states that the state of observable system may be determined by observing the initial condition x(0), followed by constructing the state response at any time t.

Clearly, Definition 2-3.1 reduces to observability in linear system theory when it is used for normal systems.

Let

$$R_s = \text{Ker}[C_1/C_1A_1/ \ \dots \ /C_1A_1^{n_1-1}] \text{ and } R_f = \text{Ker}[C_2/C_2N/ \ \dots \ /C_2N^{h-1}].$$

Then the following is true.

Theorem 2-3.1. Consider system (2-1.2).

(1). Let $u(t) \equiv 0$. Then $y(t) \equiv 0$, $t \geqslant 0$ iff $x(0) \in R_s \oplus R_f$.

(2). Its slow subsystem (2-2.1a) is observable iff

$$\text{rank}[sE-A/C] = n, \quad \forall\ s \in \mathbb{C},\ s \text{ finite}.$$

(3). The following statements are equivalent.

(a) Its fast subsystem (2-2.1b) is observable.

(b) $\text{rank}[C_2/C_2N/\ \dots\ /C_2N^{h-1}] = n_2$.

(c) $\text{Ker}[N/C_2] = \{0\}$.

(d) $\text{rank}[N/C_2] = n_2$.

(e) $\text{rank}[E/C] = n$.

(f) For any two nonsingular matrices Q_1, P_1 satisfying $Q_1EP_1 = \text{diag}(I_q, 0)$, $CP_1 = [\bar{C}_1, \bar{C}_2]$, $\bar{C}_2$ is of full column rank, $\text{rank}\bar{C}_2 = n - \text{rank}E$.

(4). The following statements are equivalent.

(i) System (2-1.2) is observable.

(ii) Both its slow and fast subsystems are observable.

(iii) $\text{rank}[sE-A/C] = n$, $\forall\ s \in \mathbb{C}$, s finite, and $\text{rank}[E/C] = n$.

(iv) The matrix

$$\Theta_1 = \begin{bmatrix} -A & E & & & \\ & -A & E & & \\ & & \ddots & & \\ & & & -A & E \\ C & & & & \\ & C & & & \\ & & C & & \\ & & & \ddots & \\ & & & & C \end{bmatrix}_{(n+r-1)n \times n^2}$$

is of full column rank n^2.

<u>Proof.</u>

(1). When $u(t) \equiv 0 \in \mathbb{C}_p^{h-1}$, state response (1-4.10) gives

$$x_1(t) = e^{A_1t}x_1(0), \qquad y_1 = C_1x_1$$

$$x_2(t) = -\sum_{i=1}^{h-1} \delta^{(i-1)}(t)N^i x_2(0), \qquad y_2 = C_2x_2 \tag{2-3.1}$$

$$y = y_1 + y_2 .$$

Note the special form of y_1 and y_2; from Theorem A-1 in Appendix A we know $y(t) \equiv 0$ if and only if $y_1(t) \equiv 0$ and $y_2(t) \equiv 0$.

If $y_1(t) = C_1x_1 = C_1e^{A_1t}x_1(0) \equiv 0$, taking differentials on both sides at $t = 0$, we have

$$[C_1/C_1A_1/\ \dots\ /C_1A_1^{n_1-1}]x_1(0) = 0,$$

i.e., $x_1(0) \in R_s$. On the other hand,

$$y_2(t) = C_2x_2(t) = -\sum_{i=1}^{h-1} \delta^{(i-1)}(t)C_2N^ix_2(0), \qquad y_2(0) = C_2x_2(0).$$

Combining it with Theorem A-1 in Appendix A yields $y_2(t) \equiv 0$ iff

$$C_2N^ix_2(0) = 0, \quad i = 0, 1, 2, \dots, h-1,$$

i.e.,

$$[C_2/C_2N/ \dots /C_2N^{h-1}]x_2(0) = 0.$$

Hence, $x_2(0) \in R_f$, $x(0) \in R_s \oplus R_f$. This is the conclusion.

(2). Slow subsystem (2-2.1a) is normal. Definition 2-3.1 becomes the classic observability definition. Thus, slow subsystem (2-2.1a) is observable iff (A_1, C_1) is observable, i.e.,

$$\operatorname{rank}[sI-A_1/C_1] = n_1, \quad \forall\ s \in \mathbb{C},\ s \text{ finite}.$$

On the other hand,

$$\operatorname{rank}\begin{bmatrix} sE-A \\ \\ C \end{bmatrix} = \operatorname{rank}\begin{bmatrix} sQEP-QAP \\ \\ CP \end{bmatrix} = \operatorname{rank}\begin{bmatrix} sI-A_1 & 0 \\ 0 & sN-I \\ C_1 & C_2 \end{bmatrix}.$$

Noticing the fact that $sN-I$ is invertible for any finite $s \in \mathbb{C}$, we have

$$\operatorname{rank}[sE-A/C] = n_2 + \operatorname{rank}[sI-A_1/C_1], \quad \forall\ s \in \mathbb{C},\ s \text{ finite},$$

which indicates that $\operatorname{rank}[sE-A/C] = n = n_1 + n_2$ holds if and only if $\operatorname{rank}[sI-A_1/C_1] = n_1$, $\forall\ s \in \mathbb{C}$, s finite, which in turn gives the conclusion.

(3). By definition, observability for a fast subsystem means that $x_2(0) = 0$ if $y_2(t) \equiv 0$, $t \geqslant 0$ when $u(t) \equiv 0$. As indicated by (1), this is equivalent to $R_f = \{0\}$. Combining these results, we know that (a) and (b) are equivalent.

Furthermore, (b) is the same as

$$\operatorname{rank}[C_2^\tau, N^\tau C_2^\tau, \dots, (N^\tau)^{h-1}C_2^\tau] = n_2,$$

i.e., (N^τ, C_2^τ) is controllable. By Theorem 2-2.1 (2),(b)-(d), (a), and (f) are all equivalent to each other.

(4). Equivalence among (i)-(iii) is a direct result of (1), and equivalence between (ii) and (iv) may be proven by the same method in (3) noticing the result in Theorem 2-2.1 part (3). Q.E.D.

Example 2-3.1. Consider the standard decomposition (2-1.8) for the circuit network in Example 1-3.2, which is not controllable by Example 2-2.1.

Since $n_1 = n_2 = 2$ and $\text{rank}[C_1/C_1A_1] = 2$ and $\text{rank}[C_2/C_2N] = 0 < 2$, its slow subsystem is observable but its fast subsystem is not. Therefore, system (2-1.8) is not observable.

Example 2-3.2. For system (2-2.5)-(2-2.6), direct computation gives that

$$\text{rank}\begin{bmatrix} sE-A \\ C \end{bmatrix} = \text{rank}\begin{bmatrix} sC_1 & 0 & 0 & -1 \\ 0 & sC_2 & -1 & 0 \\ 1 & -1 & -sL & 0 \\ -1 & 0 & 0 & -R \\ 0 & 1 & 0 & 0 \end{bmatrix} = 4$$

and

$$\text{rank}[E/C] = \text{rank}\begin{bmatrix} C_1 & 0 & 0 & 0 \\ 0 & C_2 & 0 & 0 \\ 0 & 0 & -L & 0 \\ 0 & 0 & 0 & 0 \\ 0 & 1 & 0 & 0 \end{bmatrix} = 3 < 4,$$

which indicates that its slow subsystem is observable and its fast one is not. Hence, system (2-2.5)-(2-2.6) is not observable.

Let the measure output be

$$y = Cx, \qquad C = [0 \;\; 0 \;\; 1 \;\; 1] \tag{2-3.2}$$

instead of as in (2-2.6), i.e., $y = U_{c2} + I_1$, for the system composed by (2-2.5) and (2-3.2) we have

$$\text{rank}\begin{bmatrix} sE - A \\ C \end{bmatrix} = \text{rank}\begin{bmatrix} sC_1 & 0 & 0 & -1 \\ 0 & sC_2 & -1 & 0 \\ 1 & -1 & -sL & 0 \\ -1 & 0 & 0 & -R \\ 0 & 0 & 1 & 1 \end{bmatrix} = 2 + \text{rank}\begin{bmatrix} sC_2 & sC_1 \\ -1 & 1+s^2LC_1 \\ 0 & -1-sRC_1 \end{bmatrix}.$$

In a circuit network, L, C_1, C_2 are positive scalars, thus

$$\text{rank}\begin{bmatrix} sC_2 & sC_1 \\ -1 & 1+s^2LC_1 \\ 0 & -1-sRC_1 \end{bmatrix} = 2, \quad \forall\, s \in \mathbb{C}.$$

and consequently, $\text{rank}[sE-A/C] = 4$. Furthermore, $\text{rank}[E/C] = 4$. Therefore, this system is observable by Theorem 2-3.1 (3).

Definition 2-3.2. System (2-1.2) is R-observable if it is observable in reachable set, or, in other words, any state in the reachable set may be uniquely determined by $y(t)$ and $u(\tau)$, $0 \leqslant \tau \leqslant t$.

While observability reflects the reconstruction ability of the whole state (impulse terms are included) from measure output, together with control input, R-observability characterizes the ability to reconstruct only the reachable state (impulse portions are not included, thus it is partial state) from same information source. Obviously, from the definition, system (2-1.2) is R-observable if it is observable.

Theorem 2-3.2. System (2-1.2) is R-observable iff its slow subsystem is observable, i.e.,

$$\mathrm{rank}[sE-A/C] = n, \quad \forall\ s \in \mathbb{C},\ s \text{ finite}. \tag{2-3.3}$$

Proof. According to the state representation, any reachable state has the form

$$\begin{aligned} x(t) &= [x_1(t)/x_2(t)] \\ \dot{x}_1(t) &= A_1 x_1(t) + B_1 u(t) \\ x_2(t) &= -\sum_{i=0}^{h-1} N^i B_2 u^{(i)}(t), \qquad t > 0 \\ y &= y_1 + y_2 = C_1 x_1 + C_2 x_2 \ . \end{aligned} \tag{2-3.4}$$

Note that in (2-3.4), $x_2(t)$ is uniquely determined by $u(t)$, and $y_1(t) \triangleq C_1 x_1(t) = y - C_2 x_2(t)$ is uniquely determined by $y(t)$ and $u(t)$. Thus, reconstructing the reachable state $x(t)$ from $y(t)$ and $u(t)$ is equivalent to that $x_1(t)$ may be uniquely determined by $y_1(t)$ and $u(t)$, which is, in turn, equivalent to the observability of (A_1, C_1). The conclusion is the same by Theorem 2-3.1. Q.E.D.

Theorem 2-3.3. For system (2-1.2), the following statements are equivalent.

(a). System (2-1.2) is R-observable.

(b). Its slow subsystem is observable.

(c). The matrix

$$\theta_2 = \begin{bmatrix} -A & E & & & \\ & -A & E & & \\ & & \ddots & & \\ & & & -A & E \\ & & & & -A \\ C & & & & \\ & C & & & \\ & & C & \ddots & \\ & & & & C \end{bmatrix}_{(n+r)k \times nk}$$

is of full column rank nk. Here $k \geqslant n_1$.

Proof. Equivalence between (a) and (b) is obvious by Theorem 2-3.2.

Theorem 2-3.2 states that system (2-1.2) is R-observable if and only if (2-3.3) holds, which is equivalent to

$$\mathrm{rank}[sE^{\tau} - A^{\tau},\ C^{\tau}] = n, \quad \forall\ s \in \mathbb{C},\ s \text{ finite}.$$

By Theorem 2-2.2, it in turn is equivalent to the full row rank of

$$D_3 = \begin{bmatrix} -A^\tau & & & & C^\tau & & & & \\ E^\tau & -A^\tau & & & & C^\tau & & & \\ & E^\tau & \ddots & & & & C^\tau & & \\ & & \ddots & & & & & \ddots & \\ & & & -A^\tau & & & & & C^\tau \\ & & & E^\tau & -A^\tau & & & & & C^\tau \end{bmatrix}_{nk \times (n+r)k}$$

for any $k \geqslant n_1$. Thus Θ_2 is of full column rank by $\Theta_2 = D_3^\tau$. Q.E.D.

Example 2-3.3. We noticed in Examples 2-3.1 and 2-3.2 that systems (2-1.8) and (2-2.5)-(2-2.6) are R-observable.

Example 2-3.4. By definition, the fast system

$$\begin{aligned} N\dot{x} &= x + Bu \\ y &= Cx \end{aligned} \tag{2-3.5}$$

where N is nilpotent, is always R-observable.

Definition 2-3.3. If $x_\tau(t)$ may be uniquely determined by $y_\tau(t)$ and $\Delta_\tau u(t)$ for any $\tau \geqslant 0$, system (2-1.2) will be called impulse observable.

The definition is pointwise since x_τ and y_τ have nonzero values at τ only. Note that jump behavior in input contributes to the impulse terms in x(t). Impulse observability guarantees the ability to uniquely determine the impulse behavior in x(t) from information of the impulse behavior in output and jump behavior in input. The possibility is positive under impulse observability and negative otherwise. Compared with observability and R-observability, the latter two concepts are on the finite-value terms in state response, while impulse observability focuses on the impulse terms that take infinite value. Thus these concepts have different meanings.

Lemma 2-3.1. For system (2-1.2), $y_\tau(t) \equiv 0$ when $u(t) \equiv 0$ if and only if $x_2(0) \in NR_f = \mathrm{Ker}[C_2N/C_2N^2/ \ldots /C_2N^{h-1}]$.

Proof. Since $y_\tau = Cx_\tau = C_2x_{2\tau}$ from (2-2.9) we have

$$y_\tau(t) = -\sum_{i=1}^{h-1} \delta^{(i-1)}(t)C_2N^i x_2(0).$$

Thus, by using Theorem A-1 in Appendix A, $y_\tau(t) \equiv 0$ if and only if

$$C_2N^i x_2(0) = 0, \qquad i = 1, 2, \ldots, h-1,$$

i.e.,

$$x_2(0) \in \mathrm{Ker}[C_2N/C_2N^2/ \ldots /C_2N^{h-1}]. \qquad \text{Q.E.D.}$$

Theorem 2-3.4. Consider system (2-1.2). The following statements are equivalent.

(i). System (2-1.2) is impulse observable.

(ii). Its fast subsystem (2-2.1b) is impulse observable.

(iii). $R_f \cap \mathrm{Im}N = \{0\}$.

(iv). $NR_f = \mathrm{Ker}N$.

(v). $\mathrm{Ker}N \cap \mathrm{Ker}C_2 \cap \mathrm{Im}N = \{0\}$.

(vi). Let

$$\left(\begin{bmatrix} N_{11} & 0 \\ N_{21} & N_{22} \end{bmatrix}, \; [C_{21} \quad 0] \right)$$

be the observability decomposition of (N, C_2). Here (N_{11}, C_{21}) is observable. Then it must be that either N_{22} doesn't exist, in this case (N,C_2) is observable; or $N_{22} = 0$ and $\mathrm{rank}N_{11} = \mathrm{rank}[N_{11}/N_{21}]$ holds.

(vii).
$$\mathrm{rank} \begin{bmatrix} E & A \\ 0 & E \\ 0 & C \end{bmatrix} = n + \mathrm{rank}E. \tag{2-3.6}$$

Proof. Let (2-2.1) be the standard decomposition for system (2-1.2). Assume that when $u(t) \in \mathbf{C}_p^{h-1}$, $x_{1\tau} = 0$. From

$$x_\tau(t) = [x_{1\tau}(t)/x_{2\tau}(t)] = [0/x_{2\tau}(t)]$$

$$y_\tau(t) = C_1x_{1\tau}(t) + C_2x_{2\tau}(t) = C_2x_{2\tau}(t) = y_{2\tau},$$

we know that (i) and (ii) are equivalent.

Equation (2-2.9) shows that, when $u(t) \equiv 0$,

$$x_{2\tau}\Big|_{\tau=0} = -\sum_{i=1}^{h-1} \delta^{(i-1)}(t)N^i x_2(0).$$

Thus $x_{2\tau}\Big|_{\tau=0} = 0$ if and only if $Nx_2(0) = 0$. Combining it with Lemma 2-3.1, it is easy to know (ii) is true iff $\mathrm{Ker}N = NR_f$, which estabilishes the equivalence between (ii) and (iv).

Now we will prove the equivalence between (iii) and (iv).

Assume that (iv) is true. Then for any $\alpha \in R_f \cap \mathrm{Im}N$, there exists a β such that $\alpha = N\beta$. Moreover, $\alpha \in R_f$ indicates $\beta \in NR_f = \mathrm{Ker}N$, i.e., $N\beta = 0$. Thus $\alpha = N\beta = 0$, which is (iii).

Conversely, if (iii) holds, for any $\alpha \in NR_f$, $N\alpha \in R_f \cap \mathrm{Im}N$, showing $N\alpha = 0$, $\alpha \in \mathrm{Ker}N$, or, in other words, $NR_f \subseteq \mathrm{Ker}N$. The combination of it with the obvious result $\mathrm{Ker}N \subseteq NR_f$ readily results in $NR_f = \mathrm{Ker}N$. Thus, (iii) and (iv) are equivalent.

Equivalence between (iii) and (v) may be proven in a similar way.

Further, (iv) shows

$$\mathrm{Ker}N = \mathrm{Ker}[C_2N/C_2N^2/ \; \ldots \; /C_2N^{h-1}]. \tag{2-3.7}$$

According to the laws in matrix theory, $\mathrm{Ker}N + \mathrm{Im}N^\tau = \mathbb{R}^{n_2}$. We have from (2-3.7) that

$$\mathrm{Im}N^{\tau} = N^{\tau}\mathrm{Im}[C_2^{\tau},\ N^{\tau}C_2^{\tau},\ \dots\ ,\ (N^{\tau})^{h-1}C_2^{\tau}]. \tag{2-3.8}$$

Starting from this point and Theorem 2-2.3, equivalence among (iv), (v), and (vi) may be proven. Q.E.D.

For a given system, its observability, R-observability, and impulse observability characterize its state reconstruction ability from different aspects. Apparantly, a system is impulse observable if it is observable, but its inverse is not true. Relationships among observability, R-observability, and impulse observability may be shown by the following diagram.

observability → R-observability

observability → impulse observability

Example 2-3.5. By definition, the slow subsystem (or normal system)

$$\dot{x}_1 = A_1x_1 + B_1u$$
$$y_1 = C_1x_1$$

is always impulse observable.

Example 2-3.6. For system (2-2.5)-(2-2.6), we have

$$E = \begin{bmatrix} C_1 & 0 & 0 & 0 \\ 0 & C_2 & 0 & 0 \\ 0 & 0 & -L & 0 \\ 0 & 0 & 0 & 0 \end{bmatrix},\quad A = \begin{bmatrix} 0 & 0 & 0 & 1 \\ 0 & 0 & 1 & 0 \\ -1 & 1 & 0 & 0 \\ 1 & 0 & 0 & R \end{bmatrix},\quad B = \begin{bmatrix} 0 \\ 0 \\ 0 \\ -1 \end{bmatrix},\quad C = [0\ \ 1\ \ 0\ \ 0].$$

It has been verified in Example 2-3.2 that it is not observable. But on the other hand,

$$\mathrm{rank}\begin{bmatrix} E & A \\ 0 & E \\ 0 & C \end{bmatrix} = \mathrm{rank}\left[\begin{array}{cccc|cccc} C_1 & 0 & 0 & 0 & 0 & 0 & 0 & 1 \\ 0 & C_2 & 0 & 0 & 0 & 0 & 1 & 0 \\ 0 & 0 & -L & 0 & -1 & 1 & 0 & 0 \\ 0 & 0 & 0 & 0 & 1 & 0 & 0 & R \\ \hline & & & & C_1 & 0 & 0 & 0 \\ & & & & 0 & C_2 & 0 & 0 \\ & & 0 & & 0 & 0 & -L & 0 \\ & & & & 0 & 0 & 0 & 0 \\ \cline{5-8} & & & & 0 & 1 & 0 & 0 \end{array}\right] = 7 = n + \mathrm{rank}E.$$

Therefore, it is impulse observable.

2-4. Dual Principle

A careful examination of aspects of controllabilities and observabilities would reveal the important fact that they are very similar. For example, the necessary and sufficient conditions of controllability for system (2-1.2) are

$$\begin{aligned} &\mathrm{rank}[B_1,\ A_1B_1,\ \ldots\ ,\ A_1^{n_1-1}B_1] = n_1 \\ &\mathrm{rank}[B_2,\ NB_2,\ \ldots\ ,\ N^{h-1}B_2] = n_2. \end{aligned} \tag{2-4.1}$$

Transposing these matrices, we know (2-4.1) is equivalent to

$$\mathrm{rank}[B_1^\tau /\ B_1^\tau A_1^\tau /\ \ldots\ /\ B_1^\tau (A_1^\tau)^{n_1-1}] = n_1$$

$$\mathrm{rank}[B_2^\tau / B_2^\tau N^\tau /\ \ldots\ / B_2^\tau (N^\tau)^{h-1}] = n_2,$$

which are the necessary and sufficient conditions of observability for system ($E^\tau, A^\tau, C^\tau, B^\tau$).

Summing up these arguments, system (E,A,B,C) is controllable if and only if ($E^\tau, A^\tau, C^\tau, B^\tau$) is observable; conversely, system (E,A,B,C) is observable iff ($E^\tau, A^\tau, C^\tau, B^\tau$) is controllable. This is the so-called dual principle.

Consider the following singular system:

$$\begin{aligned} E\dot{x} &= Ax + Bu \\ y &= Cx\ . \end{aligned} \tag{2-4.2}$$

The system

$$\begin{aligned} E^\tau \dot{z} &= A^\tau z + C^\tau v \\ y &= B^\tau z \end{aligned} \tag{2-4.3}$$

is called its dual system.

By using earlier results on the controllabilities and observabilities, we see clearly that the following theorem holds.

Theorem 2-4.1 (Dual Principle). Consider system (2-4.2).

1. System (2-4.2) is controllable (observable) iff its dual system (2-4.3) is observable (controllable).

2. System (2-4.2) is R-controllable (R-observable) iff its dual system (2-4.3) is R-observable (R-controllable).

3. System (2-4.2) is impulse controllable (impulse observable) iff its dual system (2-4.3) is impulse observable (impulse controllable).

The dual principle plays an important role in system theory. For this law, there is no need to study the controllabilities and observabilities separately. We may study the system itself or its dual system at our convenience.

Theorem 2-4.2. Let (1-3.21) be an EF2 of system (2-4.2) with transformation matrices Q_1 and P_1. Then the following hold.

A. System (2-4.2) (or (1-3.21)) is controllable if and only if the following hold.

(1) $\text{rank}[sE-A,\ B] = n,\ \forall\ s \in \mathbb{C}$, s finite.

(2) B_2 is of full row rank, $\text{rank}B_2 = n - \text{rank}E$.

B. System (2-4.2) is impulse controllable iff $\text{rank}[A_{22}, B_2] = n - \text{rank}E$.

Proof.

A. From Theorem 2-2.1, the necessary and sufficient conditions of controllability for system (2-4.2) are (1) and $\text{rank}[E\ \ B] = n$. But, on the other hand,

$$\text{rank}[E\ \ B] = \text{rank}[Q_1EP_1\ \ Q_1B]$$

$$= \text{rank}\begin{bmatrix} I_q & 0 & B_1 \\ 0 & 0 & B_2 \end{bmatrix} = \text{rank}E + \text{rank}B_2.$$

Thus, $\text{rank}[E\ \ B] = n$ is equivalent to $\text{rank}B_2 = n - \text{rank}E$, i.e., B_2 is of full row rank. A. is proven.

B. Moreover, by Theorem 2-2.3 (g), system (2-4.2) is impulse controllable iff

$$\text{rank}\begin{bmatrix} E & 0 & 0 \\ A & E & B \end{bmatrix} = n + \text{rank}E. \tag{2-4.4}$$

Since

$$\text{rank}\begin{bmatrix} E & 0 & 0 \\ A & E & B \end{bmatrix} = \text{rank}\begin{bmatrix} Q_1EP_1 & 0 & 0 \\ Q_1AP_1 & Q_1EP_1 & Q_1B \end{bmatrix}$$

$$= \text{rank}\begin{bmatrix} I_q & 0 & 0 & 0 & 0 \\ 0 & 0 & 0 & 0 & 0 \\ A_{11} & A_{12} & I_q & 0 & B_1 \\ A_{21} & A_{22} & 0 & 0 & B_2 \end{bmatrix} = 2\text{rank}E + \text{rank}[A_{22}\ \ B_2].$$

Equation (2-4.4) holds if and only if $\text{rank}[A_{22}, B_2] = n - \text{rank}E$, which is B. Q.E.D.

From the dual principle, we immediately have the following theorem.

Theorem 2-4.3. Let (1-3.21) be an EF2 for system (2-4.2). Then

A. System (2-4.2) is observable iff

(1) $\text{rank}[sE-A/C] = n,\ \forall\ s \in \mathbb{C}$, s finite.

(2) C_2 is of full column rank, $\text{rank}C_2 = n - \text{rank}E$.

B. System (2-4.2) is impulse observable iff $\text{rank}[A_{22}/C_2] = n - \text{rank}E$.

Example 2-4.1. Consider the system (1-3.19) with EF2 in Example 1-3.3, whose R-controllability has been proven in Example 2-2.1, i.e.,

$$\operatorname{rank}[sE-A,\ B] = n, \quad \forall\ s \in \mathbb{C},\ s \text{ finite}.$$

Since

$$\operatorname{rank}B_2 = \operatorname{rank}[0/-1] = 1 < 2 = n - \operatorname{rank}E$$

$$\operatorname{rank}[A_{22}\ \ B_2] = 2 = n - \operatorname{rank}E,$$

this system is impulse controllable, but it is not controllable. This is the conclusion of Examples 2-2.1 and 2-2.7.

For systems with EF3, we have the following theorem.

Theorem 2-4.4. Let (1-3.22) be an EF3 of system (2-4.2). Then

(i). System (2-4.2) is controllable if and only if $(\hat{E},\hat{B})$ is controllable.

(ii). The necessary and sufficient condition for R-controllability is

$$\operatorname{rank}[sI-\hat{E},\ \hat{B}] = n, \quad \forall\ s \neq 0,\ s \in \mathbb{C},\ s \text{ finite},$$

or in other words $(\hat{E},\hat{B})$ has only zero uncontrollable poles.

(iii). The system is impulse controllable iff $(\hat{E},\hat{B})$ is algebraically similar (QP =I in r.s.e. relation) to

$$\left(\begin{bmatrix} \hat{E}_1 & 0 \\ 0 & 0 \end{bmatrix},\ \begin{bmatrix} \hat{B}_1 \\ 0 \end{bmatrix} \right),$$

where $(\hat{E}_1,\hat{B}_1)$ has no zero uncontrollable poles, or

$$\operatorname{rank}\ \hat{E}[\hat{E}\ \ \hat{B}] = \operatorname{rank}\hat{E}.$$

Proof. We will prove (i), (ii), and (iii) one at a time.

(i). According to notations $\hat{E}$, $\hat{B}$, we know

$$\operatorname{rank}[sI-\hat{E},\ \hat{B}] = \operatorname{rank}[(\alpha E+A)sI - E,\ B]$$

$$= \begin{cases} \operatorname{rank}[E\ \ B] & \text{when } s = 0 \\ \operatorname{rank}[(\frac{1}{s} - \alpha)E - A,\ B] & \text{when } s \neq 0 \end{cases}$$

which shows that the necessary and sufficient conditions of controllability for $(\hat{E},\hat{B})$ are $\operatorname{rank}[E\ \ B] = n$ and $\operatorname{rank}[sE-A,\ B] = n$, $\forall\ s \in \mathbb{C}$, s finite, which, in turn, is the controllability criterion for (2-4.2).

(ii). From Theorem 2-2.2, system (2-4.2) is R-controllable iff

$$\operatorname{rank}[sE-A,\ B] = \operatorname{rank}[(s+\alpha)\hat{E} - I,\ \hat{B}] = n, \quad \forall\ s \in \mathbb{C},\ s \text{ finite}. \tag{2-4.5}$$

If $s+\alpha = 0$, equation (2-4.5) is obvious; otherwise, $s+\alpha \neq 0$. Let $\lambda = 1/(s+\alpha)$, from (2-4.5) we have

$$\text{rank}[(s+\alpha)\hat{E}-I,\ \hat{B}] = \text{rank}[\ \lambda I-\hat{E},\ \hat{B}] = n, \quad \forall\ \lambda \in \mathbb{C},\ \lambda \neq 0,\ \lambda \text{ finite},$$

which is (ii).

(iii). Note the impulse controllability condition

$$\text{rank}\begin{bmatrix} E & 0 & 0 \\ A & E & B \end{bmatrix} = \text{rank}\begin{bmatrix} \hat{E} & 0 & 0 \\ I-\alpha\hat{E} & \hat{E} & \hat{B} \end{bmatrix} = n + \text{rank}E. \qquad (2\text{-}4.6)$$

We know immediately that system (2-4.2) is impulse controllable if and only if rank $\hat{E}[\hat{E}\ \ \hat{B}] = \text{rank}\hat{E}$.

To prove the other version in (iii), we make controllability decomposition on $(\hat{E},\hat{B})$. Linear system theory assures the existence of nonsingular matrix T such that

$$T\hat{E}T^{-1} = \begin{bmatrix} \hat{E}_{11} & \hat{E}_{12} & \hat{E}_{13} \\ 0 & \hat{E}_{22} & 0 \\ 0 & 0 & \hat{E}_{33} \end{bmatrix}, \qquad T\hat{B} = \begin{bmatrix} \hat{B}_{11} \\ 0 \\ 0 \end{bmatrix} \qquad (2\text{-}4.7)$$

where $(\hat{E}_{11},\ \hat{B}_{11})$ is controllable. $\hat{E}_{22}$ is invertible and $\hat{E}_{33}$ is nilpotent.

Long computations show that rank $\hat{E}[\hat{E}\ \ \hat{B}] = \text{rank}\hat{E}$ is equivalent to

$$\text{rank}\ \hat{E}[\hat{E}\ \ \hat{B}] = \text{rank}\ T\hat{E}T^{-1}[T\hat{E}T^{-1}\ \ T\hat{B}]$$

$$= \text{rank}\begin{bmatrix} \hat{E}_{11} & 0 & \hat{E}_{13}\hat{E}_{33} \\ 0 & \hat{E}_{22}^2 & 0 \\ 0 & 0 & \hat{E}_{33}^2 \end{bmatrix} = \text{rank}\begin{bmatrix} \hat{E}_{11} & \hat{E}_{12} & \hat{E}_{13} \\ 0 & \hat{E}_{22} & 0 \\ 0 & 0 & \hat{E}_{33} \end{bmatrix}. \qquad (2\text{-}4.8)$$

Noticing the nilpotent property of $\hat{E}_{33}$, equation (2-4.8) holds iff $\hat{E}_{33} = 0$ and $\text{rank}\hat{E}_{11} = \text{rank}[\hat{E}_{11},\ \hat{E}_{13}]$. Thus there exists an M such that $\hat{E}_{13} = \hat{E}_{11}M$. Let

$$\hat{T} = \begin{bmatrix} I & 0 & M \\ 0 & I & 0 \\ 0 & 0 & I \end{bmatrix}.$$

Then $\hat{T}$ is nonsingular and

$$(\hat{T}T)\hat{E}(\hat{T}T)^{-1} = \text{diag}(\hat{E}_1,\ 0), \qquad (\hat{T}T)\hat{B} = [\hat{B}_1/0]$$

where

$$\hat{E}_1 = \begin{bmatrix} \hat{E}_{11} & \hat{E}_{12} \\ 0 & \hat{E}_{22} \end{bmatrix}, \qquad \hat{B}_1 = \begin{bmatrix} \hat{B}_{11} \\ 0 \end{bmatrix}. \qquad \text{Q.E.D.}$$

By duality we have the following theorem.

Theorem 2-4.5. Assume that (1-3.22) is the EF3 of system (2-4.2). Then

(i). System (2-4.2) is observable iff $(\hat{E},\ C)$ is observable.

(ii). System (2-4.2) is R-observable iff

$$\text{rank}[sI-\hat{E}/C] = n, \qquad \forall\ s\in\mathbb{C},\ s\neq 0,\ s \text{ finite},$$

i.e., $(\hat{E},C)$ has only zero unobservable poles.

(iii). System (2-4.2) is impulse observable iff $(\hat{E}, C)$ is similar to

$$(\text{diag}(\hat{E}_1, 0),\ [\hat{C}_1, 0])$$

in which $(\hat{E}_1,\hat{C}_1)$ has no zero unobservable poles, or, equivalently,

$$\text{rank } [\hat{E}/C]\hat{E} = \text{rank}\hat{E}.$$

Example 2-4.2. Consider the EF3 (1-3.23) of system (1-3.19). We have

$$\text{rank}[\hat{B},\ \hat{E}\hat{B},\ \hat{E}^2\hat{B},\ \hat{E}^3\hat{B}] = \text{rank}\begin{bmatrix} 0 & -1 & 1 & 0 \\ 0 & 0 & -1 & 1 \\ -1 & 1 & 0 & -1 \\ 0 & -1 & 1 & 0 \end{bmatrix} = 3 < 4.$$

Thus, $(\hat{E}, \hat{B})$ is not controllable, but

$$\text{rank}[sI-\hat{E},\ \hat{B}] = \text{rank}\begin{bmatrix} s & 0 & -1 & 0 & 0 \\ -1 & s & 0 & 0 & 0 \\ 1 & 0 & s+1 & 0 & -1 \\ 0 & 0 & -1 & s & 0 \end{bmatrix} = 4, \quad \forall\ s\neq 0,\ s\in\mathbb{C},\ s \text{ finite}$$

and

$$\text{rank } \hat{E}[\hat{E}\ \ \hat{B}] = \text{rank}\begin{bmatrix} 1 & 0 & -1 & 0 & -1 \\ 0 & 0 & 1 & 0 & 0 \\ 1 & 0 & 0 & 0 & 1 \\ -1 & 0 & -1 & 0 & -1 \end{bmatrix} = 2 = \text{rank}\hat{E}.$$

According to Theorem 2-4.4, system (1-3.19) is R-controllable, and impulse controllable, but not controllable. This coincides with the already proven results in Example 2-4.1.

Theorems 2-4.2 - 2-4.5 establish the various controllability and observability criteria under different equivalent forms. These theorems show that under EF3, the zero uncontrollable poles of $(\hat{E}, \hat{B})$ are the only difference between controllability and R-controllability. Corresponding to the original system, the difference lies in the controllability at infinity, which was already known.

2-5. System Decomposition

As pointed out in Chapter 1, for any regular singular system

$$\begin{aligned} E\dot{x} &= Ax + Bu \\ y &= Cx \end{aligned} \tag{2-5.1}$$

there exist two nonsingular matrices Q and P such that system (2-5.1) is r.s.e. to EF1:

$$\begin{aligned} \dot{x}_1 &= A_1x_1 + B_1u \\ N\dot{x}_2 &= x_2 + B_2u \\ y &= C_1x_1 + C_2x_2 \end{aligned} \qquad (2\text{-}5.2)$$

where $n_1 + n_2 = n$, $A_1 \in \mathbb{R}^{n_1 \times n_1}$, $N \in \mathbb{R}^{n_2 \times n_2}$ is nilpotent, and

$$QEP = \mathrm{diag}(I_{n_1}, N), \qquad QAP = \mathrm{diag}(A_1, I_{n_2})$$

$$QB = [B_1/B_2], \qquad CP = [C_1, C_2].$$

System (2-5.2) is the standard decomposition of (2-5.1). It is in the EF1 and successfully separates the system's two intrinsically different subsystems ----- slow and fast. This is useful in analysis, but, as no thing is perfact in nature, the main shortcoming of EF1 is its failure to separate controllability and observability without further study.

According to the well-known results in linear system theory, for any triple (A_1, B_1, C_1), a nonsingular matrix T_1 may be chosen such that (see. Appendix C):

$$T_1A_1T_1^{-1} = \begin{bmatrix} A_{11} & 0 & A_{13} & 0 \\ A_{21} & A_{22} & A_{23} & A_{24} \\ 0 & 0 & A_{33} & 0 \\ 0 & 0 & A_{43} & A_{44} \end{bmatrix}, \qquad T_1B_1 = \begin{bmatrix} B_{11} \\ B_{12} \\ 0 \\ 0 \end{bmatrix} \qquad (2\text{-}5.3)$$

$$C_1T_1^{-1} = [C_{11} \quad 0 \quad C_{13} \quad 0],$$

where $A_{ij} \in \mathbb{R}^{n_{1i} \times n_{1j}}$, $i, j = 1, 2, 3, 4$, $\sum_{i=1}^{4} n_{1i} = n_1$, and the following are true.

1. The subsystem (A_{11}, B_{11}, C_{11}) is both controllable and observable.

2. The subsystem

$$\left(\begin{bmatrix} A_{11} & 0 \\ A_{21} & A_{22} \end{bmatrix}, \begin{bmatrix} B_{11} \\ B_{12} \end{bmatrix}, [C_{11} \quad 0] \right)$$

is controllable, but unobservable.

3. The subsystem

$$\left(\begin{bmatrix} A_{11} & A_{13} \\ 0 & A_{33} \end{bmatrix}, \begin{bmatrix} B_{11} \\ 0 \end{bmatrix}, [C_{11} \quad C_{13}] \right)$$

is observable, but uncontrollable.

4. The subsystem $(A_{44}, 0,0)$ is neither controllable nor observable.

Similarly, there exists a nonsingular matrix T_2 such that (N,B_2,C_2) may be decomposed into controllability and observability form:

$$T_2NT_2^{-1} = \begin{bmatrix} N_{11} & 0 & N_{13} & 0 \\ N_{21} & N_{22} & N_{23} & N_{24} \\ 0 & 0 & N_{33} & 0 \\ 0 & 0 & N_{43} & N_{44} \end{bmatrix}, \qquad T_2B_2 = \begin{bmatrix} B_{21} \\ B_{22} \\ 0 \\ 0 \end{bmatrix} \tag{2-5.4}$$

$$C_2T_2^{-1} = [C_{21}, 0, C_{23}, 0]$$

where $N_{ii} \in \mathbb{R}^{n_{2i} \times n_{2i}}$, $i = 1, 2, 3, 4$, are nilpotent, $N_{ij} \in \mathbb{R}^{n_{2i} \times n_{2j}}$, $\sum_{i=1}^{4} n_{2i} = n_2$, and the following are true.

1. The subsystem (N_{11},B_{21},C_{21}) is both controllable and observable.
2. The subsystem

$$\left(\begin{bmatrix} N_{11} & 0 \\ N_{21} & N_{22} \end{bmatrix}, \begin{bmatrix} B_{21} \\ B_{22} \end{bmatrix}, [C_{21} \quad 0] \right)$$

is controllable, but unobservable.

3. The subsystem

$$\left(\begin{bmatrix} N_{11} & N_{13} \\ 0 & N_{33} \end{bmatrix}, \begin{bmatrix} B_{21} \\ 0 \end{bmatrix}, [C_{21} \quad C_{23}] \right)$$

is observable, but uncontrollable.

4. The subsystem $(N_{44}, 0, 0)$ is neither controllable nor observable.

Denote

$$Q_1 = \operatorname{diag}(T_1, T_2)Q, \qquad P_1 = P\operatorname{diag}(T_1^{-1}, T_2^{-1}).$$

Obviously, under this transformation, system (2-5.1) is r.s.e. to

$$\begin{bmatrix} \dot{x}_{11} \\ \dot{x}_{12} \\ \dot{x}_{13} \\ \dot{x}_{14} \end{bmatrix} = \begin{bmatrix} A_{11} & 0 & A_{13} & 0 \\ A_{21} & A_{22} & A_{23} & A_{24} \\ 0 & 0 & A_{33} & 0 \\ 0 & 0 & A_{43} & A_{44} \end{bmatrix} \begin{bmatrix} x_{11} \\ x_{12} \\ x_{13} \\ x_{14} \end{bmatrix} + \begin{bmatrix} B_{11} \\ B_{12} \\ 0 \\ 0 \end{bmatrix} u$$

$$\begin{bmatrix} N_{11} & 0 & N_{13} & 0 \\ N_{21} & N_{22} & N_{23} & N_{24} \\ 0 & 0 & N_{33} & 0 \\ 0 & 0 & N_{43} & N_{44} \end{bmatrix} \begin{bmatrix} \dot{x}_{21} \\ \dot{x}_{22} \\ \dot{x}_{23} \\ \dot{x}_{24} \end{bmatrix} = \begin{bmatrix} x_{21} \\ x_{22} \\ x_{23} \\ x_{24} \end{bmatrix} + \begin{bmatrix} B_{21} \\ B_{22} \\ 0 \\ 0 \end{bmatrix} u \tag{2-5.5a}$$

$$y = [C_{11}, 0, C_{13}, 0, C_{21}, 0, C_{23}, 0][T_1^{-1}x_1/T_2^{-1}x_2] \tag{2-5.5b}$$

where

$$T_i^{-1}x_i = [x_{i1}/x_{i2}/x_{i3}/x_{i4}], \qquad i = 1, 2$$

$$x_{ij} \in \mathbb{R}^{n_{ij}}, \ i = 1, 2, \ j = 1, 2, 3, 4, \ \sum_{j=1}^{4} n_{ij} = n_i, \ i = 1, 2,$$

$n_1 + n_2 = n$, N_{ii} is nilpotent, $i = 1, 2, 3, 4$.

Suitably rewrite the coordinate vector

$$\tilde{x}_i = [x_{1i}/x_{2i}], \quad i = 1, 2, 3, 4.$$

This means the combination of corresponding state variables in the slow and fast substates. Direct computation will verify that, under such a coordinate transformation, system (2-5.5) is r.s.e. to

$$\begin{bmatrix} \tilde{E}_{11} & 0 & \tilde{E}_{13} & 0 \\ \tilde{E}_{21} & \tilde{E}_{22} & \tilde{E}_{23} & \tilde{E}_{24} \\ 0 & 0 & \tilde{E}_{33} & 0 \\ 0 & 0 & \tilde{E}_{43} & \tilde{E}_{44} \end{bmatrix} \begin{bmatrix} \dot{\tilde{x}}_1 \\ \dot{\tilde{x}}_2 \\ \dot{\tilde{x}}_3 \\ \dot{\tilde{x}}_4 \end{bmatrix} = \begin{bmatrix} \tilde{A}_{11} & 0 & \tilde{A}_{13} & 0 \\ \tilde{A}_{21} & \tilde{A}_{22} & \tilde{A}_{23} & \tilde{A}_{24} \\ 0 & 0 & \tilde{A}_{33} & 0 \\ 0 & 0 & \tilde{A}_{43} & \tilde{A}_{44} \end{bmatrix} \begin{bmatrix} \tilde{x}_1 \\ \tilde{x}_2 \\ \tilde{x}_3 \\ \tilde{x}_4 \end{bmatrix} + \begin{bmatrix} \tilde{B}_1 \\ \tilde{B}_2 \\ 0 \\ 0 \end{bmatrix} u$$

$$y = [\tilde{C}_1, 0, \tilde{C}_3, 0][\tilde{x}_1/\tilde{x}_2/\tilde{x}_3/\tilde{x}_4] \tag{2-5.6}$$

where

$$\tilde{E}_{ii} = \mathrm{diag}(I, N_{ii}); \quad \tilde{A}_{ii} = \mathrm{diag}(A_{ii}, I), \quad i = 1, 2, 3, 4$$

$$\tilde{E}_{ij} = \mathrm{diag}(0, N_{ij}); \quad \tilde{A}_{ij} = \mathrm{diag}(A_{ij}, 0), \quad i \neq j, \ i, j = 1, 2, 3, 4$$

$$\tilde{B}_i = [B_{1i}/B_{2i}], \ i = 1, 2; \quad \tilde{C}_j = [C_{1j}, C_{2j}], \ i = 1, 3,$$

are constant matrices, $\tilde{x}_i \in \mathbb{R}^{\tilde{n}_i}$, $\tilde{n}_i = n_{1i} + n_{2i}$, $i = 1, 2, 3, 4$, $\sum_{i=1}^{4} \tilde{n}_i = n$.

From the controllability and observability criteria (Theorems 2-2.1 and 2-3.1), and the decomposition of (2-5.3)-(2-5.4), it is easy to know that the system (2-5.6) has the following properties.

a. The subsystem $(\tilde{E}_{11}, \tilde{A}_{11}, \tilde{B}_1, \tilde{C}_1)$ is both controllable and observable.

b. The subsystem

$$\left(\begin{bmatrix} \tilde{E}_{11} & 0 \\ \tilde{E}_{21} & \tilde{E}_{22} \end{bmatrix}, \begin{bmatrix} \tilde{A}_{11} & 0 \\ \tilde{A}_{21} & \tilde{A}_{22} \end{bmatrix}, \begin{bmatrix} \tilde{B}_1 \\ \tilde{B}_2 \end{bmatrix}, [\tilde{C}_1 \ \ 0] \right)$$

is controllable, but unobservable.

c. The subsystem

$$\left(\begin{bmatrix}\tilde{E}_{11} & \tilde{E}_{13}\\ 0 & \tilde{E}_{33}\end{bmatrix}, \begin{bmatrix}\tilde{A}_{11} & \tilde{A}_{13}\\ 0 & \tilde{A}_{33}\end{bmatrix}, \begin{bmatrix}\tilde{B}_1\\ 0\end{bmatrix}, [\tilde{C}_1 \quad \tilde{C}_2]\right)$$

is observable, but uncontrollable.

d. The subsystem $(\tilde{E}_{44}, \tilde{A}_{44}, 0, 0)$ is neither controllable nor observable.

Both (2-5.2) and (2-5.6) are canonical forms for system (2-5.1) under restricted system equivalence.

The significance of canonical form (2-5.6) lies in its ability to separate the controllability (observability) substates from uncontrollable (unobservable) ones. Thus it characterizes the inner structure of a system.

Decomposition (2-5.6) not only provides us with a canonical form, but its constructive proof also gives us a computation approach to obtain it: We first take standard decomposition on the system to obtain the slow-fast subsystem decomposition. Then we will take controllability and observability decomposition on each subsystem, resulting in r.s.e. system (2-5.5). Finally, (2-5.6) is ready by rearranging the coordinate vector in (2-5.5).

Particularly, let $\bar{x}_1 = [\tilde{x}_1/\tilde{x}_2]$ and $\bar{x}_2 = [\tilde{x}_3/\tilde{x}_4]$. Then (2-5.6) becomes

$$\begin{bmatrix}\bar{E}_{11} & \bar{E}_{12}\\ 0 & \bar{E}_{22}\end{bmatrix}\begin{bmatrix}\dot{\bar{x}}_1\\ \dot{\bar{x}}_2\end{bmatrix} = \begin{bmatrix}\bar{A}_{11} & \bar{A}_{12}\\ 0 & \bar{A}_{22}\end{bmatrix}\begin{bmatrix}\bar{x}_1\\ \bar{x}_2\end{bmatrix} + \begin{bmatrix}\bar{B}_1\\ 0\end{bmatrix}u$$

$$y = [\bar{C}_1, \bar{C}_2][\bar{x}_1/\bar{x}_2] \tag{2-5.7}$$

where $\bar{E}_{ij}$, $\bar{A}_{ij}$, $\bar{B}_i$, $\bar{C}_i$, i, j = 1, 2, are certain constant matrices, and $(\bar{E}_{11},\bar{A}_{11},\bar{B}_1)$ is controllable. (2-5.7) is called the controllability canonical form for system (2-5.1).

If we set $\hat{x}_1 = [\tilde{x}_1/\tilde{x}_3]$ and $\hat{x}_2 = [\tilde{x}_2/\tilde{x}_4]$, system (2-5.6) may be rewritten as

$$\begin{bmatrix}\hat{E}_{11} & 0\\ \hat{E}_{21} & \hat{E}_{22}\end{bmatrix}\begin{bmatrix}\dot{\hat{x}}_1\\ \dot{\hat{x}}_2\end{bmatrix} = \begin{bmatrix}\hat{A}_{11} & 0\\ \hat{A}_{21} & \hat{A}_{22}\end{bmatrix}\begin{bmatrix}\hat{x}_1\\ \hat{x}_2\end{bmatrix} + \begin{bmatrix}\hat{B}_1\\ \hat{B}_2\end{bmatrix}u$$

$$y = [\hat{C}_1, 0][\hat{x}_1/\hat{x}_2] \tag{2-5.8}$$

where $\hat{E}_{ij}$, $\hat{A}_{ij}$, $\hat{B}_i$, $\hat{C}_i$, i, j = 1, 2, are certain constant matrices, and $(\hat{E}_{11},\hat{A}_{11},\hat{C}_1)$ is observable. This form is called the observability canonical form of system (2-5.1).

If we set $\bar{\bar{x}}_1 = [x_1/x_{21}/x_{22}]$ and $\bar{\bar{x}}_2 = [x_{23}/x_{24}]$, noticing (2-5.4), system (2-5.2) (thus system (2-5.1)) is r.s.e. to

$$\begin{bmatrix} \bar{\bar{E}}_{11} & \bar{\bar{E}}_{12} \\ 0 & \bar{\bar{E}}_{22} \end{bmatrix}\begin{bmatrix} \dot{\bar{\bar{x}}}_1 \\ \dot{\bar{\bar{x}}}_2 \end{bmatrix} = \begin{bmatrix} \bar{\bar{A}}_{11} & \bar{\bar{A}}_{12} \\ 0 & \bar{\bar{A}}_{22} \end{bmatrix}\begin{bmatrix} \bar{\bar{x}}_1 \\ \bar{\bar{x}}_2 \end{bmatrix} + \begin{bmatrix} \bar{\bar{B}}_1 \\ 0 \end{bmatrix} u$$

$$y = [\bar{\bar{C}}_1, \bar{\bar{C}}_2][\bar{\bar{x}}_1/\bar{\bar{x}}_2] \qquad (2\text{-}5.9)$$

where $\text{rank}[\bar{\bar{E}}_{11}, \bar{\bar{B}}_1] = \bar{\bar{n}}_1$, $[\bar{\bar{E}}_{11}, \bar{\bar{B}}_1]$ is of full row rank and $\bar{\bar{E}}_{22}$ is nilpotent. This decomposition is called normalizability (the definition is given in the next chapter) decomposition. It will be needed later.

Example 2-5.1. Consider system (1-3.19), whose standard decomposition is given by (1-3.20):

$$\dot{x}_1 = \begin{bmatrix} -1 & -1 \\ 1 & 0 \end{bmatrix} x_1 + \begin{bmatrix} 1 \\ 0 \end{bmatrix} V_s(t)$$

$$0 = x_2 + [-1/0]V_s(t) \qquad (2\text{-}5.10)$$

$$y = [0 \quad 1]x_1.$$

Since

$$\left(\begin{bmatrix} -1 & -1 \\ 1 & 0 \end{bmatrix}, \begin{bmatrix} 1 \\ 0 \end{bmatrix}, [0 \quad 1] \right)$$

is both controllable and observable, we need to make decomposition on only the fast subsystem. Note that if we denote $x_2 = [x_{21}/x_{22}]$ in (2-5.10), x_{21} is controllable but not observable, and x_{22} is neither. Hence, system (2-5.10) (thus (1-3.19)) has the following canonical form:

$$\left[\begin{array}{cc|c|c} 1 & 0 & 0 & 0 \\ 0 & 1 & 0 & 0 \\ \hline 0 & 0 & 0 & 0 \\ \hline 0 & 0 & 0 & 0 \end{array}\right]\begin{bmatrix} \dot{x}_1 \\ \dot{x}_{21} \\ \dot{x}_{22} \end{bmatrix} = \left[\begin{array}{cc|c|c} -1 & -1 & 0 & 0 \\ 1 & 0 & 0 & 0 \\ \hline 0 & 0 & 1 & 0 \\ \hline 0 & 0 & 0 & 1 \end{array}\right]\begin{bmatrix} x_1 \\ x_{21} \\ x_{22} \end{bmatrix} + \left[\begin{array}{c} 1 \\ 0 \\ \hline -1 \\ \hline 0 \end{array}\right] V_s(t) \qquad (2\text{-}5.11)$$

$$y = [0 \quad 1 \mid 0 \mid 0][x_1/x_{21}/x_{22}].$$

Example 2-5.2. As given in Examples 2-2.2 and 2-3.2, the circuit network (2-2.5)-(2-2.6) is controllable, but not observable.

To obtain its canonical decomposition, we begin by determining its standard decomposition, via the method given by Lemma 1-2.2. In this system, matrix A is nonsingular. Thus $\alpha = 0$ certainly satisfies $|\alpha E+A| \neq 0$. Let $Q_1 = A^{-1}$, $P_1 = I_4$. Then

$$E_1 = Q_1 E P_1 = A^{-1}E = \begin{bmatrix} -RC_1 & 0 & 0 & 0 \\ -RC_1 & 0 & -L & 0 \\ 0 & C_2 & 0 & 0 \\ C_1 & 0 & 0 & 0 \end{bmatrix} \qquad Q_1 A P_1 = I_4. \qquad (2\text{-}5.12)$$

A similarity transformation of E_1 and separation of its Jordan blocks belonging to zero eigenvalue from those belonging to nonzero eigenvalues, we know that

$$T = \begin{bmatrix} 1 & 0 & 0 & 0 \\ 0 & 1 & 0 & 0 \\ 0 & 0 & 1 & 0 \\ 1 & 0 & 0 & R \end{bmatrix}$$

will transfer E_1 into

$$TE_1T^{-1} = \begin{bmatrix} -RC_1 & 0 & 0 & 0 \\ -RC_1 & 0 & -L & 0 \\ 0 & C_2 & 0 & 0 \\ 0 & 0 & 0 & 0 \end{bmatrix}. \qquad (2\text{-}5.13)$$

Denoting $Q = TQ_1$ and $P = P_1T^{-1}$, from (2-5.12) and (2-5.13), it is easy to obtain

$$QEP = TE_1T^{-1}, \qquad QAP = I_4. \qquad (2\text{-}5.14)$$

Thus, the coordinate transformation

$$[x_1/x_2] = P^{-1}x = Tx = [u_{c1}/u_{c2}/I_2/(RI_1+u_{c1})]$$

will transfer (2-2.5)-(2-2.6) into its canonical form

$$\left[\begin{array}{ccc|c} -RC_1 & 0 & 0 & 0 \\ -RC_1 & 0 & -L & 0 \\ 0 & C & 0 & 0 \\ \hline 0 & 0 & 0 & 0 \end{array}\right]\begin{bmatrix} \dot{x}_1 \\ \dot{x}_2 \end{bmatrix} = \left[\begin{array}{ccc|c} 1 & 0 & 0 & 0 \\ 0 & 1 & 0 & 0 \\ 0 & 0 & 1 & 0 \\ \hline 0 & 0 & 0 & 1 \end{array}\right]\begin{bmatrix} x_1 \\ x_2 \end{bmatrix} + \left[\begin{array}{c} -1 \\ 0 \\ 0 \\ \hline 0 \end{array}\right]u$$

$$y = u_{c2} = [0 \quad 1 \quad 0 \mid 0][x_1/x_2]. \qquad (2\text{-}5.15)$$

Decomposition (2-5.15) is not only the canonical form but the observable canonical form as well. In this decomposition, the substate x_1 is both controllable and observable; but substate x_2 is only controllable. Corresponding to its original system, this fact refers to that $x_2 = RI_1 + u_{c1}$ is not observable, which is obvious in Figure 2-2.1, in which $RI_1 + u_{c1} = u_e$ could not be determined from measure u_{c2}.

By using EF3, we may obtain the canonical decomposition more directly. First take the coordinate transformation to obtain EF3; then take controllability and observability decomposition on $(\hat{E},\hat{B},C)$, and finally the canonical form is obtainable. The detailed process is omitted here. Interested readers may easily do this.

2-6. Transfer Matrix and Minimal Realization

Based on the state variable concept, the state space description method for system analysis and synthesis obtains its state space model (or description equation) from the physical sense of variables and their relationships. Thus, this method allows us to understand the inner structure of systems. On the other hand, the transfer matrix reflects the outer structure: input-output relationship, or transfer relationship from input to output. To compare, the state space method may characterize the inner properties but is sometimes too complicated for practical interests; while the transfer matrix is usually unable to characterize inner properties, it provides a succinct dependence relationship between input and output. We begin our discussion from the Laplace transformation.

The Laplace transformation $L[f]$ of a function $f(t)$ is defined as

$$L[f] = \int_0^{\infty} e^{-st} f(t)dt$$

which is assumed to exist. It has the basic property:

$$L[\alpha f_1 + \beta f_2] = \alpha L[f_1] + \beta L[f_2]$$

$$L[\dot{f}] = sL[f] - f(0).$$

Taking Laplace transformation on both sides of the system equation

$$E\dot{x} = Ax + Bu \qquad (2\text{-}6.1)$$
$$y = Cx$$

we have

$$sEx(s) - Ex(0) = Ax(s) + Bu(s)$$
$$y(s) = Cx(s).$$

Under the assumption of regularity $(sE-A)^{-1}$ exists. Thus

$$y(s) = C(sE-A)^{-1}(Ex(0) + Bu(s)).$$

When $x(0) = 0$, the preceding equation yields the input-output relationship:

$$y(s) = G(s)u(s) \qquad (2\text{-}6.2)$$

in which

$$G(s) = C(sE-A)^{-1}B \qquad (2\text{-}6.3)$$

is called the transfer matrix for system (2-6.1).

Example 2-6.1. Consider system (1-3.19). Its transfer function is

$$G(s) = C(sE-A)^{-1}B$$

$$= [0 \quad 0 \quad 1] \begin{bmatrix} s & -1 & 0 & 0 \\ -1 & 0 & s & 0 \\ 1 & 0 & 0 & -1 \\ 0 & -1 & -1 & -1 \end{bmatrix}^{-1} \begin{bmatrix} 0 \\ 0 \\ 0 \\ -1 \end{bmatrix} = \frac{1}{s^2+s+1}$$

which is a strictly proper function as in the normal case.

Example 2-6.2. The singular system with only a fast system

$$N\dot{x} = x + Bu$$
$$y = Cx \tag{2-6.4}$$

where N is nilpotent, has a transfer matrix

$$G(s) = C(sN-I)^{-1}B = -CB - sCNB - \dots - s^{h-1}CN^{h-1}B$$

where h is the nilpotent index of N. Therefore, the transfer matrix of system (2-6.4) is a polynomial matrix.

Let Q and P be nonsingular and $\tilde{E} = QEP$, $\tilde{A} = QAP$, $\tilde{B} = QB$, $\tilde{C} = PC$, then the transfer matrix G(s) has the form

$$G(s) = \tilde{C}(s\tilde{E}-\tilde{A})^{-1}\tilde{B} = CP(sQEP-QAP)^{-1}QB = C(sE-A)^{-1}B \tag{2-6.5}$$

which states that r.s.e. systems have the same transfer matrix. This property is a great advantage in system theory. For systems with structures that are too complex to analyze and/or design, we may need to study an r.s.e. form convenient to our discussion and find its properties in which we are interested.

As previously proven, for a given system (2-6.1), there exist two nonsingular matrices Q and P such that (2-6.1) is r.s.e. to the canonical form (2-5.6). Direct but tedious computation shows that

$$G(s) = C(sE-A)^{-1}B = \tilde{C}_1(s\tilde{E}_{11}-\tilde{A}_{11})^{-1}\tilde{B}_1. \tag{2-6.6}$$

Therefore, the transfer matrix for a system is determined by the controllable and observable subsystem, and has no relationship to other uncontrollable or unobservable parts. Accounting for this, the transfer matrix reflects only the most important part of a system: the controllable and observable part, which is the control effect of input on the measure outputs, without any knowledge of control on the state.

Further calculation tells us

$$G(s) = \tilde{C}_1(s\tilde{E}_{11}-\tilde{A}_{11})^{-1}\tilde{B}_1 = G_1(s) + G_2(s) \tag{2-6.7}$$

where $G_1(s) = C_{11}(sI-A_{11})^{-1}B_{11}$, $G_2(s) = C_{21}(sN_{11}-I)^{-1}B_{21}$.

Another distinguishing feature may be seen from (2-6.7): Different from the case of the normal system, whose transfer matrix is strictly proper. the transfer matrix of singular systems generally have two parts. One is $G_1(s)$, which is determined by the slow subsystem and is strictly proper; the other is a polynomial $G_2(s)$, which is determined by the fast subsystem. In the general case, $G_2(s) \neq 0$. Different from the strict properness in the normal case, G(s) for a singular system is rational, but not necessarilly proper. This is a special feature of singular systems.

Example 2-6.3. Consider again the system (1-3.19) with standard decomposition (1-3.20). Its transfer matrix is

$$G(s) = C_1(sI-A_1)^{-1}B_1 + C_2(sN-I)^{-1}B_2 = C_1(sI-A_1)^{-1}B_1 \quad (C_2 = 0)$$
$$= 1/(s^2+s+1)$$

which coincides with the results in Example 2-6.1, with much less computation since r.s.e. form is used.

For a single input single output system, its transfer matrix

$$G(s) = \frac{b_{n-1}s^{n-1} + \ldots + b_1 s + b_o}{s^{n_1} + a_{n_1-1}s^{n_1-1} + \ldots + a_1 s + a_o}$$

is a rational transfer function.

In general, $n_1 < n$. G(s) is not proper if $b_{n-1} \neq 0$. Thus, viewing from the point of transfer matrix, the only difference between the normal and singular systems is the strict properness of transfer function. If it is strictly proper, the input-output relationship may be described by a normal system, otherwise, the singular system is used. As for multi-input multi-output systems, the difference lies in the properness of the transfer matrix.

It is worth pointing out that a singular system may have a strictly proper function, as the case of system (1-3.19) in Example 2-6.1, if the controllable and observable subsystem is normal.

As pointed out earlier, two r.s.e. systems have the same transfer matrix. Further, we can prove the following theorem.

Theorem 2-6.1. Two controllable and observable systems (E,A,B,C) and $(\tilde{E},\tilde{A},\tilde{B},\tilde{C})$ have the same transfer matrix if and only if they are r.s.e.

Proof. By noticing (2-6.7) the proof is easy. Q.E.D.

This theorem assures that the transfer matrix is not only determined by controllable and observable subsystem but is determined uniquely. If controllability and observability assumptions are not guaranteed, this theorem may be not true. For example, as shown in Examples 2-6.1 and 2-6.3, system (1-3.19) and the system

$$\dot{x} = \begin{bmatrix} -1 & -1 \\ 1 & 0 \end{bmatrix} x + \begin{bmatrix} 1 \\ 0 \end{bmatrix} u$$

$$y = [0 \quad 1]x \tag{2-6.8}$$

have the same transfer function $1/(s^2+s+1)$. But they are not r.s.e. to each other since they are of different orders.

As mentioned earlier, for a given singular system, we may obtain its transfer matrix. Conversely, we will be interested in that for a given transfer matrix G(s), a certain singular system may be found whose transfer matrix is G(s). This is the so-called realization theory. To be precise, we define the following.

Definition 2-6.1. Assume that $G(s) \in \mathbb{R}^{r\times m}$ is a rational matrices. If there exist matrices E, A, B, C such that

$$G(s) = C(sE-A)^{-1}B$$

where $E, A \in \mathbb{R}^{n\times n}$, $B \in \mathbb{R}^{n\times m}$, $C \in \mathbb{R}^{r\times n}$ are constant matrices, the system

$$E\dot{x} = Ax + Bu$$

$$y = Cx \tag{2-6.9}$$

will be called a singular system realization of G(s), simply, a realization of G(s).

Furthermore, if any other realization has an order greater than n, system (2-6.9) will be called the minimal order realization or, simply, minimal realization.

Example 2-6.4. Both systems (1-3.19) and (2-6.8) are the realization of transfer matrix $G(s) = 1/(s^2+s+1)$. However, system (2-6.8) is of lower order than system (1-3.19).

Lemma 2-6.1. Let g(s) be a rational function $g(s) = b(s)/a(s)$. Then there always exists a strictly proper rational function $g_1(s)$ and a polynomial $g_2(s)$ such that $g(s) = g_1(s) + g_2(s)$.

Proof. Let $d = \deg(b(s))$, $\ell = \deg(a(s))$. Here deg(.) represents the degree of a polynomial. If $d < \ell$, g(s) is strictly proper. In this case, the lemma is true for $g_1(s) = g(s)$, $g_2(s) = 0$; otherwise, $d \geq \ell$, from the polynomial division theorem, we know that there exist polynomials q(s) and r(s) such that $b(s) = q(s)a(s) + r(s)$ and $\deg(r(s)) < \deg(a(s))$. Therefore,

$$g(s) = \frac{b(s)}{a(s)} = \frac{r(s)}{a(s)} + q(s).$$

Let $g_1(s) = r(s)/a(s)$ and $g_2(s) = q(s)$, then $g_1(s)$ is strictly proper, $g_2(s)$ is a polynomial, and $g(s) = g_1(s) + g_2(s)$. Q.E.D.

Theorem 2-6.2. Any r×m rational matrix G(s) may be represented as

$$G(s) = G_1(s) + G_2(s) \tag{2-6.10}$$

where $G_1(s)$ is a strictly proper rational matrix and $G_2(s)$ is a polynomial matrix.

Proof. Let $G(s) = (g_{ij}(s))_{rxm}$, where $g_{ij}(s)$ is the element at the ith row and jth column, which is rational. From Lemma 2-6.1 there exist strictly proper function $g^1_{ij}(s)$ and polynomial $g^2_{ij}(s)$ such that

$$g_{ij}(s) = g^1_{ij}(s) + g^2_{ij}(s), \quad i = 1, 2, \dots, r, \; j = 1, 2, \dots, m.$$

Thus, the theorem is true by setting

$$G_1(s) = (g^1_{ij}(s)), \qquad G_2(s) = (g^2_{ij}(s)). \qquad \text{Q.E.D.}$$

Example 2-6.5. Consider the transfer matrix

$$G(s) = [\frac{1}{s^2+s+1} \;/\; \frac{s^4+s^3-s}{s^2+s+1}]. \tag{2-6.11}$$

Since

$$\frac{s^4+s^3-s}{s^2+s+1} = s^2 - 1 + \frac{1}{s^2+s+1},$$

$G(s)$ may be rewritten in the form of $G(s) = G_1(s) + G_2(s)$, with

$$G_1(s) = \frac{1}{s^2+s+1}\begin{bmatrix} 1 \\ 1 \end{bmatrix}, \qquad G_2(s) = \begin{bmatrix} 0 \\ s^2-1 \end{bmatrix}.$$

Lemma 2-6.2. For any polynomial matrix $P(s)$, there always exist matrices N, B_2, C_2, N is nilpotent, such that

$$P(s) = C_2(sN-I)^{-1}B_2.$$

Proof. Consider the strictly proper matrix $\bar{G}(s) = -\frac{1}{s}P(\frac{1}{s})$, which may be viewed as the transfer matrix of a normal system. Thus, according to linear system theory, there exist matrices N, B_2, C_2, such that

$$\bar{G}(s) = -\frac{1}{s}P(\frac{1}{s}) = C_2(sI-N)^{-1}B_2.$$

Noticing that $\bar{G}(s)$ has only zero poles, N has only zero eigenvalues. Thus, N is nilpotent. Direct computation will results in $P(s) = C_2(sN-I)^{-1}B_2$. Q.E.D.

Hence, system

$$N\dot{x} = x + B_2u, \qquad y = C_2x$$

may be viewed as a singular system realization for $P(s)$. Speaking from this point,

Lemma 2-6.2 shows that any polynomial matrix P(s) has a singular system realization with only a fast subsystem.

Theorem 2-6.3. Any rational matrix may have a realization of (2-6.9), which satisfies $G(s) = C(sE-A)^{-1}B$; furthermore, the realization is minimal if and only if the system is both controllable and observable.

Proof. According to Lemma 2-6.2, any rational matrix G(s) may have a decomposition $G(s) = G_1(s) + G_2(s)$, in which $G_1(s)$ is strictly proper and $G_2(s)$ is polynomial. For the strict properness of $G_1(s)$, from linear system theory, it always has a realization A_1, B_1, C_1, satisfying $G_1(s) = C_1(sI-A_1)^{-1}B_1$.

Also, Lemma 2-6.2 shows the existence of nilpotent N, B_2, C_2, such that $G_2(s) = C_2(sN-I)^{-1}B_2$.

Let

$$E = \operatorname{diag}(I,N),\quad A = \operatorname{diag}(A_1,\ I),\ B = [B_1/B_2],\quad C = [C_1,\ C_2]. \tag{2-6.12}$$

It is easy to verify that

$$G(s) = G_1(s) + G_2(s) = C(sE-A)^{-1}B.$$

The the system determined by (2-6.12) is a realization of G(s).

As for the second conclusion, we note that any transfer matrix G(s) may be decomposed into the sum of a strictly proper $G_1(s)$ and a polynomial $G_2(s)$, and the order of its realization is the sum of those of $G_1(s)$ and $G_2(s)$. Thus, system (2-6.9) is a minimal realization iff (A_1,B_1,C_1) and (N,B_2,C_2) are minimal, or equivalently, iff (A_1,B_1,C_1) ans (N,B_2,C_2) are minimal realizations of strictly proper matrices $G_1(s)$ and $\bar{G}_2(s)$, respectively, indicating both (A_1,B_1,C_1) and (N,B_2,C_2) are controllable and observable. This in turn shows the minimal realization (2-6.12) is controllable and observable. Q.E.D.

The first and second portions of Theorem 2-6.3, respectively, show the existence of realizations and minimal realizations. Meanwhile, the proof also provides us with a method to find realizations. First decompose matrix G(s) into two parts $G = G_1+G_2$, in which G_1 is strictly proper and G_2 is polynomial; then find the minimal realizations of $G_1(s)$ and $G_2(s)$. Combining the matrices in the way of (2-6.1), and we will be ready to get the minimal realization (2-6.9).

A direct result of the combination of Theorems 2-6.2 and 2-6.3 is the following corollary.

Corollary 2-6.1. Any two minimal realizations for a rational matrix are r.s.e.

Example 2-6.6. Both systems (1-3.19) and (2-6.8) are realizations of $G(s) = 1/(s^2+s+1)$. Since system (2-6.8) is controllable and observable, it is minimal. But

(1-3.19) is not.

For a given polynomial matrix,

$$P(s) = P_o + P_1 s + \dots + P_{k-1} s^{k-1} \in \mathbb{R}^{r \times m} \tag{2-6.13}$$

Lemma 2-6.2 assures the existence of $B_2 \in \mathbb{R}^{n_2 \times m}$, $C_2 \in \mathbb{R}^{r \times n_2}$ and the nilpotent matrix $N \in \mathbb{R}^{n_2 \times n_2}$ such that

$$P(s) = C_2(sN-I)^{-1}B_2. \tag{2-6.14}$$

We now introduce the Silverman-Ho algorithm to search the minimal realization. Define

$$M_o = \begin{bmatrix} -P_o & -P_1 & \dots & -P_{k-2} & -P_{k-1} \\ -P_1 & -P_2 & \dots & -P_{k-1} & 0 \\ & \dots & \dots & & \dots \\ -P_{k-2} & -P_{k-1} & \dots & \dots & 0 \\ -P_{k-1} & 0 & \dots & \dots & 0 \end{bmatrix} \in \mathbb{R}^{kr \times km}$$

$$M_1 = \begin{bmatrix} -P_1 & -P_2 & \dots & -P_{k-1} & 0 \\ -P_2 & -P_3 & \dots & 0 & 0 \\ & \dots & \dots & & \dots \\ -P_{k-1} & 0 & \dots & \dots & 0 \\ 0 & 0 & \dots & \dots & 0 \end{bmatrix} \in \mathbb{R}^{kr \times km}$$

and set $\tilde{n}_2 = \operatorname{rank} M_o$. Then take full rank decomposition

$$M_o = L_1 \cdot L_2 \tag{2-6.15}$$

where $L_1 \in \mathbb{R}^{kr \times \tilde{n}_2}$, $L_2 \in \mathbb{R}^{\tilde{n}_2 \times km}$ are of full column and row rank, respectively.

Let $\tilde{B}_2$ and $\tilde{C}_2$, respectively, be the first m columns of L_2 and the first r rows of L_1, and

$$\tilde{N} = (L_1^{\tau} L_1)^{-1} L_1^{\tau} M_1 L_2^{\tau} (L_2 L_2^{\tau})^{-1}. \tag{2-6.16}$$

Then matrices $\tilde{N}$, $\tilde{B}_2$ and $\tilde{C}_2$ form a minimal realization for $P(s)$.

<u>Proof</u>. Existence of minimal realization is proven in Lemma 2-6.2 and Theorem 2-6.3. There exists controllable and observable triple (N, B_2, C_2), N is nilpotent, such that (2-6.14) holds, i.e.,

$$P_o + P_1 s + \dots + P_{k-1} s^{k-1} = - C_2 B_2 - C_2 N B_2 s - \dots - C_2 N^{k-1} B_2 s^{k-1}.$$

Equalling the coefficients of the same order on both sides of this equation, we have

$$-P_i = C_2N^iB_2, \quad i = 0, 1, 2, \ldots, k-1.$$

(If k is less than the nilpotent index h of N, the coefficients of the higher order terms will be considered zero). Thus

$$M_o = H_1 \cdot H_2, \qquad M_1 = H_1NH_2 \tag{2-6.17}$$

where

$$H_1 = [C_2/C_2N/\ldots/C_2N^{k-1}], \qquad H_2 = [B_2, NB_2, \ldots, N^{k-1}B_2].$$

Recalling our assumption (N,B_2,C_2) is minimal. Therefore, H_1 and H_2 are of full column and row rank, respectively.

Noticing n_2 = rankH_1 = rankH_2, we have $\tilde{n}_2$ = rankM_o = rankH_1 = n_2. Furthermore, from (2-6.15) and (2-6.17) it is easy to see

$$L_1L_2 = H_1H_2 . \tag{2-6.18}$$

Left multiplying both sides of (2-6.18) by L_1^τ and paying attention to the invertibility of $L_1^\tau L_1$, we obtain $L_2 = (L_1^\tau L_1)^{-1}L_1^\tau H_1H_2$. By selection, n_2 = rankL_2 = rankH_2, and if

$$T \triangleq (L_1^\tau L_1)^{-1}L_1^\tau H_1 \in \mathbb{R}^{n_2 x n_2}$$

we know that T is nonsingular and

$$L_2 = TH_2 = [TB_2, TNB_2, \ldots, TN^{k-1}B_2]. \tag{2-6.19}$$

According to the selection of $\tilde{B}_2$, from (2-6.19) we have

$$\tilde{B}_2 = TB_2. \tag{2-6.20}$$

The substitution of (2-6.20) into (2-6.18) results in $L_1TH_2 = H_1H_2$. Right multiplying this equation by H_2^τ and noticing that $H_2H_2^\tau$ is nonsingular, we know that

$$L_1 = H_1T^{-1} = [C_2T^{-1}/C_2NT^{-1}/\ldots/C_2N^{k-1}T^{-1}].$$

Thus

$$\tilde{C}_2 = C_2T^{-1}. \tag{2-6.21}$$

Furthermore, the selection method of $\tilde{N}$ in (2-6.16), together with $L_2 = TH_2$, $L_1 = H_1T^{-1}$, and (2-6.17), yields

$$\tilde{N} = (L_1^\tau L_1)^{-1}L_1^\tau H_1NH_2L_2^\tau(L_2L_2^\tau)^{-1} = TNT^{-1}. \tag{2-6.22}$$

Equations (2-6.20) - (2-6.22) show that the realization has the same order with the minimal one (N,B_2,C_2), and the two realizations are similar. Thus (N,B_2,C_2) is a minimal realization for P(s). Q.E.D.

The method for finding the minimal realization given here is relatively simple.

Example 2-6.7. Consider the polynomial matrix $G_2(s)$ in Example 2-6.5:

$$G_2(s) = [0 \; / \; s^2-1].$$

Clearly, its minimal realization is ready when the minimal realization of polynomial $P(s) = s^2-1$ is obtained. Next, we find its minimal realization using the method mentioned earlier.

For $P(s)$, we have

$$M_o = \begin{bmatrix} 1 & 0 & -1 \\ 0 & -1 & 0 \\ -1 & 0 & 0 \end{bmatrix}, \qquad M_1 = \begin{bmatrix} 0 & -1 & 0 \\ -1 & 0 & 0 \\ 0 & 0 & 0 \end{bmatrix}.$$

Note that M_o is nonsingular. Thus we may choose $L_1 = M_o$ and $L_2 = I_3$. Therefore, $\tilde{B}_2 = [1 \;\; 0 \;\; 0]$, $\tilde{C}_2 = [1 \;\; 0 \;\; -1]$, and according to (2-6.16),

$$\tilde{N} = (L_1^\tau L_1)^{-1} L_1^\tau M_1 L_2^\tau (L_2 L_2^\tau)^{-1} = M_o^{-1} M_1 = \begin{bmatrix} 0 & 0 & 0 \\ 1 & 0 & 0 \\ 0 & 1 & 0 \end{bmatrix}.$$

Hence, $(\tilde{N}, \tilde{B}_2, \tilde{C}_2)$ is a minimal realization of $P(s) = s^2 - 1$. And the singular system

$$\begin{bmatrix} 0 & 0 & 0 \\ 1 & 0 & 0 \\ 0 & 1 & 0 \end{bmatrix} \dot{x} = x + \begin{bmatrix} 1 \\ 0 \\ 0 \end{bmatrix} u$$

$$y = \begin{bmatrix} 0 & 0 & 0 \\ 1 & 0 & -1 \end{bmatrix} x$$

is a minimal realization of $G_2(s)$.

2-7. Notes and References

Section 2-1 is due to Yip and Sincovec (1981) and the theory of various controllabilities, observabilities and duality is mainly from Cobb (1984). The matrix criteria under three equivalent forms are mainly from Dai (1987b, 1988d) and Zhou et al. (1987). There are other definitions or terms of controllability and observability for singular systems, such as structural controllability (Aoki et al, 1983; Murota, 1983; Yamada and Luenberger, 1985a, b), D_i-controllability (Pandolfi, 1980); strong controllability (Gontian and Tarn, 1982; Verghese et al, 1981), C-controllability (Yip and Sincovec, 1981). A partial list of other papers on controllability and observability includes Bender and Laub (1985), Campbell (1982b), Christodoulou and Paraskevopoulos (1985), Cullen (1986a), Kalyuthnaya (1978), Lewis and Ozcaldiran (1984), Lewis (1985a), Zhou et al. (1987).

It needs to be pointed out that under the strong restricted system equivalence (Verghese et al. 1981), the impulse controllability, controllability at infinity (Verghese et al. 1981) (which is the same as impulse controllability here), and the controllability of the fast subsystem become the same concept.

Sections 2-5 and 2-6 are from Verghese et al. (1981), Dai (1988c) and Cullen (1986b).

CHAPTER 3

FEEDBACK CONTROL

We use the term feedback control to refer to state feedback and static output feedback control, one of the most commonly used methods to change the systems's dynamic or static properties. In practical system analysis and design, the evaluation of a system's properties is based on some features that characterize the system, such as stability, static behavior, and response speed. But generally, all of the characteristics are not fulfilled at the same time. This inspires the study of adjustment. The improvement of the properties is the content of system design and forms many topics of control engineering and of control theory.

3-1. State Feedback

Consider the following singular system:

$$\begin{aligned} E\dot{x} &= Ax + Bu \\ y &= Cx \end{aligned} \tag{3-1.1}$$

where $x \in \mathbb{R}^n$, $u \in \mathbb{R}^m$, and $y \in \mathbb{R}^r$ are its state, control input, and measure output, respectively; E, $A \in \mathbb{R}^{n\times n}$, $B \in \mathbb{R}^{n\times m}$, and $C \in \mathbb{R}^{r\times n}$ are constant matrices. It is also assumed that $q \triangleq \operatorname{rank}E < n$ and system (3-1.1) is regular.

Let the state feedback control have the form of

$$u = K_1 x - K_2 \dot{x} + v \tag{3-1.2}$$

where the first term is proportional feedback and the second is derivative feedback, which may be viewed as the "speed" feedback. $v(t)$ is the new input control; and K_1, $K_2 \in \mathbb{R}^{m\times n}$ are constant matrices. Control (3-1.2) is the proportional and derivative (P-D) state feedback.

In singular systems, control input in the form of (3-1.2) is often used.

The closed-loop system formed by (3-1.1) and (3-1.2) is

$$\begin{aligned} (E + BK_2)\dot{x} &= (A + BK_1)x + Bv \\ y &= Cx \end{aligned} \tag{3-1.3}$$

Particularly, when $K_2 = 0$, the feedback control (3-1.2) becomes

$$u = K_1 x + v \tag{3-1.4}$$

which is a pure proportional (P-) state feedback, as in the normal system case. The corresponding closed-loop system is

$$\begin{aligned} E\dot{x} &= (A + BK_1)x + Bu \\ y &= Cx. \end{aligned} \tag{3-1.5}$$

To guarantee the existence and uniqueness of solution for any control input, we will hereafter assume that feedback control is confined to the set such that both (3-1.3) and (3-1.5) are regular.

3-1.1. Stability

It is wellknown that a practical system should be stable, otherwise, it could not work properly or be destroyed. Thus, stability is the most important property in a system. Therefore, we are most concerned with the stability of closed-loop systems.

Definition 3-1.1. Singular system (3-1.1) is called stable if there exist scalars α, $\beta > 0$ such that when $u(t) \equiv 0$ for $t > 0$ its state $x(t)$ satisfies

$$\|x(t)\|_2 \leqslant \alpha e^{-\beta t}\|x(0)\|_2, \qquad t > 0. \tag{3-1.6}$$

By definition, we stress stability under the Lyapunov sense, or asymptotic stability. When system (3-1.1) is stable and $u(t) \equiv 0$, $t > 0$, $\lim_{t\to\infty} x(t) = 0$.

We call the polynomial

$$f(s) = |sE-A| = a_{n_1}s^{n_1} + \ldots + a_1 s + a_0$$

the characteristic polynomial for system (3-1.1). The finite s's satisfying $f(s) = |sE-A| = 0$ are called finite poles for the system, or poles for simplicity. Since E is singular, the number of finite poles is always smaller than n for singular systems. We will use $\sigma(E,A) = \{s \mid s \in \mathbb{C},\ s \text{ finite},\ |sE-A| = 0\}$ to denote the finite pole set for the system, and $\sigma(I,A)$ will be specified as $\sigma(A)$.

Example 3-1.1. In circuit network (1-3.19),

$$E = \begin{bmatrix} 1 & 0 & 0 & 0 \\ 0 & 0 & 1 & 0 \\ 0 & 0 & 0 & 0 \\ 0 & 0 & 0 & 0 \end{bmatrix}, \qquad A = \begin{bmatrix} 0 & 1 & 0 & 0 \\ 1 & 0 & 0 & 0 \\ -1 & 0 & 0 & 1 \\ 0 & 1 & 1 & 1 \end{bmatrix}.$$

The characteristic polynomial of this system is

$$f(s) = |sE-A|$$

$$= \begin{vmatrix} s & -1 & 0 & 0 \\ -1 & 0 & s & 0 \\ 1 & 0 & 0 & -1 \\ 0 & -1 & -1 & -1 \end{vmatrix} = s^2 + s + 1 = (s+\frac{1}{2}+\frac{\sqrt{3}}{2}i)(s+\frac{1}{2}-\frac{\sqrt{3}}{2}i).$$

Thus the finite pole set is $\sigma(E,A) = \{-\frac{1}{2}-\frac{\sqrt{3}}{2}i,\ -\frac{1}{2}+\frac{\sqrt{3}}{2}i\}$.

Theorem 3-1.1. System (3-1.1) is stable if and only if

$$\sigma(E,A) \subset \mathbb{C}^-,$$

where $\mathbb{C}^- = \{s \mid s \in \mathbb{C},\ \mathrm{Re}(s) < 0\}$ represents the open left half complex plane.

Proof. When $u(t) \equiv 0$, $t > 0$, the state response $x(t)$ of system (3-1.1) satisfies

$$E\dot{x} = Ax, \quad x(0) = x_o. \tag{3-1.7}$$

Note that system (3-1.1) (thus (3-1.7)) is regular. There exist two nonsingular matrices Q and P such that (3-1.7) is r.s.e. to EF1:

$$\dot{x}_1(t) = A_1x_1(t), \qquad x_1(0) = x_{1o} \tag{3-1.8a}$$

$$N\dot{x}_2(t) = x_2(t), \qquad x_2(0) = x_{2o} \tag{3-1.8b}$$

where $QEP = \mathrm{diag}(I, N)$, $QAP = \mathrm{diag}(A_1, I)$, $x(t) = P[x_1(t)/x_2(t)]$, and $x_o = P[x_{1o}/x_{2o}]$.

Thus, system (3-1.8) has the solution

$$\left.\begin{aligned} x_1(t) &= e^{A_1t}x_{1o} \\ x_2(t) &= 0 \end{aligned}\right. \qquad t > 0.$$

Since $x_2(t) = 0$, $t > 0$, inequality (3-1.6) holds in and only if $x_1(t)$ satisfies

$$\| x_1(t)\|_2 < \| P \|_2^{-1} \alpha e^{-\beta t}\|x(0)\|_2, \qquad t > 0,$$

which is equivalent to $\sigma(A_1) \subset \mathbb{C}^-$.

On the other hand, $\sigma(E,A) = \sigma(QEP,QAP) = \sigma(\mathrm{diag}(I, N), \mathrm{diag}(A_1, I)) = \sigma(A_1)$, which means that (3-1.6) is true if and only if $\sigma(E,A) = \sigma(A_1) \subset \mathbb{C}^-$. Q.E.D.

Example 3-1.2. From Example 3-1.1, system (1-3.19) has the pole set

$$\sigma(E,A) = \{-\frac{1}{2}-\frac{3}{2}i,\ -\frac{1}{2}+\frac{3}{2}i\}.$$

Thus, it is stable by Theorem 3-1.1.

Example 3-1.3. Consider system (2-2.5). Here

$$E = \begin{bmatrix} C_1 & 0 & 0 & 0 \\ 0 & C_2 & 0 & 0 \\ 0 & 0 & -L & 0 \\ 0 & 0 & 0 & 0 \end{bmatrix}, \qquad A = \begin{bmatrix} 0 & 0 & 0 & 1 \\ 0 & 0 & 1 & 0 \\ -1 & 1 & 0 & 0 \\ 1 & 0 & 0 & R \end{bmatrix}$$

and direct computation shows its characteristic polynomial is

$$|sE-A| = (RC_1s+1)(s^2C_2L+1).$$

Since R, C_1, C_2, L > 0 for a real system, we have

$$\sigma(E,A) = \{-\frac{1}{RC_1}, \frac{1}{C_2L}i, \frac{-1}{C_2L}i\}.$$

From Theorem 3-1.1 it is unstable. In fact, it is an oscillation system whose output u_{c_2} (obtained from direct computation from (2-2.5)-(2-2.6)) satisfies

$$C_1C_2RLu_{c_2}^{(3)}(t) + C_2Lu_{c_2}^{(2)}(t) + C_1R\dot{u}_{c_2}(t) + u_{c_2}(t) = u_e(t). \quad (3\text{-}1.9)$$

Assume that the system is static at beginning, $u_{c_2}(0) = \dot{u}_{c_2}(0) = u_{c_2}^{(2)}(0) = 0$, and $u_e(t) = \delta(t)$ is the unit impulse term at the starting point. To solve $u_{c_2}(t)$, we take the Laplace transformation on both sides of (3-1.9). Then

$$u_{c_2}(s) = \frac{1}{(s^2C_2L+1)(C_1Rs+1)}$$

$$= \frac{(C_1R)^2}{(C_1R)^2+C_2L}\frac{1}{sC_1R+1} - \frac{C_2L}{(C_1R)^2+C_2L}\left[\frac{C_1R}{2\sqrt{C_2L}}\left(\frac{1}{s\sqrt{C_2L}+i}+\frac{1}{s\sqrt{C_2L}-i}\right)\right.$$

$$\left. + \frac{1}{2i}\left(\frac{1}{s\sqrt{C_2L}+i}-\frac{1}{s\sqrt{C_2L}-i}\right)\right].$$

The inverse Laplace transformation yields the represention of $u_{c_2}(t)$:

$$u_{c_2}(t) = \frac{C_1R}{(C_1R)^2+C_2L}e^{-\frac{t}{C_1R}} - \frac{C_2L}{(C_1R)^2+C_2L}\left(\frac{C_1R}{2\sqrt{C_2L}}\cos\frac{t}{\sqrt{C_2L}} - \sin\frac{t}{\sqrt{C_2L}}\right).$$

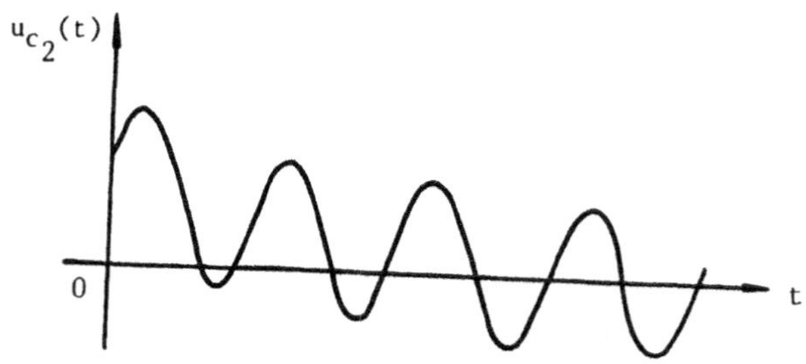

Figure 3-1.1.

Figure 3-1.1 shows the locus of this oscillation system.

3-1.2. Stabilizability and detectability

Definition 3-1.2. System (3-1.1) is called stabilizable if there exists a state feedback (3-1.4) such that the closed-loop system (3-1.5) is stable.

From Definition 3-1.2 and Theorem 3-1.1, we can prove the following theorem.

Theorem 3-1.2. System (3-1.1) is stabilizable if and only if

$$\mathrm{rank}[sE-A,\ B] = n, \quad \forall\ s \in \bar{\mathbb{C}}^+, \quad s \text{ finite} \tag{3-1.10}$$

where $\bar{\mathbb{C}}^+$ represents the closed right half complex plane; $\bar{\mathbb{C}}^+ = \{s \mid s \in \mathbb{C},\ \mathrm{Re}(s) \geqslant 0\}$, which, in turn, is equivalent to the stabilizability of the slow subsystem.

Proof. Necessity: According to Definition 3-1.2 and Theorem 3-1.1, system (3-1.1) is stabilizable iff there exists a $K_1 \in \mathbb{R}^{m\times n}$ such that $\sigma(E, A+BK_1) \subset \mathbb{C}^-$, i.e.,

$$\mathrm{rank}[sE-(A+BK_1)] = n, \quad \forall\ s \in \bar{\mathbb{C}}^+, \quad s \text{ finite}.$$

Noticing $\mathrm{rank}[sE-(A+BK_1)] = \mathrm{rank}[sE-A,\ B][I/K_1]$, we have $\mathrm{rank}[sE-A,\ B] \geqslant n$. On the other hand $[sE-A,\ B]$ is of n rows. Thus, equality holds. It is just (3-1.10).

Sufficiency: Assume that (3-1.10) holds. From Lemma 1-2.2 there must exist nonsingular matrices Q and P, such that system (3-1.1) is r.s.e. to

$$\dot{x}_1 = A_1x_1 + B_1u \tag{3-1.11a}$$

$$N\dot{x}_2 = x_2 + B_2u \tag{3-1.11b}$$

$$y = C_1x_1 + C_2x_2$$

where

$$QEP = \mathrm{diag}(I,\ N); \quad QAP = \mathrm{diag}(A_1,\ I); \quad QB = [B_1/B_2]; \quad CP = [C_1,\ C_2];$$

$$x = P[x_1/x_2], \quad x_1 \in \mathbb{R}^{n_1}, \quad x_2 \in \mathbb{R}^{n_2}; \quad n_1 + n_2 = n.$$

Paying attention to the nilpotent property of N, from (3-1.10) and (3-1.11) we have

$$\mathrm{rank}[sE-A,\ B] = \mathrm{rank}[sQEP-QAP,\ QB] = n_2 + \mathrm{rank}[sI-A_1,\ B_1] = n,$$

$$\forall\ s \in \bar{\mathbb{C}}^+, \quad s \text{ finite}.$$

Thus $\mathrm{rank}[sI-A_1,\ B_1] = n - n_2 = n_1$, $\forall\ s \in \bar{\mathbb{C}}^+$, and s is finite, indicating the stabilizability of the slow subsystem. Therefore, a matrix $\bar{K}_1 \in \mathbb{R}^{m\times n_1}$ may be chosen such that $\sigma(A_1+B_1\bar{K}_1) \subset \mathbb{C}^-$. Let $K_1 = [\bar{K}_1,\ 0]P^{-1} \in \mathbb{R}^{m\times n}$. Direct computation results in

$$\sigma(E, A+BK_1) = \sigma(QEP, Q(A+BK_1)P) = \sigma(A_1+B_1\bar{K}_1) \subset \mathbb{C}^-.$$

The system is stabilizable by definition and Theorem 3-1.1. Q.E.D.

(3-1.10) gives the stabilization criterion, which shows that the stabilization property is determined uniquely by its slow subsystem. The system is stabilizable if its slow subsystem is stabilizable, otherwise it is not. Combining it with Theorem 3-1.2, we have the following corollary.

Corollary 3-1.1. System (3-1.1) is stabilizable if it is R-controllable, but the inverse is not true.

Example 3-1.4. System (1-3.19) is stable. Thus it is stabilizable; The unstable system (2-2.5) is also stabilizable because it is R-controllable by Example 2-2.3. In fact, for this system we need only to set

$$K_1 = [1-3RC_1,\ sRC_1-LRC_1C_2,\ \frac{RC_1}{C_2} -3LRC_1,\ 0]. \tag{3-1.12}$$

The feedback control would be

$$\begin{aligned} u_e(t) &= K_1x + v \\ &= (1-3RC_1)u_{c_1} + (3RC_1-LRC_1C_2)u_{c_2} + (\frac{RC_1}{C_2} -3LRC_1)I_2 + v \end{aligned} \tag{3-1.13}$$

where v(t) is the new voltage source. Direct computation shows

$$\sigma(E,A+BK_1) = \{-1,\ -1,\ -1\} \subset \mathbb{C}^-. \tag{3-1.14}$$

Thus, the closed-loop system is stable.

As has been pointed out in Section 2-5, for regular system (3-1.1) there exist nonsingular matrices Q_1 and P_1 such that (3-1.1) is r.s.e. to the controllability canonical form (2-5.8)

$$\begin{aligned} \begin{bmatrix} E_{11} & E_{12} \\ 0 & E_{22} \end{bmatrix}\begin{bmatrix} \dot{x}_1 \\ \dot{x}_2 \end{bmatrix} &= \begin{bmatrix} A_{11} & A_{12} \\ 0 & A_{22} \end{bmatrix}\begin{bmatrix} x_1 \\ x_2 \end{bmatrix} + \begin{bmatrix} B_1 \\ 0 \end{bmatrix} u \\ y &= [C_1,\ C_2][x_1/x_2] \end{aligned} \tag{3-1.15}$$

where

$$Q_1EP_1 = \begin{bmatrix} E_{11} & E_{12} \\ 0 & E_{22} \end{bmatrix}, \quad Q_1AP_1 = \begin{bmatrix} A_{11} & A_{12} \\ 0 & A_{22} \end{bmatrix}, \quad Q_1B = \begin{bmatrix} B_1 \\ 0 \end{bmatrix}$$

$$CP_1 = [C_1,\ C_2], \qquad x = P_1[x_1/x_2],$$

and (E_{11},A_{11},B_1) is controllable.

It is easy to verify from (3-1.10) that the following corollary holds.

Corollary 3-1.2. System (3-1.1) is stabilizable if and only if in its canonical form (3-1.15) A_{22} is a stable matrix, $\sigma(A_{22}) \subset \mathbb{C}^-$.

This corollary shows that the stabilizable system has only stable uncontrollable poles. It also shows that there exists a feedback control

$$u(t) = f(x(t), t) \tag{3-1.16}$$

where $f(x(t), t)$ is a vector function of $x(t)$ and t, such that the closed-loop system formed by (3-1.1) and (3-1.16) is stable if and only if it is stabilizable. Thus any feedback in the form of (3-1.16) cannot stabilize system (3-1.1) if it is not stabilizable. This indicates the equivalence of the existence of (3-1.2) stabilizing (3-1.3) and that of (3-1.4) stabilizing (3-1.5).

The stabilizability characterizes the controllability of the system's stability. The dual concept of stabilizability is detectability which is defined as follows.

Definition 3-1.3. System (3-1.1) is detectable if its dual system $(E^\tau, A^\tau, B^\tau, C^\tau)$ is stabilizable.

This is a mathematical definition. In the next chapter, we will give another definition. A direct result of Definition 3-1.3 and Theorem 3-1.2 is the following theorem.

Theorem 3-1.3. System (3-1.1) is detectable if and only if

$$\operatorname{rank}[sE-A/C] = n, \quad \forall\ s \in \bar{\mathbb{C}}^+,\ s \text{ finite},$$

which is equivalent to the detectability of its slow subsystem.

Thus stabilizability and detectability are determined by the slow subsystem and by alone. This indicates the fact that they really only characterize properties of slow subsystem although they are defined for the whole system.

The R-observability, along with Theorem 3-1.3, shows that detectability may be deduced from R-observability, but its inverse is not true.

Example 3-1.5. Consider the system

$$\begin{bmatrix} 1 & 0 & 0 \\ 1 & -1 & 0 \\ 0 & 0 & 0 \end{bmatrix} \dot{x} = \begin{bmatrix} 3 & -2 & 0 \\ -1 & 1 & 0 \\ 0 & 2 & 1 \end{bmatrix} x + \begin{bmatrix} 1 \\ 0 \\ 0 \end{bmatrix} u$$

$$y = [1 \quad 0 \quad \tfrac{1}{4}]x.$$

We have

$$\operatorname{rank}\begin{bmatrix} sE - A \\ \\ C \end{bmatrix} = \operatorname{rank}\begin{bmatrix} s-3 & 2 & 0 \\ s+1 & -s-1 & 0 \\ 0 & -2 & -1 \\ 1 & 0 & \tfrac{1}{4} \end{bmatrix} = 3 = n, \quad \forall\ s \in \bar{\mathbb{C}}^+,\ s \text{ finite}$$

It is detectable by Theorem 3-1.3, but when $s = -1$,

$$\text{rank}\begin{bmatrix} sE - A \\ C \end{bmatrix}_{s=-1} = \text{rank}\begin{bmatrix} -4 & 2 & 0 \\ 0 & 0 & 0 \\ 0 & -2 & -1 \\ 1 & 0 & \frac{1}{4} \end{bmatrix} = 2 < n.$$

Thus it is not R-observable.

From Corollary 3-1.2 we could also obtain a further corollary.

Corollary 3-1.3. Let

$$\left(\begin{bmatrix} E_{11} & 0 \\ E_{21} & E_{22} \end{bmatrix}, \begin{bmatrix} A_{11} & 0 \\ A_{21} & A_{22} \end{bmatrix}, \begin{bmatrix} B_1 \\ B_2 \end{bmatrix}, [C_1, 0]\right)$$

be an observability canonical form for system (3-1.7) under r.s.e. equivalence, where $(E_{11}, A_{11}, B_1, C_1)$ is observable. Then, the necessary and sufficient condition of detectability is that the system has only stable unobservable poles, i.e., $\sigma(E_{11}, A_{11}) \subset \mathbb{C}^-$.

Example 3-1.6. It is easy to verify from Corollary 3-1.3 that

$$\begin{bmatrix} 1 & 0 \\ 0 & 0 \end{bmatrix}\dot{x} = \begin{bmatrix} 1 & 0 \\ 0 & 1 \end{bmatrix}x + \begin{bmatrix} 1 \\ 0 \end{bmatrix}u, \qquad y = [0 \quad 1]x$$

is not detectable.

3-1.3. Normalizability

Definition 3-1.4. System (3-1.1) is called normalizable if a feedback control (3-1.2) may be chosen such that its closed-loop (3-1.3) is normal, i.e.,

$$|E+BK_2| \neq 0. \tag{3-1.17}$$

While for any matrix $K_1 \in \mathbb{R}^{m\times n}$, the closed-loop system under (3-1.4) is always singular. Under the assumption of normalizability for system (3-1.1), we may choose appropriate feedback control (3-1.2) such that its closed-loop system is normal, enabling the application of results in normal system theory to singular ones. As long as condition (3-1.17) is satisfied, the closed-loop system (3-1.3) would become

$$\begin{aligned} \dot{x} &= (E+BK_2)^{-1}(A+BK_1)x + (E+BK_2)^{-1}Bv \\ y &= Cx \end{aligned} \tag{3-1.18}$$

and its plain feature is its n finite poles, without any infinite poles. Under the normalizability assumption, a singular system's infinite poles may be set to finite via feedback (3-1.2). Thus, no impulse terms appear in the closed-loop state response.

Theorem 3-1.4. The following statements are equivalent.

(a). System (3-1.1) is normalizable.

(b). $\text{rank}[E \quad B] = n$.

(c). The fast subsystem (1-3.11b) is controllable.

(d). In (3-1.11), $\text{rank}[N, B_2] = n_2$.

(e). For any nonsingular matrices Q_1 and P_1 satisfying $Q_1EP_1 = \text{diag}(I, 0)$, $Q_1B = [\bar{B}_1/\bar{B}_2]$, $\bar{B}_2$ is of full row rank n-rankE.

Proof. Equivalence among (b)-(e) has been shown in Theorem 2-2.1. We need only to prove the equivalence of (a) and (b).

Assume that (a) holds. By definition there exists a matrix $K_2 \in \mathbb{R}^{m\times n}$ such that $|E+BK_2| \neq 0$, i.e.,

$$\text{rank}[E+BK_2] = \text{rank}[E \quad B][I/K_2] = n.$$

Thus, $\text{rank}[E \quad B] = n$, which is (b).

Conversely, if (b) holds, from the matrix rank decomposition $Q_1EP_1 = \text{diag}(I, 0)$, where Q_1 and P_1 are nonsingular, and denoting $Q_1B = [\bar{B}_1/\bar{B}_2]$, we know from (e) that $\bar{B}_2$ must be of full row rank n-rankE, indicating $\bar{B}_2\bar{B}_2^{\tau} > 0$ is symmetric positive definite. Let $K_2 = [0 \quad \bar{B}_2^{\tau}]P_1^{-1} \in \mathbb{R}^{m\times n}$. Then

$$|E+BK| = |Q_1^{-1}||P_1^{-1}||\bar{B}_2\bar{B}_2^{\tau}| \neq 0.$$

System (3-1.1) is normalizable, i.e., (a) is true. Q.E.D.

As previously pointed out, for a given singular system, stabilizability and detectability characterize its slow subsystem properties, while normalizability is the opposite, reflecting only its fast subsystem properties. As for the whole system, normalizability clarifies the difference between singular and normal systems ---- a normalizable singular system may be changed into a normal one via suitable selection of feedback (3-1.2).

Example 3-1.7. Consider system (1-3.19). Since

$$\text{rank}[E \quad B] = \text{rank}\begin{bmatrix} 1 & 0 & 0 & 0 & 0 \\ 0 & 0 & 1 & 0 & 0 \\ 0 & 0 & 0 & 0 & 0 \\ 0 & 0 & 0 & 0 & -1 \end{bmatrix} = 3 < n = 4$$

this system is not normalizable according to Theorem 3-1.4.

Example 3-1.8. In system (2-2.5)-(2-2.6),

$$E = \begin{bmatrix} C_1 & 0 & 0 & 0 \\ 0 & C_2 & 0 & 0 \\ 0 & 0 & -L & 0 \\ 0 & 0 & 0 & 0 \end{bmatrix}, \qquad B = \begin{bmatrix} 0 \\ 0 \\ 0 \\ -1 \end{bmatrix}$$

and

$$\mathrm{rank}[E \quad B] = 4.$$

Thus it is normalizable. Let

$$K_2 = [0 \quad 0 \quad 0 \quad -1].$$

We have

$$|E+BK_2| = \begin{vmatrix} C_1 & 0 & 0 & 0 \\ 0 & C_2 & 0 & 0 \\ 0 & 0 & -L & 0 \\ 0 & 0 & 0 & 1 \end{vmatrix} = -LC_1C_2 \neq 0.$$

Example 3-1.9. By definition, the singular system with only slow subsystem (a normal system)

$$\dot{x} = Ax + Bu, \qquad y = Cx$$

is normalizable.

From Theorem 3-1.4 and controllability criterion, we see that a system is normalizable if it is controllable. Thus, if (3-1.15) is the controllability canonical form under r.s.e., the necessary and sufficient condition for the normalizability of system (3-1.1) is that E_{22} is nonsingular in (3-1.15).

Decomposition (2-5.9) is the canonical normalizability decomposition.

A system is impulse controllable if it is normalizable; the inverse is false. For example, it has been pointed out in Example 2-2.7 that system (2-1.8) (thus system (1-3.19)) is impulse controllable. But it is not normalizable, as proven in Example 3-1.7.

3-1.4. Dual normalizability

Definition 3-1.5. System (3-1.1) is dual normalizable if its dual system is normalizable.

We have the following theorem.

Theorem 3-1.5. System (3-1.1) is dual normalizable if and only if rank[E/C] = n.

It has been shown in Section 1-3 that any regular system (3-1.1) is r.s.e. to EF3:

$$\begin{aligned} \hat{E}\dot{x} &= (I - \alpha\hat{E})x + \hat{B}u \\ y &= Cx \end{aligned} \tag{3-1.9}$$

where α satisfies $|\alpha E+A| \neq 0$. Under this equivalent form we can prove the following theorem.

Theorem 3-1.7. Consider system (3-1.19).

(1). It is stabilizable if and only if

$$\operatorname{rank}[sI-\hat{E},\ \hat{B}] = n, \quad \forall\ s \neq 0,\ s \in \bar{\mathbb{C}}^{+}, \quad s \text{ finite}.$$

Or equivalently, the uncontrollable poles of $(\hat{E},\hat{B})$ must be stable or zero.

(2). It is normalizable iff $\operatorname{rank}[\hat{E}\ \ \hat{B}] = n$.

(3). It is detectable iff $\operatorname{rank}[sI-\hat{E}/C] = n$, $\forall\ s \neq 0$, $s \in \bar{\mathbb{C}}^{+}$, s finite.

(4). It is dual normalizable if and only if $\operatorname{rank}[\hat{E}/C] = n$.

Proof. We need only to prove (1) and (2) since (3) and (4) may be obtained by the dual principle.

(1). According to Theorem 3-1.2, the necessary and sufficient condition of stabilizability for system (3-1.19) is

$$\operatorname{rank}[s\hat{E}-(I-\alpha\hat{E}),\ \hat{B}] = \operatorname{rank}[(s+\alpha)\hat{E}-I,\ \hat{B}] = n, \quad \forall\ s \in \bar{\mathbb{C}}^{+}, \quad s \text{ finite}.$$

This holds if $s+\alpha = 0$; otherwise, by setting $\lambda = 1/(s+\alpha)$, it may be rewritten as

$$\operatorname{rank}[\lambda I-\hat{E},\ \hat{B}] = n, \quad \forall\ \lambda \neq 0, \quad \lambda \in \bar{\mathbb{C}}^{+}, \quad \lambda \text{ finite},$$

which is (1).

(2). Statement is obvious. Q.E.D.

So far, we have introduced various concepts in singular system theory, including controllability, R-controllability, impulse controllability, observability, R-observability, impulse observability, normalizability, dual normalizability, stabilizability, and detectability. Among these concepts there exist both similarities and differences. Their relationships are complicated, but they characterize the singular system from different points of view. As for their intrinsic properties, while controllability and observability characterize the properties for the whole system, R-controllability, R-observability, stabilizability, and detectability characterize properties for slow subsystem; and impulse controllability, impulse observability, normalizability, and dual normalizability for the fast subsystem. These concepts cannot be substituted for each other.

To sum up, the relationships among these concepts may be described by the following diagram:

controllability ⟶ R-controllability ⟶ stabilizability

controllability ⟶ normalizability ⟶ impulse controllability

controllability ⟷ R-controllability + normalizability

observability → R-observability → stabilizability

observability → dual normalizability → impulse observability

observability ⇌ R-observability + dual normalizability

where A → B represents that B may be deduced from A, and A ⇌ B indicates the equivalence of A and B.

3-2. Infinite Pole Placement

Practically, it is expected to change a system's properties by control inputs so that the closed-loop system possesses the expected properties. Many facts show that pole placement is important in affecting the dynamic and static properties of a linear time-invariant system. Thus, pole placement is a basic probelm in system design, which studies the selection of feedback control so that poles of the closed-loop system lie at the expected places in the complex plane. For singular systems, not only finite poles, but infinite ones also affect the response properties of a system. In this and the next sections, we will study the pole placement problem for infinite and finite poles, respectively.

3-2.1. Infinite pole structure in singular systems

First, we will briefly outline the concept of infinite pole structure in singular systems.

Consider singular system (3-1.1):

$$\begin{aligned} E\dot{x} &= Ax + Bu \\ y &= Cx \end{aligned} \qquad (3\text{-}2.1)$$

For system (3-2.1) the finite pole structure is determined by the zero structure of $sE-A$ at finite s and its infinite pole structure is defined as isomorphic to the zero structure of $sE-A$ at infinity, which in turn is isomorphic to that of $\frac{1}{s}E - A$ at $s = 0$.

For any regular system (3-2.1) there always exist two nonsingular matrices Q and P such that

$$QEP = \mathrm{diag}(I_{n_1}, N), \qquad QAP = \mathrm{diag}(A_1, I_{n_2})$$

where $n_1 + n_2 = n$, N nilpotent. Note that

$$Q(sE-A)P = \mathrm{diag}(sI-A_1,\ sN-I).$$

Then, the finite pole structure of system (3-2.1) is completely determined by A_1, while the infinite pole structure is isomorphic to the zero structure of

$$\frac{1}{s}N - I \tag{3-2.2}$$

at $s = 0$.

Therefore, if $N = 0$, from (3-2.2) we know that system (3-2.1) has no infinite poles, and vice versa. When $N \neq 0$, let

$$N = GF \tag{3-2.3}$$

be the full rank decomposition of N, in which G and F are of full column and row rank, respectively. Thus, an irreducible realization of $\frac{1}{s}N - I$ is (following the notation of Verghese et al., 1979):

$$R(s) = \left[\begin{array}{c|c} sI & F \\ \hline G & -I \end{array}\right] = \left[\begin{array}{c|c} T(s) & U(s) \\ \hline V(s) & W(s) \end{array}\right], \tag{3-2.4}$$

i.e.,

$$\frac{1}{s}N - I = V(s)T^{-1}(s)U(s) + W(s).$$

According to the Smith-McMillan definition of the pole-zero structure of a rational matrix (Rosenbrock, 1970; Verghese et al., 1979), (3-2.4) has the same finite zero structure as (3-2.2). Since the zero structure of (3-2.4) is determined by

$$\begin{bmatrix} sI + FG & 0 \\ 0 & -I \end{bmatrix},$$

or $sI + FG$, we have the following proposition.

Proposition 3-2.1. If $N = 0$, system (3-2.1) has no infinite poles, otherwise, its infinite pole structure is siomorphic to the zero structure of FG.

Note that N is nilpotent. Assume that

$$N = \begin{bmatrix} 0 & I_{n_2-1} \\ 0 & 0 \end{bmatrix} \in \mathbb{R}^{n_2 \times n_2}.$$

Then it has the full rank decomposition

$$N = GF, \qquad G = \begin{bmatrix} I_{n_2-1} \\ 0 \end{bmatrix}, \qquad F = [0 \ \ I_{n_2-1}].$$

Thus

$$FG = \begin{bmatrix} 0 & I_{n_2-2} \\ 0 & 0 \end{bmatrix} \in \mathbb{R}^{(n_2-1)\times(n_2-1)}.$$

This shows that when N is a Jordan block matrix, the infinite pole structure of system (3-2.1) is determined by 1-order lower Jordan block.

Note that N = 0 is the necessary and sufficient condition for the nonexistence of both infinite poles and impulse terms in the state response of system (3-2.1).

Proposition 3-2.2. No impulse terms exist in the state response of system (3-2.1) if and only if this system has no infinite poles.

From the preceding discussion, the (geometric) multiplicity of infinite poles is that of zero in N but minus one, i.e., lower than the order of Jordan blocks (belonging to zero eigenvalue) by one number. In such senses, and from the state response structure of singular systems, the impulse terms in state response of (3-2.1) are determined by the pole multiplicity at infinity. If the system has no infinite poles, there are no impulse terms in state response; conversely, if there exist infinite poles, there exist impulse terms in state response. The higher the multiplicity, the higher the order of derivatives in impulse terms. Its highest order is equal to the highest multiplicity of infinite poles.

In a parctical system, the impulse term and its derivatives are usually not expected in the state response, otherwise, strong impulse behavior may saturate the system so that it could not work or may even destroy it. Thus, except for a few cases (such as on impulse generating mechanism), impulse terms are always avoided in the state response. This probelm is equivalent to selecting feedback controls such that the closed-loop system has no infinite poles, or N = 0. this is the main problem studied in this section.

Example 3-2.1. In the standard decomposition (1-3.20), N = 0. Thus, system (1-3.19) has no infinite poles or impulse terms in its state response.

Example 3-2.2. Consider the system

$$\begin{bmatrix} 0 & 0 & 0 \\ 1 & 0 & 0 \\ 0 & 1 & 0 \end{bmatrix} \dot{x} = x + \begin{bmatrix} 1 \\ 0 \\ 0 \end{bmatrix} u$$

$$y = Cx.$$

In this system, N is a third-order Jordan block belonging to zero eigenvalue. Thus, it has an infinite pole of multiplicity 2. Impulse terms exist in its state response.

3-2.2. Infinite pole placement

Consider the pure proportional (P-) state feedback control

$$u = Kx + Bv \tag{3-2.5}$$

where v(t) is the new control input. Then (3-2.1) and (3-2.5) form the closed-loop system

$$\begin{aligned} E\dot{x} &= (A+BK)x + Bv \\ y &= Cx \end{aligned} \qquad (3\text{-}2.6)$$

Note that the infinite pole structure is determined by the fast subsystem of a system. Thus, feedback control achieves its control goal by changing the slow-fast subsystem structure.

To begin, we have the following lemma.

Lemma 3-2.1. The closed-loop system (3-2.6) has no infinite poles if and only if

$$\deg(|sE - (A+BK)|) = \text{rank}E.$$

Proof. For singular system (3-2.6), there exist nonsingular matrices Q and P such that it is r.s.e. to EF1

$$\begin{aligned} \dot{x}_1 &= A_1x_1 + B_1v \\ N\dot{x}_2 &= x_2 + B_2v \end{aligned}$$

where N is nilpotent. From the preceding discussion, system (2-3.6) has no infinite poles if and only if N = 0. Thus,

$$\deg(|sE - (A+BK)|) = \deg(|sI - A_1|) = \text{rank}E. \qquad \text{Q.E.D.}$$

Theorem 3-2.1. For system (3-2.1), there exists a feedback control (3-2.5) so that the closed-loop system has no infinite poles if and only if it is impulse controllable.

Proof. Lemma 3-2.1 shows the existence of feedback (3-2.5) such that the closed-loop system (3-2.6) has no infinite poles iff there exists a matrix K satisfying

$$\deg(|sE - (A+BK)|) = \text{rank}E. \qquad (3\text{-}2.7)$$

Necessity: Let (3-2.7) hold. Then there exist nonsingular matrices T_1 and T_2 such that

$$T_1ET_2 = \text{diag}(I_q, 0), \qquad q = \text{rank}E.$$

Denoting

$$T_1AT_2 = \begin{bmatrix} A_{11} & A_{12} \\ A_{21} & A_{22} \end{bmatrix}, \qquad T_1B = \begin{bmatrix} B_1 \\ B_2 \end{bmatrix}, \qquad KT = [K_1, K_2],$$

from (3-1.5) we have

$$\begin{aligned} \text{rank}E &= \deg(|sE - (A+BK)|) = \deg(|sT_1ET_2 - T_1(A+BK)T_2|) \\ &= \deg\left(\begin{vmatrix} sI_q - A_{11} - B_1K_1 & -A_{12} - B_1K_2 \\ -A_{21} - B_2K_1 & -A_{22} - B_2K_2 \end{vmatrix}\right) \leqslant q. \end{aligned}$$

The equality holds if and only if $|-A_{22}-B_2K_2| \neq 0$, i.e., $|A_{22}+B_2K_2| \neq 0$, indicating the normalizability of (A_{22},B_2), $\text{rank}[A_{22}, B_2] = n-\text{rank}E$. Thus, system (3-2.1) is impulse controllable.

Sufficiency: Assume that system (3-2.1) is impulse controllable. For the preceding decomposition, by Theorem 2-4.2 we know

$$\text{rank}[A_{22}, B_2] = n - \text{rank}E$$

(A_{22},B_2) is normalizable. Therefore, there exists a matrix K_2 such that

$$|A_{22}+B_2K_2| \neq 0.$$

Setting $K = [0 \quad K_2]T_2^{-1}$, direct computation yields

$$\deg(|sE-(A+BK)|) = \deg\left(\begin{vmatrix} sI-A_{11} & -A_{12}-B_1K_2 \\ -A_{21} & -A_{22}-B_2K_2 \end{vmatrix}\right) = \text{rank}E,$$

which assures the authenticity of (3-2.7). Q.E.D.

This theorem further reveals the relationship among impulse controllability, infinite poles, and impulse terms. Impulse controllability guarantees the ability to eliminate the impulse term in the state response in the closed-loop system via P-state feedback control. In this process the infinite poles are placed to the finite ones, indicating the closed-loop system has at most rankE finite poles and no infinite poles. Thus, the other infinite poles are driven to finite positions.

<u>Example 3-2.3.</u> Consider the following singular system

$$\begin{bmatrix} 0 & 0 & 0 \\ 1 & 0 & 0 \\ 0 & 1 & 0 \end{bmatrix}\dot{x} = x + \begin{bmatrix} 1 \\ 0 \\ 0 \end{bmatrix}u$$
$$y = [1 \quad 0 \quad -1]x. \tag{3-2.8}$$

Here we have

$$E = \begin{bmatrix} 0 & 0 & 0 \\ 1 & 0 & 0 \\ 0 & 1 & 0 \end{bmatrix}, \quad A = \begin{bmatrix} 1 & 0 & 0 \\ 0 & 1 & 0 \\ 0 & 0 & 1 \end{bmatrix}, \quad B = \begin{bmatrix} 1 \\ 0 \\ 0 \end{bmatrix}.$$

This system has only infinite poles, with pole set $\{\infty, \infty\}$. The multiplicity of the infinite pole is two. Thus, impulse terms are certain to appear in the state response.

It may be verified that system (3-2.8) is impulse controllable. There exists K, for example,

$$K = [0 \quad 1 \quad 1]$$

that satisfies $\text{rank}E = \deg(|sE-(A+BK)|) = 2$. For such Ks the closed-loop system

$$\begin{bmatrix} 0 & 0 & 0 \\ 1 & 0 & 0 \\ 0 & 1 & 0 \end{bmatrix} \dot{x} = \begin{bmatrix} 1 & 1 & 1 \\ 0 & 1 & 0 \\ 0 & 0 & 1 \end{bmatrix} x + \begin{bmatrix} 1 \\ 0 \\ 0 \end{bmatrix} v \tag{3-2.9}$$

$$y = [1 \quad 0 \quad -1] x$$

has no infinite poles. And the closed-loop (3-2.9) has the characteristic polynomial of

$$|sE - (A+BK)| = -(s^2 + s + 1),$$

which implies $\sigma(E, A+BK) = \{-\frac{1}{2} - \frac{\sqrt{3}}{2}i, -\frac{1}{2} + \frac{\sqrt{3}}{2}i\}$. No infinite poles appear in (3-2.9) Thus feedback control drives the two infinite poles in (3-2.8) to the finite positions.

Moreover, according to the proof, under the impulse controllability assumption, there exists a matrix K satisfying $\deg(|sE-(A+BK)|) = \text{rank}E$ if and only if

$$|A_{22} + B_2K_2| \neq 0. \tag{3-2.10}$$

Inspired by this point, we have the following algorithm for finding matrix K. We first take rank decomposition:

$$T_3A_{22}T_4 = \text{diag}(I_\mu, 0), \qquad \mu = \text{rank}A_{22}$$

where T_3 and T_4 are nonsingular. And then denote

$$T_3B_2 = \begin{bmatrix} B_{21} \\ B_{22} \end{bmatrix}.$$

Under the assumption of impulse controllability, $\text{rank}[A_{22}, B_2] = n - \text{rank}E$. B_{22} is of full row rank. Let

$$K_2 = [0 \quad B_{22}^\tau] T_4^{-1}.$$

We have $|A_{22} + B_2K_2| = |T_3^{-1}||T_4^{-1}||B_{22}B_{22}^\tau| \neq 0$. Thus, matrix $K = [0 \quad K_2]T_2^{-1}$ satisfies (3-2.7).

<u>Example 3-2.4.</u> Consider system (3-2.8). The nonsingular matrices

$$T_1 = \begin{bmatrix} 0 & 1 & 0 \\ 0 & 0 & 1 \\ 1 & 0 & 0 \end{bmatrix}, \qquad T_2 = I_3$$

transform

$$T_1ET_2 = \text{diag}(I_2, 0), \qquad T_1B = [0 \quad 0 \quad 1]^\tau, \quad T_1AT_2 = T_1$$

in which $A_{22} = 0$, $B_2 = 1$. The scalar $K_2 = 1$ satisfies $|A_{22} + B_2K_2| = 1 \neq 0$. Thus matrix

$$K = [0 \quad K_2]T_2^{-1} = [0 \quad 0 \quad 1]$$

would satisfy $\deg(|sE-(A+BK)|) = \deg(s^2+1) = 2 = \text{rank}E$.

In standard decomposition, there exist two nonsingular matrices Q and P, such that regular system (3-2.1) is r.s.e. to EF1:

$$\begin{aligned} \dot{x}_1 &= A_1x_1 + B_1u \\ N\dot{x}_2 &= x_2 + B_2u \\ y &= C_1x_1 + C_2x_2 \end{aligned} \qquad (3\text{-}2.11)$$

where $x_1 \in \mathbb{R}^{n_1}$, $x_2 \in \mathbb{R}^{n_2}$, $n_1 + n_2 = n$, and

$$QEP = \text{diag}(I, N), \quad QAP = \text{diag}(A_1, I), \quad CP = [C_1, C_2]$$

$$QB = \begin{bmatrix} B_1 \\ B_2 \end{bmatrix}, \qquad x = P\begin{bmatrix} x_1 \\ x_2 \end{bmatrix}.$$

Under decomposition (3-2.11), the P-state feedback $u = Kx + v$ becomes

$$u = \bar{K}_1x_1 + \bar{K}_2x_2 + v \qquad (3\text{-}2.12)$$

where $KP = [\bar{K}_1, \bar{K}_2]$, $\bar{K}_1 \in \mathbb{R}^{m\times n_1}$, $\bar{K}_2 \in \mathbb{R}^{m\times n_2}$.

A. When $\bar{K}_2 = 0$, feedback control (3-2.12) becomes

$$u = \bar{K}_1x_1 + v. \qquad (3\text{-}2.13)$$

In this feedback form, only substate x_1 is used for feedback purposes. We thus call it the slow substate feedback control, slow feedback hereafter.

B. If $\bar{K}_1 = 0$, feedback control (3-2.12) becomes

$$u = \bar{K}_2x_2 + v \qquad (3\text{-}2.14)$$

in which only the fast substate is derived. We will call it the fast substate feedback control, fast feedback hereafter.

C. In general, feedback control (3-2.12) includes both slow and fast substates. We will call it the combined slow-fast substate feedback, state feedback hereafter.

While the standard decomposition characterizes the system's inner structure, the slow and fast feedbacks show the effect of partial state feedback on the whole system structure.

Theorem 3-2.2. Given system (3-2.1), there exists a state feedback control (3-2.12) such that the closed-loop system has no infinite poles if and only if there exists a fast feedback (3-2.14) such that the closed-loop system (3-2.6) has no infinite poles.

Proof. By Theorem 3-2.1, the impulse controllability for (3-2.1) is the necessary and sufficient condition for existing state feedback (3-2.6) such that no infinite poles exist in its closed-loop system. The impulse controllability for (3-2.1) is equivalent to the impulse controllability of the fast subsystem. Thus, from Theorem 3-2.1 there exists a matrix $\bar{K}_2$ such that

$$\deg(|sN - (I+B_2\bar{K}_2)|) = \text{rank}N$$

and vice versa. For such a $\bar{K}_2$, we have

$$\deg(|sE - (A+BK)|) = \deg(|sQEP - Q(A+BK)P|)$$
$$= n_1 + \deg(|sN - (I+B_2\bar{K}_2)|) = \text{rank}N + n_1 = \text{rank}E$$

indicating the conclusion. Q.E.D.

This theorem shows that the pole structure at infinity is determined only by the structure of the fast subsystem, revealing that the fast subsystem reflects the system's pole structure at infinity.

Previously, we studied the distribution of infinite poles under P-state feedback. For the more general P-D state feedback control

$$u = K_1x - K_2\dot{x} + v \qquad (3\text{-}2.15)$$

where $K_1, K_2 \in \mathbb{R}^{m\times n}$, the closed-loop system formed by (3-2.1) and (3-2.15) is

$$\begin{aligned}(E+BK_2)\dot{x} &= (A+BK_1)x + Bv\\ y &= Cx.\end{aligned} \qquad (3\text{-}2.16)$$

Under the assumption of normalizability, we know that matrix $K_2 \in \mathbb{R}^{m\times n}$ may be chosen such that $|E+BK_2| \neq 0$. Therefore

Theorem 3-2.3. Assume that system (3-2.1) is normalizable. Then there exists feedback control (3-2.15) such that the closed-loop system (3-2.16) is normal, and thus has no infinite poles.

We noticed in the proof process of Theorem 3-2.1 that for any matrix K, we always have the inequality:

$$\deg(|sE - (A+BK)|) \leqslant \text{rank}E.$$

Therefore, under feedback control (3-2.5) the closed-loop system (3-2.6) is always a singular system, no matter what conditions are imposed on the system. On the other hand, we see from Theorem 3-2.3 that under the assumption of normalizability the closed-loop system is normal (with n finite poles) by suitable selection of feedback (3-2.15), which is impossible for P-state feedback (3-2.5).

Since feedback control (3-2.5) is a special form of (3-2.15), if a feedback control (3-2.5) may be chosen such that the closed-loop system (3-2.6) has no infinite poles, we may always be able to choose a suitable (3-2.15) such that (3-2.16) has no infinite poles (thus impulse terms in state response are eliminated). But the inverse is false. For instance:

Example 3-2.5. Consider the system

$$\begin{bmatrix} 0 & 1 & 0 \\ 0 & 0 & 1 \\ 0 & 0 & 0 \end{bmatrix} \dot{x} = \begin{bmatrix} 1 & 0 & 0 \\ 0 & 1 & 0 \\ 0 & 0 & 1 \end{bmatrix} x + \begin{bmatrix} 0 \\ 1 \\ 0 \end{bmatrix} u \qquad (3\text{-}2.17)$$

in which

$$E = \begin{bmatrix} 0 & 1 & 0 \\ 0 & 0 & 1 \\ 0 & 0 & 0 \end{bmatrix}, \qquad A = \begin{bmatrix} 1 & 0 & 0 \\ 0 & 1 & 0 \\ 0 & 0 & 1 \end{bmatrix}, \qquad B = \begin{bmatrix} 0 \\ 1 \\ 0 \end{bmatrix},$$

$n = 3$, rank$E = 2$, and

$$\text{rank}\begin{bmatrix} E & 0 & 0 \\ A & E & B \end{bmatrix} = 4 < 5 = n + \text{rank}E.$$

Thus, system (3-2.17) is not impulse controllable. For any P-state feedback $u = Kx + v$, there always exist impulse terms in its closed-loop system.

But, if we choose

$$u = [-1 \quad 0 \quad 0]\dot{x} + v = -K_2\dot{x} + v,$$

the closed-loop system would be

$$\begin{bmatrix} 0 & 1 & 0 \\ 1 & 0 & 1 \\ 0 & 0 & 0 \end{bmatrix} \dot{x} = \begin{bmatrix} 1 & 0 & 0 \\ 0 & 1 & 0 \\ 0 & 0 & 1 \end{bmatrix} x + \begin{bmatrix} 0 \\ 1 \\ 0 \end{bmatrix} v. \qquad (3\text{-}2.18)$$

Since the characteristic polynomial of closed-loop system (3-2.18) is

$$\deg(|s(E+BK_2) - A|) = \deg(s^2 + 1) = 2 = \text{rank}(E+BK_2),$$

closed-loop system (3-2.18) has no infinite poles; thus there are no impulse terms in its state response. This shows that impulse terms of a singular system in state response may be eliminated by derivative feedback even in the case when the system is not impulse controllable.

Theorem 3-2.4 (Dai, 1988j). There exists a P-D state feedback (3-2.15) such that the closed-loop system (3-2.16) has no infinite poles if and only if

$$\text{rank}\begin{bmatrix} E & 0 & 0 & B \\ A & E & B & 0 \end{bmatrix} = n + \text{rank}E.$$

3-3. Finite Pole Placement

Distribution of poles, especially finite poles, has a profound effect on the dynamic and static properties of a system.

Consider the system

$$\begin{aligned} E\dot{x} &= Ax + Bu \\ y &= Cx \end{aligned} \qquad (3\text{-}3.1)$$

where $x \in \mathbb{R}^n$, $u \in \mathbb{R}^m$, $y \in \mathbb{R}^r$ are its state, control input, and measure output, respectively, and the P-state feedback control

$$u = Kx + v \qquad (3\text{-}3.2)$$

where $v \in \mathbb{R}^m$. The closed-loop system is

$$\begin{aligned} E\dot{x} &= (A+BK)x + Bv \\ y &= Cx. \end{aligned} \qquad (3\text{-}3.3)$$

Lemma 3-3.1. System (3-3.1) is R-controllable if and only if (3-3.3) is R-controllable for any feedback gain matrix K satisfying $|sE-(A+BK)| \neq 0$.

Theorem 3-3.1. If system (3-3.1) is R-controllable (stabilizable), for any symmetric set Λ with n_1 elements on the complex plane, there always exists a gain matrix K such that the closed-loop finite pole set $\sigma(E,A+BK) = \Lambda\ (\subset \mathbb{C}^-)$. Here the symmetric set is the set in which complex scalars occur in conjugate pairs, and n_1 is the order of the slow subsystem in (3-3.1), i.e., $n_1 = \deg(|sE-A|)$.

Proof. By Lemma 1-2.2, for any regular system (3-3.1) there always exist two nonsingular matrices Q and P such that it is r.s.e. to EF1:

$$\dot{x}_1 = A_1x_1 + B_1u \qquad (3\text{-}3.4a)$$

$$N\dot{x}_2 = x_2 + B_2u \qquad (3\text{-}3.4b)$$

$$y = C_1x_1 + C_2x_2 \qquad (3\text{-}3.4c)$$

where $x_1 \in \mathbb{R}^{n_1}$, $x_2 \in \mathbb{R}^{n_2}$, $n_1 + n_2 = n$, N is nilpotent, and

$$\begin{aligned} &QEP = \mathrm{diag}(I,\ N), \quad QAP = \mathrm{diag}(A_1,\ I), \quad CP = [C_1,\ C_2] \\ &QB = \begin{bmatrix} B_1 \\ B_2 \end{bmatrix}, \quad P^{-1}x = \begin{bmatrix} x_1 \\ x_2 \end{bmatrix}. \end{aligned} \qquad (3\text{-}3.5)$$

Under the assumption of R-controllability (stabilizability), slow subsystem (3-3.4a) has the same property. Thus, for any Λ there exists a matrix K_1 that satisfies $\sigma(A_1+ B_1K_1) = \Lambda\ (\subset \mathbb{C}^-)$.

Let the feedback control be

$$u = K_1x_1 + v = [K_1,\ 0]P^{-1}x + v.$$

The closed-loop system is

$$\begin{aligned} \dot{x}_1 &= (A_1+B_1K_1)x_1 + B_1v \\ N\dot{x}_2 &= x_2 + B_2K_1x_1 + B_2v \\ y &= C_1x_1 + C_2x_2. \end{aligned} \qquad (3\text{-}3.6)$$

Paying attention to the nilpotent property of N, the closed-loop pole set of (3-3.6) (thus (3-3.3)) is

$$\sigma(E,A+BK) = \sigma(QEP,Q(A+BK)P) = \sigma(A_1+B_1K_1) = \Lambda \quad (\subset \mathbb{C}^-)$$

where $K = [K_1,\ 0]P^{-1}$. Q.E.D.

In the proof process we see that feedback control is a slow feedback. Therefore, if our purpose in imposing feedback control is to stabilize a system or pole placement, it is sufficient to finish this task by applying slow feedback.

However, the main disadvantage of such feedbacks is that since $n_1 \leqslant \text{rank}E$, there will exist impulse terms in the state response in the closed-loop system (3-3.3) if $N \neq 0$, which is often unexpected in practice.

Example 3-3.1. In the system

$$\begin{bmatrix} 1 & 0 & 0 \\ 0 & 0 & 1 \\ 0 & 0 & 0 \end{bmatrix}\dot{x} = \begin{bmatrix} 1 & 0 & 0 \\ 0 & 1 & 0 \\ 0 & 0 & 1 \end{bmatrix}x + \begin{bmatrix} 1 \\ 0 \\ 1 \end{bmatrix}u \qquad (3\text{-}3.7)$$

$n_1 = 1$, $n_2 = 2$, and

$$N = \begin{bmatrix} 0 & 1 \\ 0 & 0 \end{bmatrix}, \quad A_1 = 1, \quad B_1 = 1, \quad B_2 = \begin{bmatrix} 0 \\ 1 \end{bmatrix}.$$

This system is controllable. Thus it is R-controllable. Assume that $\Lambda = \{-1\}$. The $\sigma(A_1+B_1K_1) = \{-1\}$ holds by choosing $K_1 = -2$. In this case

$$u = K_1x_1 + v = [-2 \quad 0 \quad 0]x + v$$

and the corresponding closed-loop system is

$$\begin{bmatrix} 1 & 0 & 0 \\ 0 & 0 & 1 \\ 0 & 0 & 0 \end{bmatrix}\dot{x} = \begin{bmatrix} -1 & 0 & 0 \\ 0 & 1 & 0 \\ -2 & 0 & 1 \end{bmatrix}x + \begin{bmatrix} 1 \\ 0 \\ 1 \end{bmatrix}v. \qquad (3\text{-}3.8)$$

Clearly, impulse terms appear in the state response.

Therefore, our aim to impose the feedback control should place both finite and infinite poles. This requires that not only are the finite poles placed on the arbitrarily assigned positions, but the impulse terms are also eliminated in the state response in closed-loop systems. Thus, both finite and infinite pole placement should be considered at the same time, in other words, the closed-loop system should have rankE finite poles (thus no infinite poles) that lie at arbitrarily assigned positions.

Theorem 3-3.2. Let system (3-3.1) be controllable. Then for any symmetric set Λ with rankE elements on the complex plane, there always exists a gain matrix K such that the closed-loop system (3-3.3) has the finite pole set $\sigma(E, A+BK) = \Lambda$.

Proof. The proof is constructive. Let Q_1, P_1 be nonsingular matrices satisfying

$$Q_1EP_1 = \mathrm{diag}(I_q, 0), \qquad q = \mathrm{rank}E.$$

By denoting

$$Q_1AP_1 = \begin{bmatrix} A_{11} & A_{12} \\ A_{21} & A_{22} \end{bmatrix}, \quad Q_1B = \begin{bmatrix} B_1 \\ B_2 \end{bmatrix}, \quad \bar{x} = P_1^{-1}x,$$

system (3-3.1) is r.s.e. to EF2 (only state equation is needed here):

$$\begin{bmatrix} I_q & 0 \\ 0 & 0 \end{bmatrix}\dot{\bar{x}} = \begin{bmatrix} A_{11} & A_{12} \\ A_{21} & A_{22} \end{bmatrix}\bar{x} + \begin{bmatrix} B_1 \\ B_2 \end{bmatrix}u. \tag{3-3.9}$$

Accounting for the controllability assumption, system (3-3.1) is impulse controllable. By Theorem 2-4.2, $\mathrm{rank}[A_{22}, B_2] = n-q$, (A_{22}, B_2) is normalizable. Thus, a matrix $K_2 \in \mathbb{R}^{m\times(n-q)}$ may be chosen such that

$$|A_{22} + B_2K_2| \neq 0. \tag{3-3.10}$$

Setting $\bar{K} = [0 \;\; K_2]P_1^{-1}$ and

$$u = \bar{K}x + u_1 = [0 \;\; K_2]\bar{x} + u_1 \tag{3-3.11}$$

in which $u_1(t)$ is the new input for the system, the closed-loop system formed by (3-3.11) and (3-3.9) is

$$\begin{bmatrix} I_q & 0 \\ 0 & 0 \end{bmatrix}\dot{\bar{x}} = \begin{bmatrix} A_{11} & A_{12}+B_1K_2 \\ A_{21} & A_{22}+B_2K_2 \end{bmatrix}\bar{x} + \begin{bmatrix} B_1 \\ B_2 \end{bmatrix}u_1. \tag{3-3.12}$$

From (3-3.10) $(A_{22}+B_2K_2)^{-1}$ exists, and the two matrices Q_2, P_2 defined as

$$Q_2 = \begin{bmatrix} I_q & -(A_{12}+B_1K_2)(A_{22}+B_2K_2)^{-1} \\ 0 & I_{n-q} \end{bmatrix}$$

$$P_2 = \begin{bmatrix} I_q & 0 \\ -(A_{22}+B_2K_2)^{-1}A_{21} & (A_{22}+B_2K_2)^{-1} \end{bmatrix}$$

are nonsingular. Direct computation shows

$$Q_2 \mathrm{diag}(I_q, 0) P_2 = \mathrm{diag}(I_q, 0)$$

$$Q_2 \begin{bmatrix} A_{11} & A_{12}+B_1K_2 \\ A_{21} & A_{22}+B_2K_2 \end{bmatrix} P_2 = \mathrm{diag}(\tilde{A}_1, I) \qquad (3\text{-}3.13)$$

$$Q_2 \begin{bmatrix} B_1 \\ B_2 \end{bmatrix} = \begin{bmatrix} \tilde{B}_1 \\ \tilde{B}_2 \end{bmatrix}$$

where

$$\tilde{A}_1 = A_{11} - (A_{12} + B_1K_2)(A_{22} + B_2K_2)^{-1}A_{21}$$

$$\tilde{B}_1 = B_1 - (A_{12} + B_1K_2)(A_{22} + B_2K_2)^{-1}B_2, \qquad \tilde{B}_2 = B_2.$$

By defining

$$\tilde{x} = P_2^{-1}\bar{x} = P_2^{-1}P_1^{-1}x$$

from (3-3.13), system (3-3.12) is r.s.e. to

$$\mathrm{diag}(I_q, 0)\dot{\tilde{x}} = \mathrm{diag}(\tilde{A}_1, I)\tilde{x} + [\tilde{B}_1/\tilde{B}_2]u_1, \qquad (3\text{-}3.14)$$

which is R-controllable according to Lemma 3-3.1. The pair $(\tilde{A}_1, \tilde{B}_1)$ is controllable. Since $\tilde{A}_1 \in \mathbb{R}^{q\times q}$, for any symmetric set Λ with q = rankE elements, a matrix $\tilde{K}_2$ may be chosen to satisfy

$$\sigma(\tilde{A}_1 + \tilde{B}_1\tilde{K}_1) = \Lambda.$$

Define

$$u_1 = [\tilde{K}_1, 0]\tilde{x} + v.$$

The overall feedback control is

$$u = \bar{K}x + u_1 = Kx + v$$

in which $K = \bar{K} + [\tilde{K}_1 \;\; 0]P_2^{-1}P_1^{-1}$. For such a feedback gain matrix K, it is easy to verify that the closed-loop system (3-3.3) has the finite pole set of

$$\sigma(E, A+BK) = \sigma(\tilde{A}_1 + \tilde{B}_1\tilde{K}_1) = \Lambda$$

or system (3-3.3) has Λ as its finite pole set. Q.E.D.

The constructive proof also provides us with a design process to find gain matrix K.

Example 3-3.2. System (3-3.7) in Example 3-3.1 is controllable. Under the transformation matrices

$$Q_1 = \begin{bmatrix} 1 & 0 & 0 \\ 0 & 1 & 0 \\ 0 & 0 & 1 \end{bmatrix}, \qquad P_1 = \begin{bmatrix} 1 & 0 & 0 \\ 0 & 0 & 1 \\ 0 & 1 & 0 \end{bmatrix},$$

it is r.s.e. to

$$\left[\begin{array}{cc|c} 1 & 0 & 0 \\ 0 & 1 & 0 \\ \hline 0 & 0 & 0 \end{array}\right] \dot{\bar{x}} = \left[\begin{array}{cc|c} 1 & 0 & 0 \\ 0 & 0 & 1 \\ \hline 0 & 1 & 0 \end{array}\right] \bar{x} + \left[\begin{array}{c} 1 \\ 0 \\ \hline 1 \end{array}\right] u. \tag{3-3.15}$$

Thus, we have

$$A_{11} = \begin{bmatrix} 1 & 0 \\ 0 & 0 \end{bmatrix}, \quad A_{12} = \begin{bmatrix} 0 \\ 1 \end{bmatrix}, \quad B_1 = \begin{bmatrix} 1 \\ 0 \end{bmatrix},$$

$$A_{22} = 0, \quad A_{21} = [0 \quad 1], \quad B_1 = 1.$$

Let $K_2 = 1$. Then $|A_{22} + B_2K_2| = 1 \neq 0$ and

$$\tilde{A}_1 = A_{11} - (A_{12}+ B_1K_2)(A_{22}+ B_2K_2)^{-1}A_{21} = \begin{bmatrix} 1 & -1 \\ 0 & -1 \end{bmatrix}$$

$$\tilde{B}_1 = B_1 - (A_{12}+ B_1K_2)(A_{22}+ B_2K_2)^{-1}B_2 = \begin{bmatrix} 0 \\ -1 \end{bmatrix}$$

$$P_2 = \begin{bmatrix} I_q & 0 \\ -(A_{22}+ B_2K_2)^{-1}A_{21} & (A_{22}+ B_2K_2)^{-1} \end{bmatrix} = \begin{bmatrix} 1 & 0 & 0 \\ 0 & 1 & 0 \\ 0 & -1 & 1 \end{bmatrix}.$$

Note that

$$\text{rank}[\tilde{B}_1, \tilde{A}_1\tilde{B}_1] = \text{rank} \begin{bmatrix} 0 & 1 \\ -1 & 1 \end{bmatrix} = 2,$$

therefore $(\tilde{A}_1, \tilde{B}_1)$ is indeed controllable. The closed-loop system is assumed to have rankE = 2 stable finite poles -1, -1, or $\Lambda = \{-1,-1\}$. By setting

$$\tilde{K}_1 = [-4, 2]$$

we have

$$|sI - (\tilde{A}_1+\tilde{B}_1\tilde{K}_1)| = (s+1)^2.$$

Thus, $\sigma(\tilde{A}_1+\tilde{B}_1\tilde{K}_1) = \Lambda = \{-1, -1\}$.

In this case, the feedback gain matrix K is

$$K = \bar{K} + [\tilde{K}_1, 0]P_2^{-1}P_1^{-1} = [-4 \quad 1 \quad 2]$$

and the corresponding closed-loop system (3-3.3) for this special system becomes

$$\begin{bmatrix} 1 & 0 & 0 \\ 0 & 0 & 1 \\ 0 & 0 & 0 \end{bmatrix} \dot{x} = \begin{bmatrix} -3 & 1 & 2 \\ 0 & 1 & 0 \\ -4 & 1 & 3 \end{bmatrix} x + \begin{bmatrix} 1 \\ 0 \\ 1 \end{bmatrix} v \tag{3-3.16}$$

It is easy to verify that the closed-loop pole set is the expected $\{-1, -1\}$ and no infinite poles (thus no impulse portions) exist in the closed-loop system.

Summing up the results in this and last sections,we have proven the following main result of pole placement for singular systems.

Theorem 3-3.3. Consider system (3-3.1).

a. Closed-loop system (3-3.3) has at most rankE finite poles for any feedback control (3-3.2).

b. There exists a gain matrix K such that (3-3.3) has no infinite poles (thus has the most rankE finite poles) if and only if system (3-3.1) is impulse controllable.

c. If system (3-3.1) is R-controllable and impulse controllable, a matrix K may be chosen so that closed-loop system (3-3.3) has rankE arbitrarily assigned finite poles.

Under the controllability assumption, Theorem 3-3.2 states that (3-3.3) may have rankE arbitrarily assigned poles by suitably chosen feedback. This theorem further shows that the number of most possible finite poles is rankE.

Similarly we have the following corollary.

Corollary 3-3.1. There exists a matrix K satisfying both $\sigma(E,A+BK) \subset \mathbb{C}^-$ and $\deg(|sE-(A+BK)|) = \text{rank}E$ if and only if system (3-3.1) is stabilizable and impulse controllable.

In other words, stabilizability and impulse controllability are necessary and sufficient conditions for the existence of K such that the closed-loop system (3-3.3) not only is stable but also has no infinite poles (thus eliminates impulse terms in the state response).

Corollary 3-3.2. There exists a matrix G such that both $\sigma(E,A+GC) \subset \mathbb{C}^-$ and $\deg(|sE - (A+GC)|) = \text{rank}E$ are satisfied if and only if system (3-3.1) is detectable and impulse observable.

This conclusion will be needed later.

We next consider the combined proportional and derivative (P-D) state feedback control

$$u(t) = K_1x(t) - K_2\dot{x}(t) + v(t) \tag{3-3.17}$$

under which the closed-loop system is

$$\begin{aligned} (E+BK_2)\dot{x} &= (A+BK_1)x + Bv \\ y &= Cx. \end{aligned} \tag{3-3.18}$$

Lemma 3-3.2. Assume that system (3-3.1) is R-controllable (stabilizable). Then

$(E+BK_2,A,B,C)$ is R-controllable (stabilizable) for any $K_2 \in \mathbb{R}^{m\times n}$ that satisfies $|s(E+BK_2)-A| \neq 0$.

Remark. As shown in Example 3-2.5, derivative state feedback may change the impulse controllability of a system; this lemma clarifies that: Neither proportional nor derivative state feedback control can change the R-controllability (stabilizability) of a system.

Proof. By definition of R-controllability and Theorems 2-2.2 - 3-1.2, system (3-3.1) is R-controllable (stabilizable) iff

$$\operatorname{rank}[sE-A,\ B] = n, \quad \forall\ s\in\mathbb{C}, \quad s \text{ finite}$$

$$(\ \operatorname{rank}[sE-A,\ B] = n, \quad \forall\ s\in\bar{\mathbb{C}}^{+}, \quad s \text{ finite}).$$

On the other hand,

$$\operatorname{rank}[s(E+BK_2)-A,\ B] = \operatorname{rank}[sE-A,\ B]\begin{bmatrix} I & 0 \\ K_2 s & I \end{bmatrix}$$

$$= \operatorname{rank}[sE-A,\ B], \quad \forall\ s\in\mathbb{C}, \quad s \text{ finite}.$$

Thus, system $(E+BK_2,A,B,C)$ is also R-controllable (stabilizable). Q.E.D.

Theorem 3-3.4. Let Λ be an arbitrary symmetric set with n elements on the complex plane. Then controllability is the necessary and sufficient condition for the existence of (3-3.17) such that the closed-loop system (3-3.18) has the finite pole set:

$$\sigma(E+BK_2,A+BK_1) = \Lambda.$$

Proof. Sufficiency: Under the assumption of controllability, system (3-3.1) is R-controllable and normalizable. Matrix K_2 exists so that $|E+BK_2| \neq 0$. System $(E+BK_2, A,B,C)$ is R-controllable by Lemma 3-3.2. Noticing that it is a normal system and $((E+BK_2)^{-1}A,\ (E+BK_2)^{-1}B)$ is controllable, for any Λ, a matrix K_1 may be selected that satisfies

$$\sigma((E+BK_2)^{-1}A + (E+BK_2)^{-1}BK_1) = \sigma((E+BK_2)^{-1}(A+BK_1)) = \Lambda.$$

Hence, the closed-loop system (3-3.18) determined by K_1 and K_2 has the finite pole set:

$$\sigma(E+BK_2,\ A+BK_1) = \sigma((E+BK_2)^{-1}(A+BK_1)) = \Lambda.$$

Necessity: The existence of n finite poles in (3-3.18) shows $|E+BK_2| \neq 0$. Thus system (3-3.1) is normalizable; R-controllability is a direct result of arbitrariness in pole placement. Q.E.D.

Therefore, under the controllability assumption, feedback control (3-3.17) may not only drive all infinite poles to finite positions, but further make the closed-loop system normal. This feature could not be achieved via proportional feedback (3-3.2).

Example 3-3.3. System (3-3.7) is controllable. Let $\Lambda = \{-1,-1,-1\}$ be its expected closed-loop pole set, which could be achieved by derivative feedback (3-3.17) only.

Let $K_2 = [0 \;\; 1 \;\; 0]$. Then $|E+BK_2| \neq 0$. The feedback

$$u = -K_2\dot{x} + u_1 \tag{3-3.19}$$

and system (3-3.7) form the closed-loop system

$$\begin{bmatrix} 1 & 1 & 0 \\ 0 & 0 & 1 \\ 0 & 1 & 0 \end{bmatrix}\dot{x} = \begin{bmatrix} 1 & 0 & 0 \\ 0 & 1 & 0 \\ 0 & 0 & 1 \end{bmatrix}x + \begin{bmatrix} 1 \\ 0 \\ 1 \end{bmatrix}u_1. \tag{3-3.20}$$

Noticing that

$$E+BK_2 = \begin{bmatrix} 1 & 1 & 0 \\ 0 & 0 & 1 \\ 0 & 1 & 0 \end{bmatrix}$$

is nonsingular, system (3-3.20) may be written as

$$\dot{x} = \begin{bmatrix} 1 & 0 & -1 \\ 0 & 0 & 1 \\ 0 & 1 & 0 \end{bmatrix}x + \begin{bmatrix} 0 \\ 1 \\ 0 \end{bmatrix}u_1. \tag{3-3.21}$$

For any matrix $K_1 = [a_1, a_2, a_3]$, the closed-loop system formed by $u_1 = K_1x + v$ and (3-3.21) has the characteristic polynomial

$$s^3 - (1+a_2)s^2 + (a_2-1-a_3)s + (1+a_1+a_3).$$

Assume that the expected closed-loop pole set is $\{-1,-1,-1\}$. If we let $a_1 = 8$, $a_2 = -4$, and $a_3 = -8$, direct computation yields the P-D state feedback:

$$u = [0 \;\; -1 \;\; 0]\dot{x} + [8 \;\; -4 \;\; -8]x + v \tag{3-3.22}$$

and the closed-loop system

$$\begin{bmatrix} 1 & 1 & 0 \\ 0 & 0 & 1 \\ 0 & 1 & 0 \end{bmatrix}\dot{x} = \begin{bmatrix} 9 & -4 & -8 \\ 0 & 1 & 0 \\ 8 & -4 & -7 \end{bmatrix}x + \begin{bmatrix} 1 \\ 0 \\ 1 \end{bmatrix}v$$

which has the finite pole set $\{-1,-1,-1\}$.

Theorem 3-3.5. Let system (3-3.1) be stabilizable and normalizable. Then there exists a feedback control (3-3.17) such that the closed-loop system (3-3.18) is normal with only stable finite poles.

3-4. Pole Structure under State Feedback Control

3-4.1 Pole structure under pure proportional (P-) state feedback

Consider system (3-3.1)

$$\begin{aligned} E\dot{x} &= Ax + Bu \\ y &= Cx \end{aligned} \tag{3-4.1}$$

$x \in \mathbb{R}^n$, and its P-state feedback (3-3.2)

$$u = Kx + v. \tag{3-4.2}$$

As pointed out in the last section, if our aim to apply feedback (3-4.2) is merely to place the poles of the closed-loop system

$$\begin{aligned} E\dot{x} &= (A+BK)x + Bv \\ y &= Cx \end{aligned} \tag{3-4.3}$$

to expected positions, the feedback gain matrix K is not unique. Much feasibility exists in its selection. We naturally think of using this feasibility to improve the closed-loop properties.

Many facts have shown that not only pole positions but also pole structure or the Jordan forms in the coefficient matrix influence the properties of system (3-4.3) especially the dynamic properties. The so-called pole structure placement problem is using the feasibility in the selection of feedback gain matrix K to place closed-loop system's Jordan structure.

Without loss of generality, in this section we assume that system (3-4.1) is controllable, and rank$B = m$, $B \in \mathbb{R}^{n\times m}$.

As stated previously, the impulse terms are harmful in a system and should be avoided. This requires elimination of the impulse terms while we are placing pole structure. The later problem is equivalent to requiring system (3-4.3) to have rankE finite poles or no infinite poles.

To begin, take the rank decomposition on matrix E. Let $q = \text{rank}E$. Then two nonsingular matrices Q_1 and P_1 exist such that

$$Q_1EP_1 = \text{diag}(I_q, 0).$$

If we denote

$$Q_1AP_1 = \begin{bmatrix} A_{11} & A_{12} \\ A_{21} & A_{22} \end{bmatrix}, \quad Q_1B = \begin{bmatrix} B_1 \\ B_2 \end{bmatrix}, \quad x = P_1\begin{bmatrix} x_1 \\ x_2 \end{bmatrix},$$

$$x_1 \in \mathbb{R}^q, \quad x_2 \in \mathbb{R}^{n-q}, \quad KP = [\bar{K}_1, \bar{K}_2],$$

the description equation in closed-loop system (3-4.3) is r.s.e. to EF2:

$$\begin{aligned} \dot{x}_1 &= (A_{11}+B_1\bar{K}_1)x_1 + (A_{12}+B_1\bar{K}_2)x_2 + B_1 v \\ 0 &= (A_{21}+B_2\bar{K}_1)x_1 + (A_{22}+B_2\bar{K}_2)x_2 + B_2 v. \end{aligned} \tag{3-4.4}$$

Noticing the results in the last section, closed-loop system (3-4.3) has no infinite poles if and only if

$$| A_{22}+ B_2\bar{K}_2 | \neq 0. \tag{3-4.5}$$

Under this assumption, the matrices

$$Q_2 = \begin{bmatrix} I_q & -(A_{12}+B_1\bar{K}_2)(A_{22}+B_2\bar{K}_2)^{-1} \\ 0 & I_{n-q} \end{bmatrix}$$

$$P_2 = \begin{bmatrix} I_q & 0 \\ -(A_{22}+B_2\bar{K}_2)^{-1}A_{21} & (A_{22}+B_2\bar{K}_2)^{-1} \end{bmatrix}$$

would transfer system (3-4.4) into

$$\begin{aligned} \dot{\bar{x}}_1 &= (\bar{A}_1 + \bar{B}_1\bar{K}_1)\bar{x}_1 + \bar{B}_1 v \\ 0 &= \bar{x}_2 + B_2\bar{K}_1\bar{x}_1 + B_2 v \end{aligned} \tag{3-4.6}$$

where

$$\begin{aligned} \bar{A}_1 &= A_{11} - (A_{12}+B_1\bar{K}_2)(A_{22}+B_2\bar{K}_2)^{-1}A_{21} \\ \bar{B}_1 &= B_1 - (A_{12}+B_1\bar{K}_2)(A_{22}+B_2\bar{K}_2)^{-1}B_2 \\ \begin{bmatrix} \bar{x}_1 \\ \bar{x}_2 \end{bmatrix} &= P_2^{-1}\begin{bmatrix} x_1 \\ x_2 \end{bmatrix} = P_2^{-1}P_1^{-1}x. \end{aligned} \tag{3-4.7}$$

If (3-4.5) is satisfied, from (3-4.6) we know that the closed-loop pole structure is determined uniquely by the matrix $\bar{A}_1 + \bar{B}_1\bar{K}_1$ for a given $\bar{K}_2$. However, the matrix $\bar{K}_2$ (thus matrix K) that satisfies (3-4.5) is not unique. For different K we may obtain different system (3-4.6). But we can prove that the controllability indices of $(\bar{A}_1, \bar{B}_1)$ are independent of the selection of K, provided (3-4.5) is satisfied.

Theorem 3-4.1 (Dai, 1988e). Assume that system (3-4.1) is controllable and rankB = m. Let $\nu_1^c \leqslant \nu_2^c \leqslant \cdots \leqslant \nu_m^c$ be the controllability indices of $(\bar{A}_1, \bar{B}_1)$. Then they are uniquely determined by system (3-4.1) and are independent of the selection of K, provided (3-4.5) is fulfilled.

We define $\nu_1^c \leqslant \nu_2^c \leqslant \cdots \leqslant \nu_m^c$ as the controllability indices for system (3-4.1) under P-state feedback. Its computation algorithm is as follows. First, find the matrix $K \in \mathbb{R}^{m\times n}$ such that

$$\deg(|sE - (A+BK)|) = \text{rank}E.$$

Then for such a K, determine the standard decomposition; and, finally, compute the controllability indices for its slow subsystem, which is what we want. Another way is to calculate the controllability indices for $(\bar{A}_1, \bar{B}_1)$ determined by (3-4.7) in the decomposition of EF2 for system (3-4.1).

The following are our main results on pole structure placement.

Theorem 3-4.2(Dai, 1988e). Assume that system (3-4.1) is controllable. $\nu_1^c \leqslant \nu_2^c \leqslant \dots \leqslant \nu_m^c$ are its controllability indices under P-state feedback. Let $p_i(s)$, $i = 1, 2, \dots, m$, be polynomials satisfying $p_{i-1} | p_i(s)$. Then there exists a matrix K such that $\{p_i(s)\}$ is the set of invariant polynomials of the slow subsystem coefficient matrix for the closed-loop system (3-4.3) if and only if

$$\sum_{i=j}^{m} \alpha_i \geqslant \sum_{i=j}^{m} \nu_i^c, \quad j = 1, 2, \dots, m,$$

where $\sum_{i=1}^{m} \alpha_i = n$, $\alpha_i = \deg(p_i(s))$, $i = 1, 2, \dots, m$.

3-4.2. Pole Structure under Proportional and Derivative (P-D) State Feedback

Now we will consider the P-D state feedback control

$$u = K_1 x - K_2 \dot{x} + v, \tag{3-4.8}$$

where $K_1, K_2 \in \mathbb{R}^{m\times n}$. System (3-4.1) and (3-4.8) form the closed-loop system

$$\begin{aligned} (E+BK_2)\dot{x} &= (A+BK_1)x + Bv \\ y &= Cx . \end{aligned} \tag{3-4.9}$$

Since we have assumed that system (3-4.1) is controllable, it is normalizable. Matrix K_2 exists to satisfy

$$| E+BK_2 | \neq 0. \tag{3-4.10}$$

In this case, system (3-4.9) becomes

$$\begin{aligned} \dot{x} &= (\hat{A} + \hat{B}K_1)x + \hat{B}v \\ y &= Cx \end{aligned} \tag{3-4.11}$$

where

$$\hat{A} = (E+BK_2)^{-1}A, \qquad \hat{B} = (E+BK_2)^{-1}B.$$

The controllability assumption of system (3-4.1) shows that $(\hat{A}, \hat{B})$ is controllable

Let $\mu_1^c \leqslant \mu_2^c \leqslant \cdots \leqslant \mu_m^c$ be the controllability indices for $(\hat{A}, \hat{B})$. They are independent of K_2 provided (3-4.10) holds (Dai, 1988e). Such indices are called controllability indices for system (3-4.1) under P-D state feedback.

By noticing the closed-loop system form (3-4.11), we easily have the following theorem according to the linear system theory.

Theorem 3-4.3. Assume that system (3-4.1) is controllable and rankB = m; $\mu_1^c \leqslant \mu_2^c \leqslant \cdots \leqslant \mu_m^c$ are controllability indices under the P-D state feedback (3-4.9). Then for any polynomials $q_i(s)$, i = 1, 2, ... , m, satisfying

$$1.\ q_{i-1}(s) \mid q_i(s), \qquad 2.\ \sum_{i=1}^{m} \beta_i = n, \qquad \beta_i = \deg(q_i(s)),$$

the conditions

$$\sum_{i=j}^{m} \beta_i \geqslant \sum_{i=j}^{m} \mu_i^c, \qquad j = 1, 2, \ldots, m$$

are necessary and sufficient for the existence of two matrices K_1 and K_2 such that $q_i(s)$, i = 1, 2, ... , m, are the invariant polynomials of the coefficient of the closed-loop system (3-4.11).

3-5. Output Feedback Control

Output feedback includes dynamic and static feedback. The former is the content of the compensator problem, which will be studied in Chapter 5. In this section we will study static feedback of the form

$$u = Ky + v \tag{3-5.1}$$

in which y(t) is the measure output and v(t) is the new control (or reference signal), $K \in \mathbb{R}^{m\times r}$. The static output feedback is easy to realize but its special form brings about more limitations than state feedback. So it is often difficult to meet with our design purposes using such control schemes.

For system (3-4.1), we consider its static output feedback (3-5.1) under which the closed-loop system is

$$\begin{aligned} E\dot{x} &= (A+BKC)x + Bv \\ y &= Cx \end{aligned} \tag{3-5.2}$$

Lemma 3-5.1. Assume that system (3-4.1) is normalizable, dual normalizable. Then, a matrix $K \in \mathbb{R}^{m\times r}$ may be chosen that satisfies

$$|E+BKC| \neq 0. \tag{3-5.3}$$

Proof. Let

$$Q_1EP_1 = \mathrm{diag}(I_q, 0), \quad q = \mathrm{rank}E \tag{3-5.4}$$

be the rank decomposition of E, where Q_1 and P_1 are nonsingular. Denoting

$$Q_1B = \begin{bmatrix} B_1 \\ B_2 \end{bmatrix}, \qquad CP_1 = [C_1, C_2] \tag{3-5.5}$$

from Theorems 3-1.4 and 3-1.5 we know that, under the assumption of normalizability and dual normalizability, matrices B_2 and C_2 are of full row and column rank, respectively, with $\mathrm{rank}B_2 = \mathrm{rank}C_2 = n-q$. Thus, both $B_2B_2^\tau$ and $C_2^\tau C_2$ are nonsingular.

Setting $K = B_2^\tau C_2^\tau$, it is easy to verify

$$|E+BKC| = |Q_1^{-1}||P_1^{-1}||Q_1EP_1 + Q_1BKCP_1|$$

$$= |Q_1^{-1}||P_1^{-1}|\begin{vmatrix} I_q + B_1B_2^\tau C_2^\tau C_1 & B_1B_2^\tau C_2^\tau C_2 \\ B_2B_2^\tau C_2^\tau C_1 & B_2B_2^\tau C_2^\tau C_2 \end{vmatrix}$$

$$= |Q_1^{-1}||P_1^{-1}||B_2B_2^\tau C_2^\tau C_2| \neq 0.$$

Q.E.D.

Starting from this theorem, we may prove the following.

Theorem 3-5.1. Impulse controllability and impulse observability are necessary and sufficient conditions for the existence of output feedback (3-5.1) such that no impulse terms exist in the state response of (3-5.2) (or it has no infinite poles).

Proof. The necessity is obvious since (3-5.1) is a special case of state feedback. To prove the sufficiency, we decompose

$$Q_1AP_1 = \begin{bmatrix} A_{11} & A_{12} \\ A_{21} & A_{22} \end{bmatrix}.$$

where Q_1 and P_1 satisfy (3-5.4). By assumption, system (3-4.1) is impulse controllable and impulse observable. Thus $\mathrm{rank}[A_{22}, B_2] = \mathrm{rank}[A_{22}/C_2] = n-q$. Lemma 3-5.1 shows that a matrix K_1 may be chosen such that

$$|A_{22} + B_2K_1C_2| \neq 0.$$

For such a matrix K_1, we have

$$\deg(|sE - (A+BK_1C)|) = \deg(|sQ_1EP_1 - Q_1(A+BK_1C)P_1|)$$

$$= \deg(\begin{vmatrix} sI_q - A_{11} - B_1K_1C_1 & -A_{12} - B_1K_1C_2 \\ -A_{21} - B_2K_1C_1 & -A_{22} - B_2K_2C_2 \end{vmatrix})$$

$$= \deg(|-A_{22}-B_2K_2C_2|s^q + \ldots \text{ lower order terms})$$

$$= q = \text{rank}E.$$

No impulse portion appears in the closed-loop state response. Q.E.D.

Example 3-5.1. Consider system (3-2.8)

$$\begin{bmatrix} 0 & 0 & 0 \\ 1 & 0 & 0 \\ 0 & 1 & 0 \end{bmatrix} \dot{x} = \begin{bmatrix} 1 & 0 & 0 \\ 0 & 1 & 0 \\ 0 & 0 & 1 \end{bmatrix} x + \begin{bmatrix} 1 \\ 0 \\ 0 \end{bmatrix} u$$

$$y = [1 \quad 0 \quad -1]x \qquad (3\text{-}5.6)$$

in which we have $n = 3$, rank$E = 2$ and

$$E = \begin{bmatrix} 0 & 0 & 0 \\ 1 & 0 & 0 \\ 0 & 1 & 0 \end{bmatrix}, \quad A = \begin{bmatrix} 1 & 0 & 0 \\ 0 & 1 & 0 \\ 0 & 0 & 1 \end{bmatrix}, \quad B = \begin{bmatrix} 1 \\ 0 \\ 0 \end{bmatrix}, \quad C = [1 \quad 0 \quad -1]. \qquad (3\text{-}5.7)$$

Since

$$\text{rank}\begin{bmatrix} E & 0 & 0 \\ A & E & B \end{bmatrix} = 5 = n + \text{rank}E$$

$$\text{rank}\begin{bmatrix} E & A \\ 0 & E \\ 0 & C \end{bmatrix} = 5 = n + \text{rank}E,$$

system (3-5.6) is impulse controllable and impulse observable. An output feedback $u = Ky + v$ exists such that its closed-loop system has no infinite poles. Next, we find matrix K.

The nonsingular matrices

$$Q_1 = \begin{bmatrix} 0 & 1 & 0 \\ 0 & 0 & 1 \\ 1 & 0 & 0 \end{bmatrix}, \qquad P_1 = \begin{bmatrix} 1 & 0 & 0 \\ 0 & 1 & 0 \\ 0 & 0 & 1 \end{bmatrix}$$

transfer

$$Q_1EP_1 = \left[\begin{array}{cc|c} 1 & 0 & 0 \\ 0 & 1 & 0 \\ \hline 0 & 0 & 0 \end{array}\right], \qquad Q_1AP_1 = \left[\begin{array}{cc|c} 0 & 1 & 0 \\ 0 & 0 & 1 \\ \hline 1 & 0 & 0 \end{array}\right], \qquad Q_1B = \left[\begin{array}{c} 0 \\ 0 \\ \hline 1 \end{array}\right]$$

$$CP = [1 \quad 0 \quad -1].$$

Thus, here we have $A_{22} = 0$, $B_2 = 1$, and $C_2 = -1$. Choose $K = 1$. It satisdies

$$|A_{22} + B_2KC_2| \neq 0.$$

Corresponding to (3-5.2) the closed-loop system is

$$\begin{bmatrix} 0 & 0 & 0 \\ 1 & 0 & 0 \\ 0 & 1 & 0 \end{bmatrix} \dot{x} = \begin{bmatrix} 2 & 0 & -1 \\ 0 & 1 & 0 \\ 0 & 0 & 1 \end{bmatrix} x + \begin{bmatrix} 1 \\ 0 \\ 0 \end{bmatrix} v$$

$$y = [1 \quad 0 \quad -1]x. \qquad (3\text{-}5.8)$$

It is easy to verify that

$$\deg(|sE-(A+BKC)|) = \deg(2-s^2) = 2 = \mathrm{rank}E.$$

No impulse terms exist in the state response of (3-5.8).

3-6. Notes and References

There are many papers on the pole placement problem for singular systems. For reference on this problem one may refer to AL-Nasr et al. (1983a), Armentano (1984), Bajic (1986), Bender (1985), Campbell (1982b), Christodoulou and Paraskevopoulou (1984), Cobb (1981), Cullen (1986), Kautsky and Nichols (1986), Lovass-Nagy, Powers and Yan (1986), Mukundan and Ayawansa (1983), Pandolfi (1980), Wang et al. (1987), and Zhou et al. (1987). Papers on pole structure placement include Dai (1988e), Dooren (1981), Fletcher et al. (1986), Lewis and Ozcaldiran (1985), and Ozcaldiran and Lewis (1987). For reference on output feedback, one may see Dai (1988e), AL-Nasr et al. (1983b).

On the infinite pole problem, in this book we adopt the definition of Verghese et al. (1979). Another definition of infinite poles, defined as the zero structure of N in the EFI (3-1.11), is given in Campbell (1982b). Typical references on infinite pole problem are Vardulakis et al. (1982) and Verghese et al. (1979, 1981).

Stability here is an asymptotic stability in the sense of Lyapunov.

Normalizability, in the term of regularity, was first used by Mukundan and Ayawansa (1983), and formly defined in Wang and Dai (1986b), and Zhou et al. (1987). Since regularity has been used to define a kind of matrix pencils (the matrix pencil (E,A) is called regular if $|sE-A| \not\equiv 0$) and the systems studied in this book by many articles, we use normalizability instead of regularity here for distinction.

CHAPTER 4

STATE OBSERVATION

As pointed out in the last chapter, state feedback control is important in system design. Under certain conditions, this method enables us to place pole structure such that the closed-loop system has a satisfactory property. Many facts have shown that for deterministic systems, the state feedback method is convenient and practical in most cases. However, this method is based on the assumption that the state is available for our use, which is unfortunately not always the case. In practice the state is usually not directly available. What we can obtain is the control input u(t) and the measure output y(t) rather than the state itself. So the state feedback control generally can not be realized directly. The problem of overcoming this difficulty must be solved. At present there are two popular approaches. In the first one, the measure output is used instead of state, but often no satisfactory results are obtained. The second method emphasizes (asymptotically) estimating (observing) state x(t) from control input u(t) and measure output y(t), in which the state estimation is used in lieu of real state in constructing the feedback control. Thus, the feedback control law is realized.

4-1. Singular State Observer

Consider the singular system

$$\begin{aligned} E\dot{x}(t) &= Ax(t) + Bu(t) \\ y(t) &= Cx(t) \end{aligned} \tag{4-1.1}$$

where $x(t) \in \mathbb{R}^n$, $u(t) \in \mathbb{R}^m$, $y(t) \in \mathbb{R}^r$ are its state, control input, and measure output, respectively; E, A, B, and C are constant matrices of appropriate dimensions. It is assumed that system (4-1.1) is regular and $q \triangleq \text{rank}E < n$.

In system (4-1.1), its initial state x(0) is usually not known in advance. And it is difficult or impossible to reconstruct exactly the state x(t). The so-called state observer ("observer" for simplicity) is reconstructing the state asymptotically. Based on such thoughts, a system Σ, which is a state observer for system (4-1.1), should audio-visually satisfy the following two necessary conditions.

1. The input of Σ should be the control input and measure output of system (4-

1.1).

2. Its output should satisfy the asymptotical condition $\lim_{t\to\infty}(w(t)-x(t)) = 0$. This equation holds for any initial condition $x(0)$.

Systems satisfying these two conditions are called state observers for system (4-1.1).

A diagram of observers is shown in Figure 4-1.1.

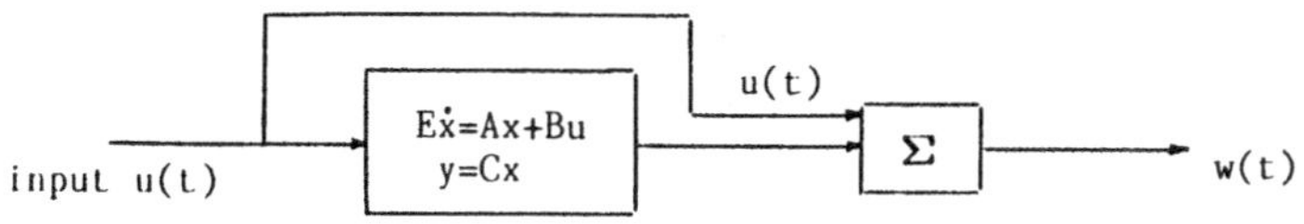

Figure 4-1.1. State Observer

Accounting for above observation we have the following definition.

Definition 4-1.1. Consider system (4-1.1). If the following system

$$\begin{aligned} E_c\dot{x}_c(t) &= A_cx_c(t) + B_cu(t) + Gy(t) \\ w(t) &= F_cx_c(t) + Fy(t) + Hu(t) \end{aligned} \tag{4-1.2}$$

where $x_c(t) \in \mathbb{R}^{n_c}$, $w(t) \in \mathbb{R}^n$, $E_c, A_c \in \mathbb{R}^{n_c \times n_c}$, B_c, G, F_c, F, and H are constant matrices of appropriate dimensions, satisfies

$$\lim_{t\to\infty}(w(t) - x(t)) = 0 \tag{4-1.3}$$

for any initial condition $x_c(0)$, $x(0)$, we call system (4-1.2) a state observer ("observer" for simplicity) for system (4-1.2). If $\mathrm{rank}E_c < n_c$, observer (4-1.2) is called the singular observer; Otherwise, $\mathrm{rank}E_c = n_c$, $E_c = I_{n_c}$ is assumed without loss of generality, and observer (4-1.2) is called a normal observer.

Obviously, the normal observer is the same as observers in linear system theory.

In this section, we are concerned with the design method of singular observers, the normal observers will be studied in the following section.

Theorem 4-1.1. Suppose that system (4-1.1) is detectable. Then it has a singular observer of the following form:

$$\begin{aligned} E\dot{x}_c &= Ax_c + Bu + G(y - Cx_c) \\ w &= x_c \end{aligned} \tag{4-1.4}$$

such that $\lim_{t\to\infty}(w-x) = 0$, $\forall\ x(0)$, $x_c(0)$.

Proof. Under the assumption of detectability, there exists a matrix $G \in \mathbb{R}^{n\times r}$ satisfying

$$\sigma(E,\ A-GC) \subset \mathbb{C}^{-}. \qquad (4\text{-}1.5)$$

Consider the singular system (4-1.4) determined by G. Let

$$e(t) = w(t) - x(t) = x_c(t) - x(t)$$

be the estimation error between real and estimated states. It is easy to know that e(t) satisfies

$$E\dot{e}(t) = (A-GC)e(t), \qquad e(0) = x_c(0) - x(0).$$

Thus, by (4-1.5) and Theorem 3-1.1, we have

$$\lim_{t\to\infty} e(t) = \lim_{t\to\infty} (w(t)-x(t)) = 0, \ \forall\ x(0),\ x_c(0).$$

Q.E.D.

In the state equation of singular observer (4-1.4), the first two terms are the same as the corresponding ones in (4-1.1), and there is the third term $G(y(t)-Cx_c(t))$ on the right side in (4-1.4), which acts as the error modification term to modify the estimating error brought about by initial conditions $x_c(0)$ and $x(0)$, such that w(t) estimates x(t) asymptotically.

Apparently, $x_c(t) \equiv x(t)$ provided $x_c(0) = x(0)$. But this case is always false in a system.

Example 4-1.1. Consider the circuit network (2-2.5)-(2-2.6):

$$\begin{bmatrix} C_1 & 0 & 0 & 0 \\ 0 & C_2 & 0 & 0 \\ 0 & 0 & -L & 0 \\ 0 & 0 & 0 & 0 \end{bmatrix} \dot{x} = \begin{bmatrix} 0 & 0 & 0 & 1 \\ 0 & 0 & 1 & 0 \\ -1 & 1 & 0 & 0 \\ 1 & 0 & 0 & R \end{bmatrix} x + \begin{bmatrix} 0 \\ 0 \\ 0 \\ -1 \end{bmatrix} u \qquad (4\text{-}1.6)$$

$$y = [0 \quad 1 \quad 0 \quad 0]x.$$

Example 2-3.3 proved its R-observability. Thus it is detectable. Denote

$$G = [g_1,\ g_2,\ g_3,\ g_4]^{\tau}.$$

We have

$$|sE-(A-GC)| = LC_1C_2Rs^3 + (LC_2+g_2LC_1R)s^2 + (g_2L-g_3C_1R)s + 1-g_3-g_4+g_1R. \qquad (4\text{-}1.7)$$

Setting

$$\begin{aligned} g_1 &= 0 \\ g_2 &= (C_2/(C_1R))(3C_1R-1) \\ g_3 &= (C_1R)^{-1}(-3LC_1C_2R+C_1R+g_2L) \\ g_4 &= 1-g_3-LC_1C_2R, \end{aligned} \qquad (4\text{-}1.8)$$

Equation (4-1.7) becomes

$$|sE-(A-GC)| = LC_1C_2R(s+1)^3.$$

Thus $\sigma(E, A-GC) = \{-1, -1, -1\} \subset \mathbb{C}^-$. For matrix G chosen as in (4-1.8), Theorem 4-1.1 gives the state observer for system (2-2.5)-(2-2.6):

$$\begin{bmatrix} C_1 & 0 & 0 & 0 \\ 0 & C_2 & 0 & 0 \\ 0 & 0 & -L & 0 \\ 0 & 0 & 0 & 0 \end{bmatrix} \dot{x}_c = \begin{bmatrix} 0 & -g_1 & 0 & 1 \\ 0 & -g_2 & 1 & 0 \\ -1 & 1-g_3 & 0 & 0 \\ 1 & -g_4 & 0 & R \end{bmatrix} x_c + \begin{bmatrix} 0 \\ 0 \\ 0 \\ -1 \end{bmatrix} u + \begin{bmatrix} g_1 \\ g_2 \\ g_3 \\ g_4 \end{bmatrix} y$$

$$w = x_c.$$

Theorem 4-1.2. System (4-1.1) has a singular observer (4-1.4) if and only if it is detectable.

Proof. Sufficiency is given by Theorem 4-1.1. We will now prove its necessity. Assume that (4-1.4) is a singular observer for system (4-1.1). Then it must be

$$\lim_{t\to\infty} e(t) = 0, \quad \forall\ x(0),\ x_c(0). \tag{4-1.9}$$

Since the error $e(t) = w(t)-x(t)$ satisfies

$$E\dot{e} = (A-GC)e, \qquad e(0) = x_c(0) - x(0) \tag{4-1.10}$$

from (4-1.9), we know the system is stable. Thus, $\sigma(E,A-GC) \subset \mathbb{C}^-$ by Theorem 3-1.1, i.e., system (4-1.1) is detectable. Q.E.D.

If $E = I_n$ is allowed, Theorems 4-1.1 and 4-1.2 become the well-known results in linear system theory.

Meanwhile, we may conclude from Theorem 4-1.2 that the detectability characterizes the ability to detect the state by a certain dynamic system. Under the detectability assumption, the state $x(t)$ may be reconstructed asymptotically by an observer from the directly obtained information $u(t)$ and $y(t)$.

Definition 4-1.2. System (4-1.1) is called detectable if it has a singular observer (4-1.4).

This definition is obviously consistent with Chapter 3.

4-2. Normal State Observer

In the last section, we studied the existence and design method for singular observers. However, as pointed out before, input derivatives are usually involved in the state of singular systems. Thus, in addition to u(t) and y(t), their derivatives are also involved in the state of singular observers. Since these terms are not usually available, it is difficult to realize singular observers, although they do exist. Furthermore, the strength of a high-frequency noise is often amplified by its derivatives. This states that the singular observer is sensitive to high frequency noises. All these arguments spur us to consider normal observers.

Consider system (4-1.1):

$$\begin{aligned} E\dot{x} &= Ax + Bu \\ y &= Cx \end{aligned} \tag{4-2.1}$$

and its observer

$$\begin{aligned} \dot{x}_c(t) &= A_c x_c(t) + B_c u(t) + Gy(t) \\ w(t) &= F_c x_c(t) + Fy(t) + Hu(t) \end{aligned} \tag{4-2.2}$$

where $x_c(t) \in \mathbb{R}^{n_c}$, A_c, B_c, G, F_c, F, and H are constant matrices of appropriate dimensions. Observer (4-2.2) is called a full-order normal observer if $n_c = n$; otherwise, $n_c < n$, and it is termed a reduced-order normal observer.

<u>Theorem 4-2.1.</u> Assume that system (4-2.1) is detectable, dual normalizable, and rankC = r. Then it has a reduced-order normal observer of order n - r of the following form:

$$\begin{aligned} \dot{x}_c &= A_c x_c + B_c u + Gy \\ w &= F_c x_c + Fy \end{aligned} \tag{4-2.3}$$

<u>Proof</u>. By assumption, rankC = r. Without loss of generality we assume $C = [C_1, C_2]$, and $\text{rank}C_1 = r$, $C_1 \in \mathbb{R}^{r\times r}$. Let

$$P = \begin{bmatrix} C_1^{-1} & -C_1^{-1}C_2 \\ 0 & I_{n-r} \end{bmatrix} .$$

Then P is nonsingular and $CP = [I_r, 0]$. By denoting

$$EP = \begin{bmatrix} \tilde{E}_{11} & \tilde{E}_{12} \\ \tilde{E}_{21} & \tilde{E}_{22} \end{bmatrix} \begin{matrix} \}r \\ \}n-r \end{matrix}$$

and also noticing the dual normalizability assumption for system (4-2.1), from rank[E /C] = n we have

$$\text{rank}\begin{bmatrix} E \\ C \end{bmatrix} = \text{rank}\begin{bmatrix} EP \\ CP \end{bmatrix} = \text{rank}\begin{bmatrix} \tilde{E}_{11} & \tilde{E}_{12} \\ \tilde{E}_{21} & \tilde{E}_{22} \\ I_r & 0 \end{bmatrix}$$

$\text{rank}[\tilde{E}_{12}/\tilde{E}_{22}] = n-r$. Thus matrix $[Q_{11}/Q_{21}] \in \mathbb{R}^{n\times r}$ exists such that

$$\text{rank}\begin{bmatrix} Q_{11} & \tilde{E}_{12} \\ Q_{21} & \tilde{E}_{22} \end{bmatrix} = n.$$

Let

$$Q = \begin{bmatrix} Q_{11} & \tilde{E}_{12} \\ Q_{21} & \tilde{E}_{22} \end{bmatrix}^{-1}.$$

Then Q is nonsingular and $Q[\tilde{E}_{12}/\tilde{E}_{22}] = [0/I_{n-r}]$. For these nonsingular matrices Q and P, we have

$$QEP = Q\begin{bmatrix} \tilde{E}_{11} & \tilde{E}_{12} \\ \tilde{E}_{21} & \tilde{E}_{22} \end{bmatrix} = \begin{bmatrix} E_{11} & 0 \\ E_{21} & I_{n-r} \end{bmatrix}, \qquad CP = [I_r, \ 0],$$

where $Q[\tilde{E}_{11}/\tilde{E}_{21}] = [E_{11}/E_{21}]$. Denote

$$QAP = \begin{bmatrix} A_{11} & A_{12} \\ A_{21} & A_{22} \end{bmatrix}, \qquad QB = \begin{bmatrix} B_1 \\ B_2 \end{bmatrix}.$$

From above discussion, under the coordinate transformation $x = P[x_1/x_2]$ system (4-2.1) is r.s.e. to

$$\begin{aligned} E_{11}\dot{x}_1 &= A_{11}x_1 + A_{12}x_2 + B_1u \\ E_{21}\dot{x}_1 + \dot{x}_2 &= A_{21}x_1 + A_{22}x_2 + B_2u \\ y &= x_1 \end{aligned} \tag{4-2.4}$$

where $x_1 \in \mathbb{R}^r$, $x_2 \in \mathbb{R}^{n-r}$.

From (4-2.4) we see that the substate x_1 may be obtained directly from measure output $y(t)$, and the substate x_2 satisfies the following dynamic equation:

$$\begin{aligned} \dot{x}_2 &= A_{22}x_2 + A_{21}y - E_{21}\dot{y} + B_2u \\ \tilde{y} &\triangleq E_{11}\dot{y} - A_{11}y - B_1u = A_{12}x_2 \end{aligned} \tag{4-2.5}$$

Moreover, since system (4-2.1) is assumed to be detectable,

$$\text{rank}\begin{bmatrix} sE-A \\ C \end{bmatrix} = \text{rank}\begin{bmatrix} sQEP-QAP \\ CP \end{bmatrix} = \text{rank}\begin{bmatrix} sE_{11}-A_{11} & -A_{12} \\ sE_{21}-A_{21} & sI-A_{22} \\ I_r & 0 \end{bmatrix} = n,$$

$$\forall \ s \in \bar{\mathbb{C}}^+, \ s \text{ finite}.$$

Therefore

$$\operatorname{rank}\begin{bmatrix} sI-A_{22} \\ A_{12} \end{bmatrix} = n-r, \quad \forall\ s\in\bar{\mathbb{C}}^{+},\ s \text{ finite},$$

showing that (A_{22},A_{12}) is detectable. Thus there exists a matrix $G_2 \in \mathbb{R}^{(n-r)xr}$ such that $\sigma(A_{22}-G_2A_{12}) \subset \mathbb{C}^{-}$. For this matrix G_2, from Theorem 4-1.1(consider the special case when $E = I_n$) we know that system (4-2.5) has the following state observer:

$$\dot{\hat{x}}_2 = (A_{22}-G_2A_{12})\hat{x}_2 + A_{21}y - E_{21}\dot{y} + B_2u + G_2\tilde{y} \tag{4-2.6}$$

such that $\lim_{t\to\infty} (\hat{x}_2-x_2) = 0,\ \forall\ x_2(0),\ \hat{x}_2(0)$.

By denoting $x_c = \hat{x}_2 + (E_{21}-G_2E_{11})y$, from (4-2.6) x_c satisfies

$$\dot{x}_c = (A_{22}-G_2A_{12})x_c + (A_{21}-G_2A_{11}-(A_{22}-G_2A_{12})(E_{21}-G_2E_{11}))y + (B_2-G_2B_1)u$$

$$x_2 = x_c - (E_{21}-G_2E_{11})y \tag{4-2.7}$$

Since $x_1=y$ and $\hat{x}_2$ is an asymptotic estimation of x_2, it is obvious that

$$\begin{bmatrix} y \\ x_2 \end{bmatrix} = P\begin{bmatrix} 0 \\ I_{n-r} \end{bmatrix}x_c + P\begin{bmatrix} I_r \\ -E_{21}+G_2E_{11} \end{bmatrix}y$$

is an asymptotic estimation of the state x for system (4-2.1). This equation, together with (4-2.7), gives the following state observer of order n-r for system (4-2.1):

$$\dot{x}_c = A_cx_c + B_cu + Gy$$
$$w = F_cx_c + Fy$$

where $x_c \in \mathbb{R}^{n-r}$, $w \in \mathbb{R}^n$, and

$$A_c = A_{22} - G_2A_{12}$$
$$B_c = B_2 - G_2B_1$$
$$G = A_{21} - G_2A_{11} - (A_{22}-G_2A_{12})(E_{21}-G_2E_{11})$$
$$F_c = P\begin{bmatrix} 0 \\ I_{n-r} \end{bmatrix},$$
$$F = P\begin{bmatrix} I_r \\ -E_{21}+G_2E_{11} \end{bmatrix}.$$

Q.E.D.

State observer (4-2.3) is characterized by its normality and its output, which is a linear combination of x_c and y. Such observers are easier to physically realize than singular observers; they are diagrammed in Figure 4-2.1.

<u>Example 4-2.1</u>. Consider the singular system

$$\begin{bmatrix} 0 & 1 & 0 \\ 0 & 0 & 1 \\ 0 & 0 & 0 \end{bmatrix}\dot{x} = \begin{bmatrix} 1 & 0 & 0 \\ 0 & 1 & 0 \\ 0 & 0 & 1 \end{bmatrix}x + \begin{bmatrix} 0 & 0 \\ 1 & 0 \\ 0 & 1 \end{bmatrix}u \tag{4-2.8a}$$

$$y = [1 \quad 0 \quad 0]x \tag{4-2.8b}$$

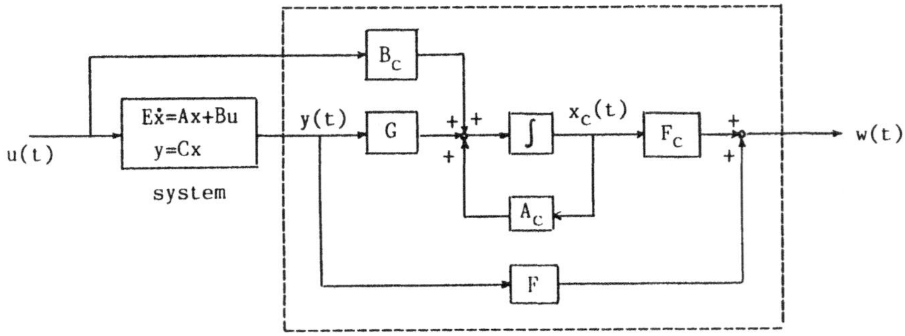

Figure 4-2.1. State Observer (4-2.3)

In system (4-2.8) $n = 3$, $r = 1$,

$$E = \begin{bmatrix} 0 & 1 & 0 \\ 0 & 0 & 1 \\ 0 & 0 & 0 \end{bmatrix}, \quad A = \begin{bmatrix} 1 & 0 & 0 \\ 0 & 1 & 0 \\ 0 & 0 & 1 \end{bmatrix}, \quad B = \begin{bmatrix} 0 & 0 \\ 1 & 0 \\ 0 & 1 \end{bmatrix}, \quad C = [1 \quad 0 \quad 0].$$

This system is observable that rank$C = 1 = r$. Thus, it has an observer of the form (4-2.3). Let

$$Q = \begin{bmatrix} 0 & 0 & 1 \\ 1 & 0 & 0 \\ 0 & 1 & 0 \end{bmatrix}, \quad P = \begin{bmatrix} 1 & 0 & 0 \\ 0 & 1 & 0 \\ 0 & 0 & 1 \end{bmatrix}, \quad x = \begin{bmatrix} x_1 \\ x_2 \end{bmatrix}$$

$x_1 \in \mathbb{R}^1$, $x_2 \in \mathbb{R}^2$. Direct computation shows that system (4-2.8) is r.s.e. to

$$\begin{bmatrix} 0 & 0 & 0 \\ 0 & 1 & 0 \\ 0 & 0 & 1 \end{bmatrix}\begin{bmatrix} \dot{x}_1 \\ \dot{x}_2 \end{bmatrix} = \begin{bmatrix} 0 & 0 & 1 \\ 1 & 0 & 0 \\ 0 & 1 & 0 \end{bmatrix}\begin{bmatrix} x_1 \\ x_2 \end{bmatrix} + \begin{bmatrix} 0 & 1 \\ 0 & 0 \\ 1 & 0 \end{bmatrix} u$$

$$y = [1 \quad 0 \quad 0][x_1/x_2],$$

i.e.,

$$\begin{aligned} y &= x_1 \\ \dot{x}_2 &= \begin{bmatrix} 0 & 0 \\ 1 & 0 \end{bmatrix} x_2 + \begin{bmatrix} 1 \\ 0 \end{bmatrix} y + \begin{bmatrix} 0 & 0 \\ 1 & 0 \end{bmatrix} u \\ \tilde{y} &\triangleq [0 \quad -1]u = [0 \quad 1]x_2 , \end{aligned} \tag{4-2.9}$$

in which

$$A_{22} = \begin{bmatrix} 0 & 0 \\ 1 & 0 \end{bmatrix}, \qquad A_{12} = [0 \quad 1].$$

(A_{22}, A_{12}) is observable, and the matrix $G_2 = [1 \quad 1]^\tau$ satisfies $\sigma(A_{22}-G_2A_{12}) = \{-\frac{1}{2}+\frac{3}{2}i, -\frac{1}{2}-\frac{3}{2}i\} \subset \mathbb{C}^-$. Using this matrix G_2 we can constructe the following observer for the substate x :

$$\begin{aligned} \dot{x}_c &= \begin{bmatrix} 0 & -1 \\ 1 & -1 \end{bmatrix} x_c + \begin{bmatrix} 1 \\ 0 \end{bmatrix} y + \begin{bmatrix} 0 & -1 \\ 1 & -1 \end{bmatrix} u \\ w &= \begin{bmatrix} 1 \\ 0 \\ 0 \end{bmatrix} y + \begin{bmatrix} 0 & 0 \\ 1 & 0 \\ 0 & 1 \end{bmatrix} x_c \end{aligned} \tag{4-2.10}$$

such that $\lim_{t\to\infty} (w-x) = 0, \quad \forall\ x(0),\ x_c(0).$

From state represention, the state x(t) of system (4-2.8) is

$$x = \begin{bmatrix} 0 & 0 \\ -1 & 0 \\ 0 & -1 \end{bmatrix} u + \begin{bmatrix} -1 & 0 \\ 0 & -1 \\ 0 & 0 \end{bmatrix} \dot{u} + \begin{bmatrix} 0 & -1 \\ 0 & 0 \\ 0 & 0 \end{bmatrix} \ddot{u}$$

which includes input derivatives of orders up to 2. However, it is implied from (4-2.10) that the state x(t) may be asymptotically approximated by a value w that does not involve visibly input derivatives, and where w is the measure of a normal system. This interesting phenomenon reveals a common fact in normal state observers for singular systems. A value apparently includes the input derivatives may be sometimes asymptotically tracked by a value that doesn't involve apparently input derivatives. The core point for this fact lies in that input derivatives are included invisibly in measure output y(t).

<u>Theorem 4-2.2</u>. Let system (4-2.1) be both detectable and impulse observable. Then it always has a normal observer of order rankE of the following form:

$$\begin{aligned} \dot{x}_c &= A_c x_c + B_c u + Gy \\ w &= F_c x_c + Fy + Hu. \end{aligned} \tag{4-2.11}$$

<u>Proof</u>. Under the assumption of detectability and impluse observability, from Corollary 3-3.2 we know there exists a matrix $\bar{G} \in \mathbb{R}^{n\times r}$ such that

$$\deg(|sE - (A-\bar{G}C)|) = \mathrm{rank}E, \qquad \sigma(E,\ A-\bar{G}C) \subset \mathbb{C}^-. \tag{4-2.12}$$

Thus, by Theorem 4-1.1, the system

$$\begin{aligned} E\dot{\bar{x}}_c &= (A-\bar{G}C)\bar{x}_c + Bu + \bar{G}y \\ w &= \bar{x}_c \end{aligned} \tag{4-2.13}$$

is a state observer for system (4-2.1), such that

$$\lim_{t\to\infty} (w - x) = 0, \quad \forall\ \bar{x}_c(0),\ x(0).$$

Noticing the equations in (4-2.12), two nonsingular matrices Q_1 and P_1 exist such that

$$Q_1EP_1 = \text{diag}(I_q, 0), \quad Q_1(A-\bar{G}C)P_1 = \text{diag}(\bar{A}_1, I), \quad q = \text{rank}E \tag{4-2.14}$$

and $\bar{A}_1 \in \mathbb{R}^{q\times q}$, $\sigma(\bar{A}_1) = \sigma(E,A-\bar{G}C) \subset \mathbb{C}^-$. Denote

$$Q_1B = \begin{bmatrix} B_1 \\ B_2 \end{bmatrix}, \qquad B_1 \in \mathbb{R}^{q\times m}, \quad B_2 \in \mathbb{R}^{(n-q)\times m},$$

$$Q_1\bar{G} = \begin{bmatrix} G_1 \\ G_2 \end{bmatrix}, \qquad G_1 \in \mathbb{R}^{q\times r}, \quad G_2 \in \mathbb{R}^{(n-q)\times r}, \tag{4-2.15}$$

$$P^{-1}\bar{x}_c = \begin{bmatrix} x_{c_1} \\ x_{c_2} \end{bmatrix} \qquad x_{c_1} \in \mathbb{R}^q, \quad x_{c_2} \in \mathbb{R}^{n-q}.$$

State observer (4-2.13) is r.s.e. to

$$\begin{aligned} \dot{x}_{c_1} &= \bar{A}_1x_{c_1} + B_1u + G_1y \\ 0 &= x_{c_2} + B_2u + G_2y \\ w &= P\,[x_{c_1}/x_{c_2}], \end{aligned}$$

i.e.,

$$\begin{aligned} \dot{x}_c &= \bar{A}_1x_c + B_1u + G_1y \\ w &= P_1\begin{bmatrix} I \\ 0 \end{bmatrix}x_c - P_1\begin{bmatrix} 0 \\ I \end{bmatrix}B_2u - P_1\begin{bmatrix} 0 \\ I \end{bmatrix}G_2y \end{aligned}$$

where $x_c = x_{c_1} \in \mathbb{R}^q$, which is a normal observer of the form (4-2.11). Q.E.D.

For any nonsingular matrices Q_1 and P_1 determined by (4-2.14), if we denote

$$CP_1 = [C_1, C_2], \quad C_1 \in \mathbb{R}^{r\times q}, \quad C_2 \in \mathbb{R}^{r\times(n-q)}, \tag{4-2.16}$$

we can prove that $\text{rank}C_1$ is independent of the selection of matrix $\bar{G}$ provided that $\bar{G}$ satisfies (4-2.12).

Lemma 4-2.1. $d \triangleq \text{rank}C_1$ is a fixed value that is independent of the selection of $\bar{G}$, provided (4-2.12) holds.

Proof. To prove that $d = \text{rank}C_1$ is independent of $\bar{G}$, we assume that $\tilde{G}$ is any other matrix satisfying (4-2.12). That is

$$\deg(|sE-(A-\tilde{G}C)|) = \text{rank}E, \quad \sigma(E, A-\tilde{G}C) \subset \mathbb{C}^-. \tag{4-2.17}$$

Then, for the matrices Q_1 and P_1 determined by (4-2.14), we have

$$Q_1(A-\tilde{G}C)P_1 = Q_1(A-\bar{G}C+(\bar{G}-\tilde{G})C)P_1 = Q_1(A-\bar{G}C)P_1 + Q_1(\bar{G}-\tilde{G})CP_1$$

$$= \begin{bmatrix} \bar{A}_1 + \tilde{G}_1C_1 & \tilde{G}_1C_2 \\ \tilde{G}_2C_1 & I+\tilde{G}_2C_2 \end{bmatrix}$$

where

$$Q_1(\bar{G}-\tilde{G}) = [\tilde{G}_1/\tilde{G}_2].$$

Since $\tilde{G}$ satisfies the first equation in (4-2.17), from the results in Section 3-2 (inequality (3-2.10)) we know that

$$|I + \tilde{G}_2C_2| \neq 0.$$

Let

$$Q_2 = \begin{bmatrix} I_q & -\tilde{G}_1C_1(I+\tilde{G}_2C_2)^{-1} \\ 0 & I_{n-q} \end{bmatrix} Q_1, \qquad P_2 = P_1\begin{bmatrix} I_q & 0 \\ -(I+\tilde{G}_2C_2)^{-1}\tilde{G}_2C_1 & (I+\tilde{G}_2C_2)^{-1} \end{bmatrix}.$$

Then Q_2 and P_2 are nonsingular and

$$Q_2EP_2 = \mathrm{diag}(I_q,\ 0), \qquad Q_2(A-\tilde{G}C)P_2 = \mathrm{diag}(\tilde{A}_1,\ I)$$
$$CP_2 = [\bar{C}_1,\ \bar{C}_2]$$

where

$$\tilde{A}_1 = \bar{A}_1 + \tilde{G}_1C_1 - \tilde{G}_1C_2(I+\tilde{G}_2C_2)^{-1}\tilde{G}_2C_1$$
$$\bar{C}_1 = C_1 - C_2(I+\tilde{G}_2C_2)^{-1}\tilde{G}_2C_1 = (I+C_2\tilde{G}_2)^{-1}C_1$$
$$\bar{C}_2 = C_2(I+\tilde{G}_2C_2)^{-1}.$$

The last line in this set of equations clearly shows that $\bar{d} \triangleq \mathrm{rank}\bar{C}_1 = \mathrm{rank}C_1 = d$. Note the arbitrariness of $\tilde{G}$ and that the standard decomposition is unique in the sense of similarity. We can conclude that $d = \mathrm{rank}C_1$ is independent of $\bar{G}$. Q.E.D.

<u>Remark</u>. It is worthy of pointing out that in the decomposition

$$QEP = \mathrm{diag}(I_q,\ 0), \qquad CP = [C_1,\ C_2]$$

$\mathrm{rank}C_1$ is dependent of the selection of Q and P. For example, let

$$E = \begin{bmatrix} 0 & 1 & 0 \\ 1 & 0 & 1 \\ 0 & 0 & 0 \end{bmatrix}, \qquad C = [1 \quad 0 \quad 2].$$

Then it may be verified that both

$$Q_1 = \begin{bmatrix} 1 & 0 & 0 \\ 0 & 1 & 0 \\ 0 & 0 & 1 \end{bmatrix}, \qquad P_1 = \begin{bmatrix} 0 & 1 & -1 \\ 1 & 0 & 0 \\ 0 & 0 & 1 \end{bmatrix}$$

and

$$Q_2 = \begin{bmatrix} 1 & 0 & 0 \\ 0 & 1 & 0 \\ 0 & 0 & 1 \end{bmatrix}, \qquad P_2 = \begin{bmatrix} 0 & 2 & -1 \\ 1 & 0 & 0 \\ 0 & -1 & 1 \end{bmatrix}$$

satisfy

$$Q_1EP_1 = Q_2EP_2 = \mathrm{diag}(I_q,\ 0).$$

However, we have

$$CP_1 = [0 \quad 1 \quad 1] \triangleq [C_1, \ C_2], \qquad CP_2 = [0 \quad 0 \quad 1] \triangleq [\bar{C}_1, \ \bar{C}_2].$$

Hence

$$\text{rank}C_1 = \text{rank}[0 \quad 1] = 1 \neq \text{rank}\bar{C}_1 = 0.$$

Theorem 4-2.3. Suppose that system (4-2.1) is detectable and impulse observable. Then it always has a normal observer of order rankE-d of the following form

$$\begin{aligned} \dot{x}_c &= A_c x_c + B_c u + Gy \\ w &= F_c x_c + Fy + Hu \end{aligned} \tag{4-2.18}$$

where $d = \text{rank}C_1$, C_1 is determined in (4-2.16).

Proof. By assumption, system (4-2.1) is detectable and impulse observable. In the proof process we know that there exists a matrix $\bar{G} \in \mathbb{R}^{n \times r}$ satisfying (4-2.12). In this case, system (4-2.1) may be rewritten as

$$\begin{aligned} E\dot{x} &= (A-\bar{G}C)x + Bu + \bar{G}y \\ y &= Cx. \end{aligned} \tag{4-2.19}$$

Note that there exist two nonsingular matrices Q_1 and P_1 such that (4-2.14) and (4-2.15) hold. Thus, if we define the coordinate transformation

$$P_1^{-1}x = \begin{bmatrix} x_1 \\ x_2 \end{bmatrix}, \qquad x_1 \in \mathbb{R}^q, \quad x_2 \in \mathbb{R}^{n-q},$$

system (4-2.19) is r.s.e. to

$$\begin{aligned} \dot{x}_1 &= \bar{A}_1 x_1 + B_1 u + G_1 y \\ 0 &= x_2 + B_2 u + G_2 y \\ y &= C_1 x_1 + C_2 x_2 \end{aligned}$$

where $CP_1 = [C_1 \quad C_2]$. This system can be further written as

$$\begin{aligned} x_2 &= -B_2 u - G_2 y \\ \dot{x}_1 &= \bar{A}_1 x_1 + B_1 u + G_1 y \\ \tilde{y} &\triangleq C_1 x_1 = (I_r + C_2 G_2) y + C_2 B_2 u. \end{aligned} \tag{4-2.20}$$

With no harm to our discussion, assume that $d = \text{rank}C_1 > 0$, which is independent of $\bar{G}$ (Lemma 4-2.1) and Q_1, P_1. Thus, a matrix M exists such that

$$MC_1 = \begin{bmatrix} \bar{C}_1 \\ 0 \end{bmatrix}$$

and $\bar{C}_1$ is of full row rank d.

On the other hand, system (4-2.1) is detectable. Hence

$$n = \text{rank}\begin{bmatrix} sE-A \\ C \end{bmatrix} = \text{rank}\begin{bmatrix} sQ_1EP_1 - Q_1(A-\bar{G}C)P_1 \\ CP_1 \end{bmatrix}$$

$$= \text{rank}\begin{bmatrix} sI_q - \bar{A}_1 & 0 \\ 0 & -I \\ C_1 & C_2 \end{bmatrix} = n - q + \text{rank}\begin{bmatrix} sI_q - \bar{A}_1 \\ \bar{C}_1 \end{bmatrix},$$

$$\forall \ s \in \bar{\mathbb{C}}^+, \ s \text{ finite},$$

which in turn shows that

$$\text{rank}\begin{bmatrix} sI_q - \bar{A}_1 \\ \bar{C}_1 \end{bmatrix} = q, \quad \forall \ s \in \bar{\mathbb{C}}^+, \ s \text{ finite},$$

i.e., $(\bar{A}_1, \bar{C}_1)$ is detectable.

Next we will construct the Luenberger observer for substate x_1. Define

$$M_1 = [I_q, \ 0]M, \qquad \bar{y} = M_1\tilde{y} = M_1(I+C_2G_2)y + M_1C_2B_2u.$$

From (4-2.20) we know x_1 satisfies the following dynamic equation:

$$\begin{aligned} \dot{x}_1 &= \bar{A}_1x_1 + B_1u + G_1y \\ \bar{y} &= M_1C_1x_1 = \bar{C}_1x_1. \end{aligned} \tag{4-2.21}$$

Since $\bar{C}_1$ is of full row rank, a matrix $\bar{C}_2$ may be chosen such that rank$[\bar{C}_1/\bar{C}_2]$ = q and $T = [\bar{C}_1/\bar{C}_2] \in \mathbb{R}^{q\times q}$ is nonsingular with the fact

$$\bar{C}_1T^{-1} = [I_d, \ 0].$$

Let

$$TA_1T^{-1} = \begin{bmatrix} \bar{A}_{11} & \bar{A}_{12} \\ \bar{A}_{21} & \bar{A}_{22} \end{bmatrix}, \qquad TB_1 = \begin{bmatrix} \bar{B}_{11} \\ \bar{B}_{12} \end{bmatrix}, \qquad TG_1 = \begin{bmatrix} \bar{G}_{11} \\ \bar{G}_{12} \end{bmatrix}$$

$$Tx_1 = \begin{bmatrix} x_{11} \\ x_{12} \end{bmatrix}, \qquad x_{11} \in \mathbb{R}^d, \quad x_{12} \in \mathbb{R}^{q-d}.$$

System (4-2.21) is similar to

$$\begin{aligned} \dot{x}_{11} &= \bar{A}_{11}x_{11} + \bar{A}_{12}x_{12} + \bar{B}_{11}u + \bar{G}_{11}y \\ \dot{x}_{12} &= \bar{A}_{21}x_{11} + \bar{A}_{22}x_{12} + \bar{B}_{12}u + \bar{G}_{12}y \\ \bar{y} &= x_{11} \end{aligned} \tag{4-2.22}$$

in which x_{11} is given directly by $\bar{y}$. Thus, we need only to construct the state observer for substate x_{12}. Noticing that x_{12} satisfies

$$\begin{aligned} \dot{x}_{12} &= \bar{A}_{22}x_{12} + \bar{B}_{12}u + \bar{G}_{12}y + \bar{A}_{21}\bar{y} \\ z &\triangleq \bar{A}_{12}x_{12} = \dot{\bar{y}} - \bar{A}_{11}\bar{y} - \bar{B}_{11}u - \bar{G}_{11}y. \end{aligned} \tag{4-2.23}$$

We have proven previously the detectability of $(\bar{A}_1, \bar{C}_1)$. Since

$$\operatorname{rank}\begin{bmatrix} sI - \bar{A}_1 \\ \bar{C}_1 \end{bmatrix} = \operatorname{rank}\begin{bmatrix} sI - \bar{A}_{11} & -\bar{A}_{12} \\ -\bar{A}_{21} & sI - \bar{A}_{22} \\ I_d & 0 \end{bmatrix} = d + \operatorname{rank}\begin{bmatrix} sI - \bar{A}_{22} \\ \bar{A}_{12} \end{bmatrix},$$

$$\forall\ s \in \bar{\mathbb{C}}^+,\ s \text{ finite}.$$

$(\bar{A}_{22}, \bar{A}_{12})$ is detectable. There exists a matrix G_3 satisfying

$$\sigma(\bar{A}_{22} - G_3\bar{A}_{12}) \subset \mathbb{C}^-.$$

Let G_3 be determined as above. The system

$$\dot{\hat{x}}_{12} = (\bar{A}_{22} - G_3\bar{A}_{12})\hat{x}_{12} + \bar{B}_{12}u + \bar{G}_{12}y + \bar{A}_{21}\bar{y} + G_3 z \tag{4-2.24}$$

will be an observer for x_{12}, such that

$$\lim_{t\to\infty} (x_{12} - \hat{x}_{12}) = 0, \quad \forall\ x_{12}(0),\ \hat{x}_{12}(0).$$

Define $x_c = \hat{x}_{12} - \bar{A}_{21}\bar{y}$. From (4-2.24) we know x_c satisfies

$$\begin{aligned} \dot{x}_c &= A_c x_c + B_c u + Gy \\ x_{12} &= x_c + \bar{A}_{21}\bar{y} \end{aligned} \tag{4-2.25}$$

where

$$\begin{aligned} A_c &= \bar{A}_{22} - G_3\bar{A}_{12} \\ B_c &= \bar{B}_{12} - G_3\bar{B}_{11} + (\bar{A}_{12} - G_3\bar{A}_{11} + (\bar{A}_{22} - G_3\bar{A}_{12})G_3)M_1C_2B_2 \\ G &= \bar{G}_{12} - G_3\bar{G}_{11} + (\bar{A}_{12} - G_3\bar{A}_{11} + (\bar{A}_{22} - G_3\bar{A}_{12})G_3)M_1(I + C_2G_2). \end{aligned}$$

On the other hand,

$$\hat{x}_1 = T^{-1}([0/I_{q-d}]x_c + [I_d/G_3]M_1(I + C_2G_2)y + [I_d/G_3]M_1C_2B_2u) \tag{4-2.26}$$

is apparently an asymptotic estimation of x_1. Thus

$$\hat{x} = P_1\begin{bmatrix} x_1 \\ -B_2u - G_2y \end{bmatrix} \tag{4-2.27}$$

is an asymptotic estimation of state $x(t)$ for system (4-2.1), such that $\lim_{t\to\infty}(x - \hat{x}) = 0$.

The combination of (4-2.24) - (4-2.27) immediately yields the reduced order normal state observer for system (4-2.1):

$$\begin{aligned} \dot{x}_c &= A_c x_c + B_c u + Gy \\ w &= F_c x_c + Fy + Hu \end{aligned}$$

which is of order rankE − d. Here $w = \hat{x}$, and

$$F_c = P_1 \begin{bmatrix} T^{-1}[0/I_{q-d}] \\ 0 \end{bmatrix}$$

$$F = P_1 \begin{bmatrix} T^{-1}[I_d/G_3]M_1(I+C_2G_2) \\ -G_2 \end{bmatrix}$$

$$H = P_1 \begin{bmatrix} T^{-1}[I_d/G_3]M_1C_2B_2 \\ -B_2 \end{bmatrix}.$$

Q.E.D.

The normal observer (4-2.11) and (4-2.18) are characterized by their measure outputs in which u(t) is visibly included.

Figure 4-2.2 is a diagram of normal observers (4-2.11) and (4-2.18).

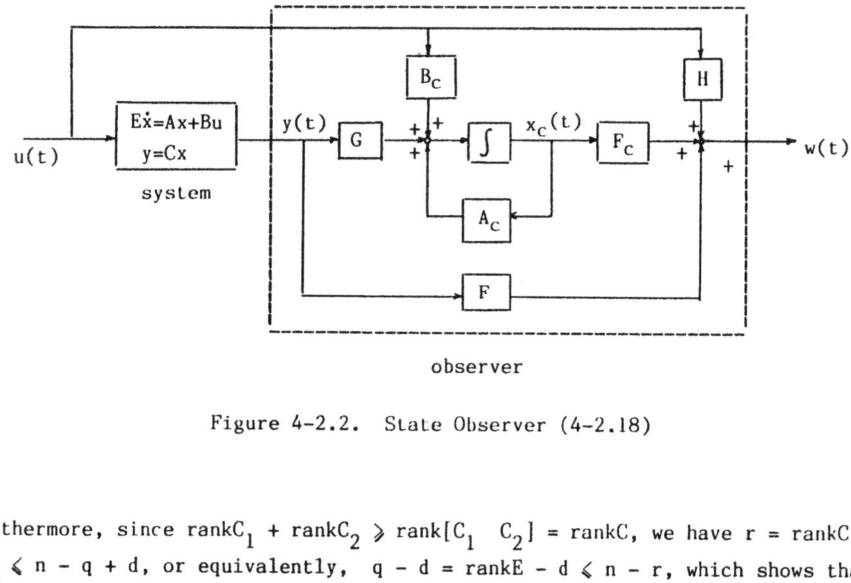

Figure 4-2.2. State Observer (4-2.18)

Furthermore, since $\mathrm{rank}C_1 + \mathrm{rank}C_2 \geqslant \mathrm{rank}[C_1 \;\; C_2] = \mathrm{rank}C$, we have $r = \mathrm{rank}C \leqslant d + \mathrm{rank}C_2 \leqslant n - q + d$, or equivalently, $q - d = \mathrm{rank}E - d \leqslant n - r$, which shows that the order of observer (4-2.18) may be lower than that of observer (4-2.3).

Thus, in the case where system (4-2.1) is dual normalizable (it is clearly impulse observable) there are two design approaches: designing a normal observer of order n−r according to Theorem 4-2.1, or designing a normal observer of lower order rankE−d according to Theorem 4-2.3. However, since input u is directly involved in measure equation (4-2.18), the measure w is sensitive to disturbance when noises are involved in u (acting as the disturbance input for original system). Thus, in the cases where high precision is required, it is better to choose observer (4-2.3) even though it is of higher order.

A direct result of the combination of Theorems 4-2.1 and 4-2.3 is the following corollary.

Corollary 4-2.1. If system (4-2.1) is observable, the poles of its state observer (4-2.3) may be arbitrarily placed, and if it is R-controllable and impulse observable, the poles of its state observer (4-2.18) may be arbitrarily placed.

Example 4-2.2. For the coefficients of the following system

$$\begin{bmatrix} 0 & 1 & 0 \\ 0 & 0 & 1 \\ 0 & 0 & 0 \end{bmatrix}\dot{x} = \begin{bmatrix} 1 & 0 & 0 \\ 0 & 1 & 0 \\ 0 & 0 & 1 \end{bmatrix}x + \begin{bmatrix} 0 & 0 \\ 1 & 0 \\ 0 & 1 \end{bmatrix}u$$
$$y = [1 \quad 0 \quad 0]x \tag{4-2.28}$$

the matrix $\bar{G} = [0 \quad 0 \quad 1]^{\tau}$ satisfies $\deg(|sE-(A-\bar{G}C)|) = \text{rank}E = 2$.

Let the coordinate transformation be

$$Q = \begin{bmatrix} 1 & 0 & 1 \\ 0 & 1 & 0 \\ 0 & 0 & -1 \end{bmatrix}, \qquad P = \begin{bmatrix} 0 & 1 & 1 \\ 1 & 0 & 0 \\ 0 & 1 & 0 \end{bmatrix},$$

$$P^{-1}x = \begin{bmatrix} x_1 \\ x_2 \end{bmatrix}, \qquad x_1 \in \mathbb{R}^2, \quad x_2 \in \mathbb{R}^2.$$

System (4-2.28) (corresponding to the rewritten form (4-2.20) of system (4-2.19)) is r.s.e. to

$$\dot{x}_1 = \begin{bmatrix} 0 & 1 \\ 1 & 0 \end{bmatrix}x_1 + \begin{bmatrix} 0 & 1 \\ 1 & 0 \end{bmatrix}u + \begin{bmatrix} 1 \\ 0 \end{bmatrix}y$$
$$0 = x_2 + [0 \quad -1]u - y \tag{4-2.29}$$
$$y = [0 \quad 1]x_1 + x_2,$$

which implies that substates x_1 and x_2 satisfy, respectively,

$$x_2 = y - [0 \quad -1]u$$
$$\dot{x}_1 = \begin{bmatrix} 0 & 1 \\ 1 & 0 \end{bmatrix}x_1 + \begin{bmatrix} 0 & 1 \\ 1 & 0 \end{bmatrix}u + \begin{bmatrix} 1 \\ 0 \end{bmatrix}y$$
$$\tilde{y} \triangleq [0 \quad 1]x_1 = [0 \quad -1]u.$$

According to the method given in Theorem 4-2.3, it is easy to construct the Luenberger observer for substate x_1:

$$\dot{x}_c = -x_c + [-1 \quad 1]u + y$$
$$x_1 = \begin{bmatrix} x_c + y \\ [0 \quad -1]u \end{bmatrix}. \tag{4-2.30}$$

Therefore, system (4-2.28) has a reduced order observer:

$$\dot{x}_c = -x_c + [-1 \quad 1]u + y$$
$$w = \begin{bmatrix} 1 \\ 0 \\ 0 \end{bmatrix} x_c + \begin{bmatrix} 0 \\ 0 \\ 1 \end{bmatrix} y + \begin{bmatrix} 0 & -1 \\ 0 & -1 \\ 0 & 1 \end{bmatrix} u \qquad (4\text{-}2.31)$$

such that

$$\lim_{t \to \infty} (w - x) = 0, \quad \forall \; x_c(0), \; x(0).$$

Both (4-2.10) and (4-2.31) are reduced-order normal observers for system (4-2.8). But (4-2.31) is of lower order (order 1). This is done at the cost that $u(t)$ is involved visibly in the measure output.

4-3. General Structure of State Observers

Consider the system

$$E\dot{x} = Ax + Bu$$
$$y = Cx. \qquad (4\text{-}3.1)$$

In previous sections we studied the design of state observers for singular systems and proposed design methods. But, as we know, any dynamic system Σ satisfying the two conditions given at the beginning of Section 4-1 would be a state observer for system (4-3.1). Thus, state observers are not unique and adopt a wide range of forms. For this reason, we now discuss the general structure of observers for singular systems.

4-3.1. General Case

As pointed out earlier, the aim of constructing an observer is to realize the state feedback control, in which not all state variables are required to be known. A certain linear combination of state would be sufficient for our use.

<u>Definition 4-3.1</u>. Consider system (4-3.1). For any matrix $K \in \mathbb{R}^{p \times n}$, if there exists a dynamic system:

$$E_c\dot{x}_c = A_c x_c + B_c u + Gy$$
$$w = F_c x_c + Fy + Hu \qquad (4\text{-}3.2)$$

where $x_c \in \mathbb{R}^{n_c}$, $w \in \mathbb{R}^p$, E_c, A_c, B_c, F_c, G, F, and H are constant matrices of appropriate dimensions, such that

$$\lim_{t \to \infty} (Kx - w) = 0, \quad \forall \; x(0), \; x_c(0).$$

we define the following.

a. If $\mathrm{rank}E_c < n_c$, system (4-3.2) is called a singular Kx observer.

b. Otherwise $\mathrm{rank}E_c = n_c$, $E_c = I$ is assumed without loss of generality, and (4-3.2) is termed a normal Kx observer.

The general (4-3.2) is the so-called function (Kx) observer.

Given regular system (4-3.1), there exist two nonsingular matrices Q and P such that

$$\begin{aligned} QEP &= \mathrm{diag}(I_{n_1}, N), \qquad QAP = \mathrm{diag}(A_1, I_{n_2}), \\ QB &= [B_1/B_2], \qquad CP = [C_1, C_2], \end{aligned} \tag{4-3.3}$$

where $A_1 \in \mathbb{R}^{n_1 \times n_1}$, $N \in \mathbb{R}^{n_2 \times n_2}$, $B_i \in \mathbb{R}^{n_i \times m}$, $C_i \in \mathbb{R}^{r \times n_i}$, $i = 1, 2$; $n_1 + n_2 = n$, and N is nilpotent with an index of h.

Theorem 4-3.1 (Wang and Dai, 1987a). Suppose that system (4-3.1) is controllable. Then for any matrix $L \in \mathbb{R}^{n_c \times n}$, $\bar{H} \in \mathbb{R}^{n_c \times n}$,

$$\lim_{t \to \infty} (Lx - x_c - \bar{H}u) = 0, \quad \forall\ x(0),\ x_c(0) \tag{4-3.4}$$

holds if and only if

a. $B_c - A_c\bar{H} = MB$.

b. $MA - A_cL = GC$.

c. $ME - E_c(L + \bar{H}B_2^{\tau}(\bar{R}\bar{R}^{\tau})^{-1}\tilde{P}_2) = 0$, $E_c\bar{H}B_2^{\tau}(\bar{R}\bar{R}^{\tau})^{-1}N = 0$, $E_c\bar{H}B_2^{\tau}(\bar{R}\bar{R}^{\tau})^{-1}B_2 = E_c\bar{H}$.

d. $\sigma(E_c, A_c) \subset \mathbb{C}^-$,

where $\bar{R} = [B_2, NB_2, \ldots, N^{h-1}B_2]$, $\tilde{P}_2 = [0 \quad I_{n_2}]P^{-1}$, B_2, N, P are determined by (4-3.3).

Its proof is lengthy and tedious, so it is omitted here. This theorem is a basic result on observer structure from which some interesting results may be obtained.

Corollary 4-3.1. Let system (4-3.1) be controllable. Then the dynamic system

$$\begin{aligned} E_c\dot{x}_c &= A_cx_c + B_cu \\ w &= x_c \end{aligned} \tag{4-3.5}$$

is its state observer if and only if a matrix $M \in \mathbb{R}^{n \times n}$ exists such that

(1). $E_c = ME$, $B_c = MB$, $A_c = MA - GC$.

(2). $\sigma(E_c, A_c) \subset \mathbb{C}^-$.

(3). $\mathrm{rank}[M \quad G] = n$, (E,A,B,C) is detectable.

Proof. (1) and (2) are direct results of Theorem 4-3.1 when $L = I_n$ and $\bar{H} = 0$.

Note that rank$[sE_c-A_c]$ = rank[sME−MA+GC] = rank[M G][sE−A/C] = n, $\forall\, s \in \bar{\mathbb{C}}^+$, s finite. It is easy to obtain (3). Q.E.D.

From this corollary we see that the observer in Theorem 4-1.1 is its special form of $M = I_n$. This shows that it is reasonable to design the observer in the form of (4-1.4).

<u>Theorem 4-3.2</u>. (Wang and Dai, 1987a). Assume that system (4-3.1) is controllable and system (4-3.2) is observable. Then for any matrix $K \in \mathbb{R}^{p\times n}$, system (4-3.2) is a function Kx observer for system (4-3.1) if and only if matrices L and $M \in \mathbb{R}^{n_c\times n}$, $\bar{H} \in \mathbb{R}^{n_c\times m}$, $J \in \mathbb{R}^{p\times n}$ may be chosen such that

1. $B_c = A_c\bar{H} + MB$.
2. $MA - A_cL = GC$.
3. $ME - E_c(L+\bar{H}B_2^\tau(\bar{R}\bar{R}^\tau)^{-1}\tilde{P}_2) = 0$, $E_c\bar{H}B_2^\tau(\bar{R}\bar{R}^\tau)^{-1}N = 0$, $E_c\bar{H}B_2^\tau(\bar{R}\bar{R}^\tau)^{-1}B_2 = E_c\bar{H}$.
4. $K = F_cL + FC + J$.
5. $H = F_c\bar{H} + JPQB$, $JPQE = 0$.
6. $\sigma(E_c, A_c) \subset \mathbb{C}^-$.

Here the items are defined by Theorem 4-3.1.

Showing the general structure of function observers, this theorem determines, for system (4-3.1), the class of Kx observers to which any Kx observer belongs without any other forms. And conversely, any dynamic system in this class would be a Kx observer for system (4-3.1). This implies the non-uniqueness of state observers.

In Theorem 4-3.2, the assumption of observability on system (4-3.2) is reasonable. In the case of unobservability, its observable subsystem may be used as a Kx observer for system (4-3.1) which is of a lower order.

4-3.2. General structure of normal observers

As proven in Section 4-2, under certain conditions the normal state or Kx observers may be designed for system (4-3.1). The observer in general is in one of the following three forms:

$$\Sigma_I: \quad \begin{aligned} \dot{x}_c &= A_cx_c + B_cu + Gy \\ w &= x_c \end{aligned}$$

$$\Sigma_{II}: \quad \begin{aligned} \dot{x}_c &= A_cx_c + B_cu + Gy \\ w &= F_cx_c + Fy \end{aligned}$$

$$\Sigma_{III}: \begin{aligned} \dot{x}_c &= A_c x_c + B_c u + Gy \\ w &= F_c x_c + Fy + Hu. \end{aligned}$$

Clearly, Σ_I and Σ_{II} are special forms of observer Σ_{III}. Now we will study their general structure as an observer for system (4-3.1).

Theorem 4-3.3. Assume that system (4-3.1) is controllable and $q = \text{rank}E < n$. Then there exists no dynamic system Σ_I that is a normal observer for this system.

Proof. According to Corollary 4-3.1, if Σ_I is a normal observer for the controllable system (4-3.1) there always exists a matrix $M \in \mathbb{R}^{n\times n}$ such that $ME = I_n$, which conflicts with the assumption $\text{rank}E < n$. Thus Σ_I doesn't exist. Q.E.D.

This theorem shows the fact that the general singular system has no normal observers in the form of Σ_I. But, as pointed out in Section 4-2, normal observers in the form of Σ_{II} or Σ_{III} often exist. Thus we are fortunate to be able to avoid such observers Σ_I.

Theorem 4-3.4 (Wang and Dai, 1987a). Assume that system (4-3.1) is controllable and system Σ_{II} is observable. Then Σ_{II} is a normal Kx observer for system (4-3.1) if and only if there exists a matrix $M \in \mathbb{R}^{n_c \times n}$ satisfying

1. $B_c = MB$.
2. $MA - A_c ME = GC$.
3. $K = F_c ME + FC$.
4. $\sigma(A_c) \subset \mathbb{C}^-$.

According to this theorem, it is easy to verify that the observer given by Theorem 4-2.1 is a special form of Σ_{II} when $M = [-G_2/I]$ where G_2 satisfies $\sigma(A_{22}-G_2A_{12}) \subset \mathbb{C}^-$.

Based on this theorem, we may prove the following result on the minimal order of observers.

Theorem 4-3.5. Let system (4-3.1) be controllable and $\text{rank}C = r$. Then if the observable system Σ_{II} is a normal observer such that

$$\lim_{t\to\infty} (x(t) - w(t)) = 0, \quad \forall\ x_c(0),\ x(0),$$

it must be

$$\min n_c = n - r.$$

Proof. Assume that Σ_{II} is a normal state observer for system (4-3.1). Then $K = I_n$. By Theorem 4-3.4 there exists a matrix $M \in \mathbb{R}^{n_c \times n}$ such that

$$I_n = F_c ME + FC = [F_c \quad F]\begin{bmatrix} ME \\ C \end{bmatrix},$$

indicating

$$\operatorname{rank}\begin{bmatrix} ME \\ C \end{bmatrix} = n.$$

Hence, rankME $\geqslant$ n-r. Therefore $n_c \geqslant$ n-r by the fact $M \in \mathbb{R}^{n_c \times n}$. In Theorem 4-2.1 we designed a normal observer Σ_{II} of order n-r for system (4-3.1). Thus min n_c = n-r. Q.E.D.

Therefore, observers designed following the process in Theorem 4-2.1 are of minimal order among observers of the form Σ_{II}.

Theorem 4-3.6. Assume that system (4-3.1) is controllable and Σ_{III} is observable. Σ_{III} is a normal observer for system (4-3.1) if and only if there exist matrices L and $M \in \mathbb{R}^{n_c \times n}$, $\bar{H} \in \mathbb{R}^{n_c \times n}$, $J \in \mathbb{R}^{n \times n}$ such that

1. $B_c = A_c\bar{\Pi} + MB$.
2. $MA - A_cL = GC$.
3. $ME - L - \bar{\Pi}B_2^{\tau}(\bar{R}\bar{R}^{\tau})^{-1}\tilde{P}_2 = 0, \quad \bar{\Pi}B_2^{\tau}(\bar{R}\bar{R}^{\tau})^{-1}N = 0, \quad \bar{H}B_2^{\tau}(\bar{R}\bar{R}^{\tau})^{-1}B_2 = \bar{\Pi}.$
4. $I = F_cL + FC + J$.
5. $H = F_c\bar{H} + JPQB, \quad JPQE = 0$.
6. $\sigma(A_c) \subset \mathbb{C}^-$.

Here every item is defined by Theorem 4-3.1.

4-4. Disturbance Decoupling State Observers

We note that the systems we have studied so far have no extra disturbance inputs, which often exist in practice. When such disturbances exist as input in a system, the asymptotical state estimation is expected not to be affected by unknown disturbances so that we may have a precise estimation of state. This is the thought of disturbance decoupling state observers (DDSO).

Consider the system with extra disturbance inputs:

$$\begin{aligned} E\dot{x} &= Ax + Bu + Mf \\ y &= Cx + Df \end{aligned} \tag{4-4.1}$$

where $x \in \mathbb{R}^n$, $u \in \mathbb{R}^m$, $y \in \mathbb{R}^r$ are its state, control input (or reference signals), and measure output; E, A, B, C, M, D are constant matrices of appropriate dimensions; $f(t) \in \mathbb{R}^d$ is extra disturbance input satisfying the following differential equation

$$\dot{f}(t) = Lf(t) \tag{4-4.2}$$

where $L \in \mathbb{R}^{d \times d}$. Since stable disturbance has no effect on state estimation, which is an asymptotical behavior, we will hereafter assume that system (4-4.2) is unstable, i.e., $\sigma(L) \subset \bar{\mathbb{C}}^+$.

Definition 4-4.1. Consider system (4-4.1) and the following dynamic system

$$\begin{aligned} E_c\dot{x}_c(t) &= A_c x_c(t) + B_c u(t) + Gy(t) \\ w(t) &= F_c x_c(t) + Fy(t) + Hu(t) \end{aligned} \tag{4-4.3}$$

where $x_c \in \mathbb{R}^{n_c}$, $w \in \mathbb{R}^n$, E_c, A_c, B_c, G, F_c, F, and H are constant matrices of appropriate dimensions, such that

$$\lim_{t\to\infty} (x(t) - w(t)) = 0$$

for any initial conditions $x(0)$, $x_c(0)$ and $f(t)$ satisfying (4-4.2).

1. When $\text{rank}E_c < n_c$, system (4-4.3) will be termed a singular DDSO for system (4-4.1).

2. Otherwise, $\text{rank}E_c = n_c$, and $E_c = I_{n_c}$ is assumed, system (4-4.3) is called a normal DDSO.

Theorem 4-4.1. Assume that system (4-4.1) is controllable when $f(t) \equiv 0$. Then the system

$$\begin{aligned} E_c\dot{x}_c &= A_c x_c + B_c u + Gy \\ w &= x_c \end{aligned} \tag{4-4.4}$$

is a singular DDSO for system (4-4.1) if and only if there exists a matrix $T \in \mathbb{R}^{n\times n}$ such that

(1). $E_c = TE$, $A_c = TA - GC$, $B_c = TB$.

(2). $\sigma(E_c, A_c) \subset \mathbb{C}^-$.

(3). $GD = TM$.

Proof. Necessity: Let (4-4.4) be a DDSO for system (4-4.1). Then for any initial condition $x(0)$, $x_c(0)$ we have

$$\lim_{t\to\infty} (x(t) - w(t)) = \lim_{t\to\infty} (x(t) - x_c(t)) = 0.$$

Particularly, when $f(0) = 0$ (in which case $f(t) \equiv 0$), system (4-4.4) would be a state observer for system

$$E\dot{x} = Ax + Bu, \quad y = Cx \tag{4-4.5}$$

Therefore, as shown in Corollary 4-3.1, there exists a matrix $T \in \mathbb{R}^{n\times n}$ satisfying (1) and (2).

Let

$$e(t) = x(t) - w(t) = x(t) - x_c(t).$$

By left multiplying (4-4.1) by T and then substracting (4-4.4), paying attention to assumption (1) and (2), we obtain

$$E_c\dot{e}(t) = A_c e(t) + (TM-GD)f(t).$$

Since $\sigma(E_c,A_c)\subset \mathbb{C}^-$, and $\lim_{t\to\infty} e(t) = 0$ holds for any initial conditions $x(0)$, $x_c(0)$ and any $f(t)$ satisfying (4-4.2), we know $\lim_{t\to\infty} \dot{e}(t) = 0$, $\forall\ x(0)$, $x_c(0)$, implying

$$\lim_{t\to\infty} (TM-GD)f(t) = 0, \quad \forall\ f(0).$$

Hence $TM-GD = 0$, which is (3).

Sufficiency is easy to obtain by the inverse process. Q.E.D.

This theorem shows the general structure of DDSOs. From the condition $E_c = TE$, a singular system doesn't have full-order DDSOs of the form

$$\dot{x}_c = A_c x_c + B_c u + Gy$$
$$w = x_c.$$

Now we will discuss the existence condition and design methods for one kind of singular (or normal) DDSOs.

Theorem 4-4.2. System (4-4.1) has a singular DDSO of the following form

$$\begin{aligned} E\dot{x}_c &= Ax_c + Bu + G(y-Cx_c) \\ w &= x_c \end{aligned} \tag{4-4.6}$$

if and only if there exists a matrix $G \in \mathbb{R}^{n\times r}$ such that

1. $\sigma(E,A-GC) \subset \mathbb{C}^-$. 2. $M = GD$.

Proof. This is a special case of Theorem 4-4.1 when $T = I_n$. Q.E.D.

Theorem 4-4.2 shows that the existence condition of singular DDSOs is independent of the equation that $f(t)$ satisfies. Therefore, if (4-4.6) is a DDSO for system (4-4.1) with respect to disturbance satisfying (4-4.2), it must be a DDSO for arbitrary disturbance $f(t)$, and vise versa. With this property, (4-4.6) is called a singular DDSO for any disturbance $f(t)$, which needs not satisfy equation (4-4.2).

In the case of $\text{rank}D = d \leqslant r$, with no loss of generality, we assume $D = [I_d/0]$. Then we can prove the following theorem.

Theorem 4-4.3. System (4-4.1) has a DDSO of the form (4-4.6) for arbitrary disturbance $f(t)$ if and only if

$$\text{rank}\begin{bmatrix} sE-A & -M \\ C & D \end{bmatrix} = d + n, \quad \forall\ s\in\bar{\mathbb{C}}^+, \quad s \text{ finite}.$$

Proof. Denote

$$C = \begin{bmatrix} C_1 \\ C_2 \end{bmatrix}, \qquad C_1 \in \mathbb{R}^{d\times n}, \quad C_2 \in \mathbb{R}^{(r-d)\times n}.$$

From $D = [I_d/0]$, we know

$$\operatorname{rank}\begin{bmatrix} sE-A & -M \\ C & D \end{bmatrix} = \operatorname{rank}\begin{bmatrix} sE-A+MC_1 & 0 \\ C_2 & 0 \\ 0 & I_d \end{bmatrix}, \quad \forall\, s \in \mathbb{C}.$$

Thus

$$\operatorname{rank}\begin{bmatrix} sE-A & -M \\ C & D \end{bmatrix} = n + d \iff \operatorname{rank}\begin{bmatrix} sE-A+MC_1 \\ C_2 \end{bmatrix} = n, \quad \forall\, s \in \bar{\mathbb{C}}^+,\ s \text{ finite} \tag{4-4.7}$$

where $\iff$ denotes if and only if.

Necessity: Assume that system (4-4.1) has a singular DDSO (4-4.6) for arbitrary disturbance $f(t)$. By Theorem 4-4.2, there exist a matrix G satisfying

$$\sigma(E, A-GC) \subset \mathbb{C}^-, \quad M = GD.$$

Denoting

$$G = [G_1, G_2], \quad G_1 \in \mathbb{R}^{n\times d}, \quad G_2 \in \mathbb{R}^{n\times(r-d)},$$

and noticing $D = [I_d/0]$, $C = [C_1/C_2]$, from $M = GD$ we have $M = G_1$, and

$$\sigma(E, A-G_1C_1-G_2C_2) = \sigma(E, A-MC_1-G_2C_2) \subset \mathbb{C}^-.$$

Thus

$$\operatorname{rank}(sE-A+MC_1+G_2C_2) = n, \quad \forall\, s \in \bar{\mathbb{C}}^+,\ s \text{ finite}.$$

Therefore we have

$$\operatorname{rank}\begin{bmatrix} sE-A+MC_1+G_2C_2 \\ C_2 \end{bmatrix} = \operatorname{rank}\begin{bmatrix} I & G_2 \\ 0 & I \end{bmatrix}\begin{bmatrix} sE-A+MC_1 \\ C_2 \end{bmatrix} = \operatorname{rank}\begin{bmatrix} sE-A+MC_1 \\ C_2 \end{bmatrix}$$

$$= n, \quad \forall\, s \in \bar{\mathbb{C}}^+,\ s \text{ finite}.$$

Combining it with (4-4.7) we immediately obtain

$$\operatorname{rank}\begin{bmatrix} sE-A & -M \\ C & D \end{bmatrix} = n + d, \quad \forall\, s \in \bar{\mathbb{C}}^+,\ s \text{ finite}.$$

Sufficiency: From (4-4.7) we have

$$\operatorname{rank}\begin{bmatrix} sE-A+MC_1 \\ C_2 \end{bmatrix} = n, \quad \forall\, s \in \bar{\mathbb{C}}^+,\ s \text{ finite},$$

i.e., $(E, A-MC_1, C_2)$ is detectable. Therefore, there exists $G_2 \in \mathbb{R}^{n\times(r-d)}$ such that

$$\sigma(E, A-MC_1-G_2C_2) \subset \mathbb{C}^-.$$

The following relationship holds

$$\sigma(E, A-GC) \subset \mathbb{C}^-, \qquad M = GD,$$

by simply setting $G = [M, G_2]$. According to Theorem 4-4.2 we know system (4-4.6) determined by such a G is a DDSO for system (4-4.1). Q.E.D.

Let us define the disturbance transition zeros of system (4-4.1) to be the complex scalars satisfying

$$\mathrm{rank}\begin{bmatrix} sE-A & -M \\ C & D \end{bmatrix} < n + \min(r, d).$$

Since $d \leq r$, the necessary and sufficient condition for the existence of singular DDSOs for arbitrary disturbance is that all disturbance transition zeros are in the open left complex plane, no unstable transition zeros exist. This is the physical sense of what is stated in Theorem 4-4.3. It is worth noticing that if rankD = d is false (such as D = 0 and $M \neq 0$), the DDSO (4-4.6) may not exist although conditions in Theorem 4-4.3 are satisfied.

Example 4-4.1. Consider the system with disturbance inputs:

$$\begin{bmatrix} 0 & 1 & 0 \\ 0 & 0 & 1 \\ 0 & 0 & 0 \end{bmatrix}\dot{x} = \begin{bmatrix} 1 & 0 & 0 \\ 0 & 1 & 0 \\ 0 & 0 & 1 \end{bmatrix}x + \begin{bmatrix} 0 & 0 \\ 1 & 0 \\ 0 & 1 \end{bmatrix}u + \begin{bmatrix} 0 \\ 2 \\ 1 \end{bmatrix}f(t)$$
$$y = [1 \quad 0 \quad 0]x + f(t) \tag{4-4.8}$$

in which

$$E = \begin{bmatrix} 0 & 1 & 0 \\ 0 & 0 & 1 \\ 0 & 0 & 0 \end{bmatrix}, \quad A = \begin{bmatrix} 1 & 0 & 0 \\ 0 & 1 & 0 \\ 0 & 0 & 1 \end{bmatrix}, \quad B = \begin{bmatrix} 0 & 0 \\ 1 & 0 \\ 0 & 1 \end{bmatrix}, \quad M = \begin{bmatrix} 0 \\ 2 \\ 1 \end{bmatrix}$$
$$C = [1 \quad 0 \quad 0], \qquad D = 1 \tag{4-4.9}$$

Since

$$\mathrm{rank}\begin{bmatrix} sE-A & -M \\ C & D \end{bmatrix} = \mathrm{rank}\begin{bmatrix} -1 & s & 0 & 0 \\ 0 & -1 & s & 2 \\ 0 & 0 & -1 & 1 \\ 1 & 0 & 0 & 1 \end{bmatrix} = 4 = n+d, \quad \forall \ s \in \bar{\mathbb{C}}^{+}, \ s \text{ finite},$$
$$\tag{4-4.10}$$

by Theorem 4-4.3 we know that system (4-4.8) has a singular DDSO of the form (4-4.6) for arbitrary disturbance f(t).

Note that D = 1. It will be G = M provided (4-4.6) exists. Thus a DDSO for arbitrary disturbance is readily given by Theorem 4-4.2:

$$\begin{bmatrix} 0 & 1 & 0 \\ 0 & 0 & 1 \\ 0 & 0 & 0 \end{bmatrix}\dot{x}_c = \begin{bmatrix} 1 & 0 & 0 \\ -2 & 1 & 0 \\ -1 & 0 & 1 \end{bmatrix}x_c + \begin{bmatrix} 0 & 0 \\ 1 & 0 \\ 0 & 1 \end{bmatrix}u + \begin{bmatrix} 0 \\ 2 \\ 1 \end{bmatrix}y$$
$$w = x_c. \tag{4-4.11}$$

Verified from (4-4.10) the transition zero set is $\{-1, -1\} \subset \mathbb{C}^-$, which is stable.

For the same reason as pointed out before, singular DDSOs are often difficult to realize. To overcome this difficulty we now consider the normal DDSOs for system (4-4.1).

Theorem 4-4.4. Consider system (4-4.1). Assume that there exists a matrix $G \in \mathbb{R}^{n\times r}$ satisfying

1. $\sigma(E, A-GC) \subset \mathbb{C}^-$.
2. $\deg(|sE-(A-GC)|) = \text{rank}E$.
3. $M = GD$.

Then this system always has a reduced-order normal DDSO of the following form:

$$\begin{aligned} \dot{x}_c &= A_c x_c + B_c u + G_c y \\ w &= F_c x_c + Fy + Hu \end{aligned} \qquad (4\text{-}4.12)$$

where $x_c \in \mathbb{R}^q$.

Proof. Under the assumption, from Theorem 4-4.2 we know that for this matrix $G \in \mathbb{R}^{n\times r}$, the system

$$\begin{aligned} E\dot{\bar{x}}_c &= A\bar{x}_c + Bu + G(y-C\bar{x}_c) \\ w &= \bar{x}_c \end{aligned} \qquad (4\text{-}4.13)$$

is a DDSO for system (4-4.1). Moreover, condition 2 shows that no impulse terms appear in the state response of (4-4.13). Thus, two nonsingular matrices Q and P exist, such that system (4-4.13) is r.s.e. to

$$\begin{aligned} \dot{x}_1 &= A_{c_1} x_1 + B_1 u + G_1 y \\ 0 &= x_2 + B_2 u + G_2 y \\ w &= P\begin{bmatrix} x_1 \\ x_2 \end{bmatrix} = P\begin{bmatrix} I_q \\ 0 \end{bmatrix} x_1 + P\begin{bmatrix} 0 \\ I_{n-q} \end{bmatrix} x_2 \end{aligned} \qquad (4\text{-}4.14)$$

where

$$QEP = \text{diag}(I_q, 0), \qquad Q(A-GC)P = \text{diag}(A_{c_1}, I_{n-q}),$$

$$QB = \begin{bmatrix} B_1 \\ B_2 \end{bmatrix}, \quad QG = \begin{bmatrix} G_1 \\ G_2 \end{bmatrix}, \quad \bar{x}_c = P\begin{bmatrix} x_1 \\ x_2 \end{bmatrix}, \quad x_1 \in \mathbb{R}^q, \; x_2 \in \mathbb{R}^{n-q}.$$

Apparently, if we choose $x_c = x_1$, $A_c = A_{c_1}$, $B_c = B_1$, $G_c = G_1$, and

$$F = P\begin{bmatrix} I_q \\ 0 \end{bmatrix}, \qquad F = -P\begin{bmatrix} 0 \\ I_{n-q} \end{bmatrix}, \qquad H = -P\begin{bmatrix} 0 \\ I_{n-q} \end{bmatrix} G_2,$$

system (4-4.14) is (4-4.12). Q.E.D.

Corollary 4-4.1. Detectability and impulse observability are necessary conditions for the existence of normal DDSOs for system (4-4.1).

Therefore, conditions in Theorem 4-4.4 are somewhat strong.

Example 4-4.2. It has been proven in Example 4-4.1 that the system

$$\begin{bmatrix} 0 & 1 & 0 \\ 0 & 0 & 1 \\ 0 & 0 & 0 \end{bmatrix} \dot{x}_c = \begin{bmatrix} 1 & 0 & 0 \\ -2 & 1 & 0 \\ -1 & 0 & 1 \end{bmatrix} x_c + \begin{bmatrix} 0 & 0 \\ 1 & 0 \\ 0 & 1 \end{bmatrix} u + \begin{bmatrix} 0 \\ 2 \\ 1 \end{bmatrix} y$$

$$w = x_c \tag{4-4.15}$$

is a DDSO for system (4-4.8). Note that there is no impulse portion in state response of this DDSO. We can choose the nonsingular matrices

$$Q = \begin{bmatrix} 1 & 0 & 1 \\ 0 & 1 & -2 \\ 0 & 0 & -1 \end{bmatrix}, \qquad P = \begin{bmatrix} 0 & 1 & 1 \\ 1 & 0 & 0 \\ 0 & 1 & 0 \end{bmatrix}$$

such that

$$QEP = \left[\begin{array}{cc|c} 1 & 0 & 0 \\ 0 & 1 & 0 \\ \hline 0 & 0 & 0 \end{array}\right], \qquad Q(A-GC)P = \left[\begin{array}{cc|c} 0 & 1 & 0 \\ 1 & -2 & 0 \\ \hline 0 & 0 & 1 \end{array}\right].$$

Let the coordinate transformation be

$$P^{-1}x_c = \begin{bmatrix} x_1 \\ x_2 \end{bmatrix}, \qquad x_2 \in \mathbb{R}^2, \quad x_2 \in \mathbb{R}^1.$$

System (4-4.15) is r.s.e. to

$$\begin{aligned} \dot{x}_1 &= \begin{bmatrix} 0 & 1 \\ 1 & -2 \end{bmatrix} x_1 + \begin{bmatrix} 0 & 1 \\ 1 & -2 \end{bmatrix} u + \begin{bmatrix} 1 \\ 0 \end{bmatrix} y \\ 0 &= x_2 + [0 \ -1]u - y \\ w &= P[x_1/x_2] \end{aligned} \tag{4-4.16}$$

which may be further rewritten as

$$\dot{x}_1 = \begin{bmatrix} 0 & 1 \\ 1 & -2 \end{bmatrix} x_1 + \begin{bmatrix} 0 & 1 \\ 1 & -2 \end{bmatrix} u + \begin{bmatrix} 1 \\ 0 \end{bmatrix} y$$

$$w = \begin{bmatrix} 0 & 1 \\ 1 & 0 \\ 0 & 1 \end{bmatrix} x_1 + \begin{bmatrix} 0 & -1 \\ 0 & 0 \\ 0 & 0 \end{bmatrix} u + \begin{bmatrix} -1 \\ 0 \\ 0 \end{bmatrix} y$$

which is just a normal DDSO in the form of (4-4.12).

4-5. State Observer with Disturbance Estimation

Previously given DDSOs give an asymptotical state estimation independent of extra disturbance. As a result, no information of disturbance is given in such DDSOs, which may be required in engineering design. For this reason, we now consider the state observer that estimates both state and disturbance at the same time. Such observers are called state observers with disturbance estimation. They are one kind of DDSOs.

Consider system (4-4.1). Let

$$\bar{E} = \begin{bmatrix} E & 0 \\ 0 & I_d \end{bmatrix}, \quad \bar{A} = \begin{bmatrix} A & M \\ 0 & L \end{bmatrix}, \quad \bar{B} = \begin{bmatrix} B \\ 0 \end{bmatrix}, \quad \bar{C} = [C \quad D], \quad z = \begin{bmatrix} x \\ f \end{bmatrix}.$$

Then $z \in \mathbb{R}^{n+d}$. By combining (4-4.1) with (4-4.2), we obtain the following system

$$\begin{aligned} \bar{E}\dot{z} &= \bar{A}z + \bar{B}u \\ y &= \bar{C}z. \end{aligned} \tag{4-5.1}$$

According to the results of Theorems 4-1.1 and 4-2.2, when the augmented system (4-5.1) is detectable, it has a singular state observer:

$$\begin{aligned} \bar{E}\dot{z}_c &= (\bar{A}-\bar{G}\bar{C})z_c + \bar{B}u + \bar{G}y \\ w &= z_c \end{aligned} \tag{4-5.2}$$

where $z_c \in \mathbb{R}^{n+d}$. When system (4-5.1) is detectable and impulse observable, it has a normal state observer of the following form

$$\begin{aligned} \dot{z}_c &= A_c z_c + B_c u + Gy \\ w &= F_c z_c + Fy + Hu \end{aligned} \tag{4-5.3}$$

where $z_c \in \mathbb{R}^{n_c}$, and $n_c \leqslant \text{rank}E + d$.

Both (4-5.2) and (4-5.3) are state observers with disturbance estimation for system (4-4.1). The estimation of disturbance is given by

$$\hat{f} = [0 \quad I_d]w$$

such that $\lim_{t\to\infty}(\hat{f} - f) = 0$, $\forall$ $f(0)$, $x(0)$, $z_c(0)$.

Meanwhile, state estimation is given by $\hat{x} = [I_n \quad 0]w$ such that $\lim_{t\to\infty}(\hat{x} - x) = 0$, $\forall$ $x(0)$, $z_c(0)$, $f(0)$.

Theorem 4-5.1. Consider system (4-4.1).

1. If $(\bar{E},\bar{A},\bar{B},\bar{C})$ is detectable, system (4-4.1) has a singular DDSO (4-5.2).
2. If $(\bar{E},\bar{A},\bar{B},\bar{C})$ is detectable and impulse observable, system (4-4.1) has a normal DDSO of the form (4-5.3).

It must be pointed out that the DDSOs given here estimate disturbance at the cost of higher order.

It is interesting to view the close relationship between DDSOs, given in the last section and the state observer with disturbance estimation here.

Assume that rank$D = d$ and the condition in Theorem 4-4.3

$$\text{rank}\begin{bmatrix} sE-A & -M \\ C & D \end{bmatrix} = n + d, \quad \forall\, s \in \bar{\mathbb{C}}^+, \ s \text{ finite},$$

is satisfied. Then system (4-4.1) has a singular DDSO (4-4.6) for arbitrary disturbance $f(t)$. Since

$$\text{rank}\begin{bmatrix} s\bar{E}-\bar{A} \\ \bar{C} \end{bmatrix} = \text{rank}\begin{bmatrix} sE-A & -M \\ 0 & sI-L \\ C & D \end{bmatrix} \geqslant \text{rank}\begin{bmatrix} sE-A & -M \\ C & D \end{bmatrix} = n + d,$$

$$\forall\ s \in \bar{\mathbb{C}}^{+},\ s \text{ finite}.$$

System (4-5.1) is detectable. As a direct result of Theorem 4-5.1, system (4-4.1) has a DDSO with state estimation, from whose output, disturbance estimation may be obtained. Sometimes, DDSO (4-5.2) may exist even though rankD = d is not satisfied (Wang and Dai, 1987b), but (4-4.6) may not.

Similarly, when rankD = d and conditions in Theorem 4-4.4 are satisfied, from Theorem 4-4.4 we know system (4-4.1) has a DDSO (4-4.12) for arbitrary disturbance. Also note that

$$\text{rank}\begin{bmatrix} s\bar{E}-\bar{A} \\ \bar{C} \end{bmatrix} = \text{rank}\begin{bmatrix} sE-A & -M \\ 0 & sI-L \\ C & D \end{bmatrix} = \text{rank}\begin{bmatrix} sE-A+GC & 0 \\ 0 & sI-L \\ C & D \end{bmatrix} = n + d,$$

$$\forall\ s \in \bar{\mathbb{C}}^{+},\ s \text{ finite}.$$

System (4-5.1) is detectable. Hence if we choose $\bar{G} = [G/0]$ it will be

$$\deg(|s\bar{E}-(\bar{A}-\bar{G}\bar{C})|) = \deg(|sI-L||sE-(A-GC)|) = \text{rank}E,$$

or in other words, system (4-5.1) is impulse observable. As a result, from Theorem 4-4.1 we can conclude that system (4-4.1) has a normal DDSO with disturbance estimation.

These arguments show an important principle. If system (4-4.1) has a singular (or normal)DDSO for arbitrary disturbance, we may choose an appropriate DDSO from which the disturbance is estimated.

<u>Example 4-5.1</u>. Consider the system

$$\begin{aligned} E\dot{x} &= Ax + Mf \\ \dot{f} &= Lf \\ y &= Cx \end{aligned} \tag{4-5.4}$$

in which

$$E = \begin{bmatrix} 1 & 0 \\ 0 & 0 \end{bmatrix},\quad A = \begin{bmatrix} 1 & -1 \\ 0 & 1 \end{bmatrix},\quad M = \begin{bmatrix} 0 & 1 \\ 1 & 2 \end{bmatrix},\quad L = \begin{bmatrix} 0 & -1 \\ 1 & 0 \end{bmatrix} \tag{4-5.5}$$

$$C = [1 \quad 0].$$

For this system D = 0, M ≠ 0, Theorem 4-4.4 is not applicable. However, since

$$\bar{E} = \begin{bmatrix} E & 0 \\ 0 & I \end{bmatrix} = \begin{bmatrix} 1 & 0 & 0 & 0 \\ 0 & 0 & 0 & 0 \\ 0 & 0 & 1 & 0 \\ 0 & 0 & 0 & 1 \end{bmatrix},\quad \bar{A} = \begin{bmatrix} A & M \\ 0 & L \end{bmatrix} = \begin{bmatrix} 1 & -1 & 0 & 1 \\ 0 & 1 & 1 & 2 \\ 0 & 0 & 0 & -1 \\ 0 & 0 & 1 & 0 \end{bmatrix},$$

$$\bar{C} = [C \quad D] = [1 \quad 0 \quad 0 \quad 0].$$

$(\bar{E},\bar{A},\bar{B},\bar{C})$ is detectable as well as impulse observable. We may construct a state observer (4-5.3) with disturbance estimation.

Therefore, the conditions given by Theorem 4-5.1 are much looser than before. But, on the other hand, the DDSO given by this theorem is not applicable to arbitrary disturbance since the model of disturbance is needed.

Finally, we would like to further explain.

Theorem 4-5.2. System (4-5.1) is impulse observable if and only if system (4-4.1) is impulse observable when $f(t) \equiv 0$.

Proof. The conclusion is easy to obtain by noticing

$$\operatorname{rank}\begin{bmatrix} \bar{E} & \bar{A} \\ 0 & \bar{E} \\ 0 & \bar{C} \end{bmatrix} = \operatorname{rank}\begin{bmatrix} E & 0 & A & M \\ 0 & I & 0 & L \\ 0 & 0 & E & 0 \\ 0 & 0 & 0 & I \\ 0 & 0 & C & D \end{bmatrix} = 2d + \operatorname{rank}\begin{bmatrix} E & A \\ 0 & E \\ 0 & C \end{bmatrix}.$$

Q.E.D.

This shows that impulse observability is independent of disturbance inputs, which satisfy a normal differential equation.

4-6. Notes and References

This chapter is based on Wang and Dai (1986b, 1987a, 1987b). The observer problem has been considered by several researchers. In EL-Tohami et al. (1983) another method based on pseudo-inverse was used to studied this problem. The other works on state observers for continuous-time singular systems are Saidahmed (1985), Shafai and Carroll (1987), and Wang (1984).

CHAPTER 5

DYNAMIC COMPENSATION FOR SINGULAR SYSTEMS

As pointed out in Chapter 3, stability is a basic property possessed by a well-working system. Therefore, it should be the first thing considered in control system design. To achieve this, dynamic compensation is the common means usually used, which constructs a dynamic compensator using the measure output of original system as its input and its output acting on the original system as a feedback control input, such that the whole closed-loop system is stable. This method also provides an approach to realize the state feedback control. Dynamic compensation has many other purposes such as tracking and model matching. Here only the compensation for stability purpose is studied, as space is limited.

Dynamic compensation is one form of dynamic feedback control of outputs.

5-1. Singular Dynamic Compensators

Consider the linear singular system

$$\begin{aligned} E\dot{x}(t) &= Ax(t) + Bu(t) \\ y(t) &= Cx(t) \end{aligned} \tag{5-1.1}$$

where $x(t) \in \mathbb{R}^n$, $u(t) \in \mathbb{R}^m$, $y(t) \in \mathbb{R}^r$ are its state, control input, and measure output; $E, A \in \mathbb{R}^{n\times n}$, $B \in \mathbb{R}^{n\times m}$, $C \in \mathbb{R}^{r\times n}$ are constant matrices. Without loss of generality, we assume that system (5-1.1) is regular and rank$E = q < n$.

The so-called singular dynamic compensator (compensator for simplicity) is a dynamic system in the following form

$$\begin{aligned} E_c\dot{x}_c(t) &= A_cx_c(t) + B_cy(t) \\ u(t) &= F_cx_c(t) + Fy(t) \end{aligned} \tag{5-1.2}$$

where $x_c(t) \in \mathbb{R}^{n_c}$, E_c, A_c, B_c, F_c, F are constant matrices of appropriate dimensions and (5-1.2) is regular, such that the closed-loop system

$$\begin{aligned} E\dot{x}(t) &= (A+BFC)x(t) + BF_cx_c(t) \\ E_c\dot{x}_c(t) &= A_cx_c(t) + B_cy(t) \end{aligned} \tag{5-1.3}$$

is stable, i.e.,

$$\sigma\left(\begin{bmatrix} E & 0 \\ 0 & E_c \end{bmatrix}, \begin{bmatrix} A+BFC & BF_c \\ B_cC & A_c \end{bmatrix}\right) \subset \mathbb{C}^-. \tag{5-1.4}$$

Figure 5-1.1 is the diagram of compensators.

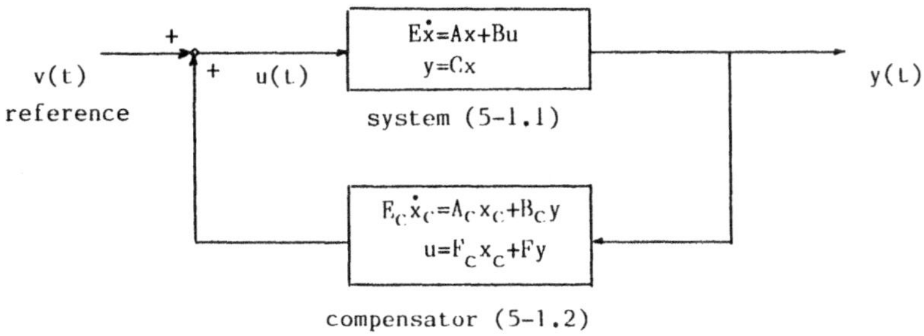

Figure 5-1.1

Theorem 5-1.1. Assume that system (5-1.1) is stabilizable and detectable. Then it has a singular compensator of the form

$$\begin{aligned} E\dot{x}_c &= (A-GC)x_c + Bu + Gy \\ u &= Kx_c \end{aligned} \tag{5-1.5}$$

where $G \in \mathbb{R}^{nxr}$ and $K \in \mathbb{R}^{mxn}$ satisfy $\sigma(E,A-GC) \subset \mathbb{C}^-$ and $\sigma(E,A+BK) \subset \mathbb{C}^-$, respectively.

Proof. By assuming the stabilizability, there exists a matrix $K \in \mathbb{R}^{mxn}$ that satisfies $\sigma(E,A+BK) \subset \mathbb{C}^-$. The state feedback determined by K:

$$u = Kx \tag{5-1.6}$$

stabilizes the closed-loop system $E\dot{x} = (A+BK)x$.

Note that state feedback control cannot be realized directly in general case. We now consider its realization via state observers.

It is assumed that system (5-1.1) is detectable, a matrix $G \in \mathbb{R}^{nxr}$ can be chosen such that $\sigma(E,A-GC) \subset \mathbb{C}^-$. Thus

$$\begin{aligned} E\dot{x}_c &= Ax_c + Bu + G(y-Cx_c) \\ w &= x_c \end{aligned}$$

is a singular state observer for system (5-1.1) such that $\lim_{t\to\infty}(w-x) = 0$. Using $u = Kx_c = Kw$ in lieu of (5-1.6), we obtain the closed-loop system

$$\begin{aligned} E\dot{x} &= Ax + BKx_c \\ E_c\dot{x}_c &= (A-GC)x_c + BKx_c + Gy \end{aligned}$$

whose pole set may be obtained by direct computation:

$$\sigma\left(\begin{bmatrix} E & 0 \\ 0 & E \end{bmatrix}, \begin{bmatrix} A & BK \\ GC & A-GC+BK \end{bmatrix}\right) = \sigma(E,A+BK) \cup \sigma(E,A-GC) \subset \mathbb{C}^-.$$

Therefore, the system determined by (5-1.5) is a singular compensator for system (5-1.1). Q.E.D.

As mentioned in Chapter 4, one aim of designing observers is to realize state feedback. From this theorem we see further that the dynamic compensator is merely a combination of state feedback and observer.

The separation principle holds in the closed-loop pole set, i.e., the pole set of the closed-loop system is the union of the pole sets under state feedback and observer that can be designed separately, which makes this method convenient for application.

Combining this discussion with the results on pole placement, we easily obtain the following corollary.

Corollary 5-1.1. If system (5-1.1) is R-controllable (controllable), R-observable (observable), a dynamic compensator (5-1.5) may be designed such that the $2n_1$ finite poles of its closed-loop system can be arbitrarily assigned to any symmetric set in the complex plane, especially in the open left half plane. Here n_1 is the order of its slow subsystem.

It has been proven in Theorem 5-1.1 that detectability and stabilizability are sufficient conditions for the existence of dynamic compensators. We can prove that it they are also necessary.

Theorem 5-1.2. System (5-1.1) has a dynamic compensator if and only if it is stabilizable and detectable.

Proof. The sufficiency is given by Theorem 5-1.1. We only need to prove its necessity.

Let (5-1.2) be a compensator for system (5-1.1). Then its closed-loop system is stable, i.e., (5-1.4) holds, or in another form

$$\text{rank}\begin{bmatrix} sE-(A+BFC) & -BF_c \\ -B_cC & sE_c-A_c \end{bmatrix} = n + n_c, \quad \forall\, s \in \bar{\mathbb{C}}^+, \text{ s finite.} \tag{5-1.7}$$

Hence

$$\text{rank}[sE-(A+BFC) \quad -BF_c] \geqslant n + n_c, \quad \forall\, s \in \bar{\mathbb{C}}^+, \text{ s finite.}$$

Noticing that the matrix $[sE-(A+BFC) \quad -BF_c]$ is of n rows, we know

$$\text{rank}[sE-(A+BFC) \quad -BF_c] = \text{rank}[sE-A \quad B]\begin{bmatrix} I & 0 \\ -FC & -F_c \end{bmatrix} = n, \quad \forall\, s \in \bar{\mathbb{C}}^+, \text{ s finite.}$$

Thus, $\text{rank}[sE-A, B] = n$, $\forall\, s \in \bar{\mathbb{C}}^+$, s finite, so system (5-1.1) is stabilizable.

In a similar way we can prove that it is also detectable. Q.E.D.

Theorem 5-1.2 is the result for the normal system when $E = I$. It is worth pointing out that the dynamic compensator itself is not necessarily stable although the overall closed-loop system (5-1.3) is stable.

Example 5-1.1. Consider the circuit network (2-2.5)-(2-2.6):

$$\begin{aligned} E\dot{x} &= Ax + Bu \\ y &= Cx \end{aligned} \tag{5-1.8}$$

with coefficient matrices

$$E = \begin{bmatrix} C_1 & 0 & 0 & 0 \\ 0 & C_2 & 0 & 0 \\ 0 & 0 & -L & 0 \\ 0 & 0 & 0 & 0 \end{bmatrix}, \quad A = \begin{bmatrix} 0 & 0 & 0 & 1 \\ 0 & 0 & 1 & 0 \\ -1 & 1 & 0 & 0 \\ 1 & 0 & 0 & R \end{bmatrix}, \quad B = \begin{bmatrix} 0 \\ 0 \\ 0 \\ -1 \end{bmatrix},$$

$$C = [0 \quad 1 \quad 0 \quad 0].$$

It has been proven in Chapter 2 that it is stabilizable and detectable. Therefore, it has a compensator (5-1.5).

We have verified in Example 3-1.4 that the matrix K

$$K = [1-3RC_1, \quad 3RC_1-LRC_1C_2, \quad \frac{RC_1}{C_2} - 3LRC_1, \quad 0]$$

satisfies

$$\sigma(E, A+BK) = \{-1, -1, -1\} \subset \mathbb{C}^-.$$

For the sake of simplicity in discussion, it is assumed that $C_1 = C_2 = R = L = 1$. Then $K = [-2 \quad 2 \quad -2 \quad 0]$. Let $G = [-9 \quad 0 \quad -9 \quad 0]^{\tau}$. We have $|sE-(A-GC)| = s^3 + s^2 + 10s + 1$, which may be verified to be a stable polynomial by Routh criterion (Fortmann and Hitz, 1977). Thus $\sigma(E, A-GC) \subset \mathbb{C}^-$.

For such matrices K and G, by Theorem 5-1.1, system (5-1.8) has a compensator:

$$\begin{aligned} \begin{bmatrix} 1 & 0 & 0 & 0 \\ 0 & 1 & 0 & 0 \\ 0 & 0 & -1 & 0 \\ 0 & 0 & 0 & 0 \end{bmatrix} \dot{x}_c &= \begin{bmatrix} 0 & 9 & 0 & 1 \\ 0 & 0 & 1 & 0 \\ -1 & 0 & 0 & 0 \\ 3 & -2 & 2 & 1 \end{bmatrix} x_c + \begin{bmatrix} -9 \\ 0 \\ -9 \\ 0 \end{bmatrix} y \\ u &= [-2 \quad 2 \quad -2 \quad 0] x_c. \end{aligned} \tag{5-1.9}$$

5-2. Full-Order Normal Compensators

Because of the difficulty in realizing singular systems, although singular compensators exist theoretically, they are often difficult to physically realize. To cancel this disadvantage, we use naturally normal compensators, which imply compensators described by the following normal system:

$$\begin{aligned} \dot{x}_c &= A_c x_c + B_c y \\ u &= F_c x_c + Fy \end{aligned} \tag{5-2.1}$$

where $x_c \in \mathbb{R}^{n_c}$, A_c, B_c, F_c, and F are constant matrices of appropriate dimensions.

Compensation function of normal compensators may be described by Figure 5-2.1.

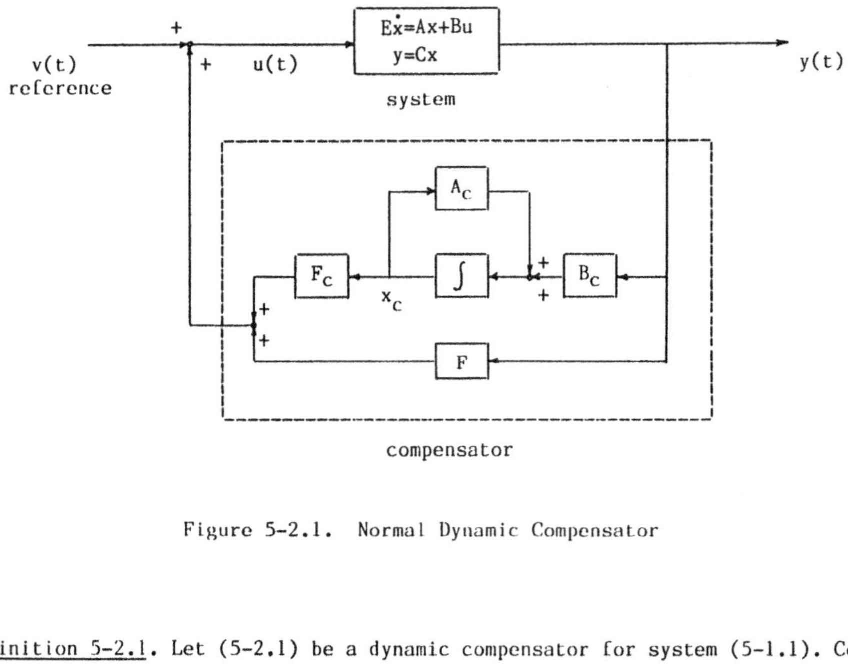

Figure 5-2.1. Normal Dynamic Compensator

Definition 5-2.1. Let (5-2.1) be a dynamic compensator for system (5-1.1). Corresponding to the two cases, $n_c = n$, $n_c < n$, system (5-2.1) is called the full-order or reduced-order, respectively, normal compensators for system (5-1.1).

Only full-order compensators are discussed in this section, reduced-order ones will be studied in the next section. To design a normal compensator for system (5-1.1), we consider the P-D state feedback

$$u = K_1 x - K_2 \dot{x} \tag{5-2.2}$$

where K_1, $K_2 \in \mathbb{R}^{mxn}$ are constant matrices.

Theorem 5-2.1. Assume that system (5-1.1) is normalizable. If it is detectable and stabilizable, a full-order compensator exists.

Proof. The proof is constructive. Under the assumption of normalizability, a matrix $K_2 \in \mathbb{R}^{mxn}$ may be chosen to satisfy

$$|E+BK_2| \neq 0. \tag{5-2.3}$$

By Lemma 3-3.2, the system $(E+BK_2,A,B,C)$ is also stabilizable, implying that a $K_1 \in \mathbb{R}^{mxn}$ exists, such that

$$\sigma(E+BK_2,A+BK_1) \subset \mathbb{C}^-. \tag{5-2.4}$$

The matrices K_1 and K_2 satisfying (5-2.3)-(5-2.4) determine feedback (5-2.2) and thus determine the closed-loop system

$$(E+BK_2)\dot{x} = (A+BK_1)x, \tag{5-2.5}$$

which is stable by (5-2.4). Since x and $\dot{x}$ are generally inaccessible, feedback (5-2.2) cannot be realized directly. Now we consider its realization via observers. Note the assumption of detectability, a matrix $G \in \mathbb{R}^{nxr}$ may be chosen that satisfies

$$\sigma(E,A-GC) \subset \mathbb{C}^-. \tag{5-2.6}$$

Therefore, according to Theorem 4-1.1,

$$E\dot{x}_c = (A-GC)x_c + Bu + Gy \tag{5-2.7}$$

is a singular observer for system (5-1.1). Let the control be

$$u = K_1x_c - K_2\dot{x}_c \tag{5-2.8}$$

instead of (5-2.2). Substituting it into (5-1.1) and (5-2.7), we obtain the closed-loop system

$$\begin{bmatrix} E & BK_2 \\ 0 & E+BK_2 \end{bmatrix}\begin{bmatrix} \dot{x} \\ \dot{x}_c \end{bmatrix} = \begin{bmatrix} A & BK_1 \\ GC & A-GC+BK_1 \end{bmatrix}\begin{bmatrix} x \\ x_c \end{bmatrix} \tag{5-2.9}$$

with finite pole set

$$\sigma\left(\begin{bmatrix} E & BK_2 \\ 0 & E+BK_2 \end{bmatrix}, \begin{bmatrix} A & BK_1 \\ GC & A-GC+BK_1 \end{bmatrix}\right) = \sigma(E+BK_2,A+BK_1) \cup \sigma(E,A-GC),$$

which shows that the closed-loop system (5-2.9) is stable with a pole set of the union of those of feedback and observer.

Combining (5-2.7) and (5-2.8), we see that if

$$A_c = (E+BK_2)^{-1}(A-GC+BK_1)$$

$$B_c = (E+BK_2)^{-1}G$$

$$F_c = K_1 - K_2(E+BK_2)^{-1}(A-GC+BK_1)$$

$$F = - K_2(E+BK_2)^{-1}G$$

it will be

$$\dot{x}_c = A_c x_c + B_c y$$
$$u = F_c x_c + Fy. \qquad (5\text{-}2.10)$$

Q.E.D.

Apparently, whenever $E = I_n$, K_2 may be taken as $K_2 = 0$. System (5-2.10) is the usual compensator for the normal systems. A direct result of Corollary 3-3.2 and Theorem 5-2.1 is the following corollary.

Corollary 5-2.1. Assume that system (5-1.1) is stabilizable and detectable. If, in addition, it is normalizable and impulse observable, a compensator can be designed such that the closed-loop system has n+rankE stable finite poles.

Thus no impulse terms exist in the state response of the closed-loop system.

Example 5-2.1. Consider the circuit network (5-1.8) which is stabilizable, detectable, and normalizable. $C_1 = C_2 = R = L = 1$ is assumed for simplicity.

Choosing $K_2 = [0 \;\; 0 \;\; 0 \; -1]$, we have $|E+BK_2| = -1 \neq 0$. And for any matrix K_1, $\sigma(E+BK_2, A+BK_1) = \sigma(\hat{A}+\hat{B}K_1)$, in which

$$\hat{A} = (E+BK_2)^{-1}A = \begin{bmatrix} 0 & 0 & 0 & 1 \\ 0 & 0 & 1 & 0 \\ 1 & -1 & 0 & 0 \\ 1 & 0 & 0 & 1 \end{bmatrix}, \qquad \hat{B} = (E+BK_2)^{-1}B = \begin{bmatrix} 0 \\ 0 \\ 0 \\ -1 \end{bmatrix}.$$

For any $K_1 = [a_1, a_2, a_3, a_4]$, direct computation shows

$$|s(E+BK_2) - (A+BK_1)| = |sI - (\hat{A}+\hat{B}K_1)|$$
$$= s^4 + (a_4-1)s^3 + a_1 s^2 + (a_3+a_4-1)s + a_1+a_2-1,$$

which is stable if we let $a_1 = 6$, $a_2 = -4$, $a_3 = 0$, and $a_4 = 5$. $\sigma(E+BK_2, A+BK_1) = \{-1, -1, -1, -1\} \subset \mathbb{C}^-$.

Similarly, when G is taken to be $G = [-9 \;\; 0 \;\; -9 \;\; 0]^\tau$. from Example 5-1.1 we know $\sigma(E, A-GC) \subset \mathbb{C}^-$.

Let K_1, K_2, and G be defined as previously, we have

$$A_c = (E+BK_2)^{-1}(A-GC+BK\;) = \begin{bmatrix} 0 & 9 & 0 & 1 \\ 0 & 0 & 1 & 0 \\ 1 & -10 & 0 & 0 \\ -5 & 4 & 0 & -4 \end{bmatrix}$$

$$B_c = (E+BK_2)^{-1}G = [-9 \;\; 0 \;\; 9 \;\; 0]^\tau$$

$$F_c = K_1 - K_2 A_c = [1 \;\; 0 \;\; 0 \;\; 1]$$

$$F = -K_2(E+BK_2)^{-1}G = -K_2B_c = 0.$$

By Theorem 5-2.1, we obtain the full-order compensator for system (5-1.8):

$$\dot{x}_c = \begin{bmatrix} 0 & 9 & 0 & 1 \\ 0 & 0 & 1 & 0 \\ 1 & -10 & 0 & 0 \\ -5 & 4 & 0 & -4 \end{bmatrix} x_c + \begin{bmatrix} -9 \\ 0 \\ 9 \\ 0 \end{bmatrix} y$$

$$u = [1 \quad 0 \quad 0 \quad 1]x_c. \tag{5-2.11}$$

The closed-loop system formed by (5-2.11) and (5-1.8) has seven finite poles, thus no impulse terms exist in its state response.

Generally speaking, Singular systems are often not normalizable, so the design method in Theorem 5-2.1 is not applicable. Now we consider designing compensators for general singular systems. To begin, we consider the normalizability decomposition (2-5.9) for system (5-1.1). First, for any regular system (5-1.1), there exist two nonsingular matrices Q and P such that (Lemma 1-2.2):

$$QEP = \operatorname{diag}(I_{n_1}, N), \qquad QAP = \operatorname{diag}(A_1, I_{n_2})$$

$$QB = [B_1/B_2], \qquad CP = [C_1 \quad C_2],$$

where $n_1 + n_2 = n$, N is nilpotent. Taking controllability decomposition of (N, B_2), from linear system theory there exists a nonsingular matrix T such that

$$TNT^{-1} = \begin{bmatrix} N_{11} & N_{12} \\ 0 & N_{22} \end{bmatrix}, \qquad TB_2 = \begin{bmatrix} B_{21} \\ 0 \end{bmatrix},$$

where (N_{11}, B_{21}) is controllable, $N_{11} \in \mathbb{R}^{n_3 x n_3}$, $N_{22} \in \mathbb{R}^{n_4 x n_4}$ are nilpotent, $n_3 + n_4 = n_2$. Let

$$Q_1 = \operatorname{diag}(I_{n_1}, T)Q, \qquad P_1 = P\operatorname{diag}(I_{n_1}, T^{-1})$$

$$P_1^{-1}x = \begin{bmatrix} \tilde{x}_1 \\ \tilde{x}_2 \end{bmatrix}, \qquad \tilde{x}_1 \in \mathbb{R}^{n_1+n_3}, \quad \tilde{x}_2 \in \mathbb{R}^{n_4}$$

Then system (5-1.1) is r.s.e. to

$$\begin{aligned} \tilde{E}_{11}\dot{\tilde{x}}_1 + \tilde{E}_{12}\dot{\tilde{x}}_1 &= \tilde{A}_{11}\tilde{x}_1 + \tilde{B}_1 u \\ N_{22}\dot{\tilde{x}}_2 &= \tilde{x}_2 \\ y &= \tilde{C}_1\tilde{x}_1 + \tilde{C}_2\tilde{x}_2 \end{aligned} \tag{5-2.12}$$

where

$$\tilde{E}_{11} = \operatorname{diag}(I_{n_1}, N_{11}), \quad \tilde{A}_1 = \operatorname{diag}(A_1, I_{n_3}), \quad \tilde{E}_{12} = [0/N_{12}],$$

$$\tilde{B}_1 = [B_1/B_{21}], \quad \tilde{C}_1 = [C_1, C_{21}], \quad C_2T^{-1} = [C_{21}, \tilde{C}_2]. \tag{5-2.13}$$

According to our decomposition, (N_{22}, B_{21}) is controllable. Thus $(\tilde{E}_{11},\tilde{A}_1,\tilde{B}_1,\tilde{C}_1)$ is normalizable. System (5-2.12) is a normalizability decomposition. Furthermore, from

$$\text{rank}[sE-A,\ B] = \text{rank}\begin{bmatrix} s\tilde{E}_{11}-\tilde{A}_1 & \tilde{B}_1 & s\tilde{E}_{12} \\ 0 & 0 & sN_{22}-I_{n_4} \end{bmatrix} = n, \quad \forall\ s \in \bar{\mathbb{C}}^+,\ s \text{ finite}$$

and the invertibility of $sN_{22}-I_{n_4}$ we know that $(\tilde{E}_{11},\tilde{A}_1,\tilde{B}_1,\tilde{C}_1)$ is stabilizable if system (5-1.1) is stabilizable.

Similarly $(\tilde{E}_{11},\tilde{A}_1,\tilde{B}_1,\tilde{C}_1)$ is detectable if system (5-1.1) is detectable.

Therefore, if system (5-1.1) is stabilizable and detectable, the subsystem

$$\begin{aligned} \tilde{E}_{11}\dot{z} &= \tilde{A}_1 z + \tilde{B}_1 u \\ \bar{y} &= \tilde{C}_1 z \end{aligned} \tag{5-2.14}$$

is normalizable, stabilizable, and detectable. Theorem 5-2.1 shows that there exists a normal compensator

$$\begin{aligned} \dot{x}_c &= A_c x_c + B_c \bar{y} \\ u &= F_c x_c + F\bar{y} \end{aligned} \tag{5-2.15}$$

where $x_c \in \mathbb{R}^{n-n_4}$, such that the closed-loop system

$$\begin{bmatrix} \tilde{E}_{11} & 0 \\ 0 & I_{n-n_4} \end{bmatrix}\begin{bmatrix} \dot{z} \\ \dot{x}_c \end{bmatrix} = \begin{bmatrix} \tilde{A}_1+\tilde{B}_1F\tilde{C}_1 & \tilde{B}_1F_c \\ B_c\tilde{C}_1 & A_c \end{bmatrix}\begin{bmatrix} z \\ x_c \end{bmatrix} \tag{5-2.16}$$

is stable.

Using the matrices defined by (5-2.15), we construct the following system

$$\begin{aligned} \dot{x}_c &= A_c x_c + B_c y \\ u &= F_c x_c + Fy. \end{aligned} \tag{5-2.17}$$

By direct computation, we may verify that (5-2.17) is a compensator for system (5-1.1) and the closed-loop systems (5-1.1)-(5-2.17) and (5-2.16) have the same slow subsystem structure (Dai and Wang, 1987b).

Summing up this discussion, we thus prove the following theorem.

Theorem 5-2.2. Designing a normal compensator for system (5-1.1) is equivalent to designing its normalizable subsystem (5-2.14) in the sense that their closed-loop systems have the same slow subsystem.

A direct result of Theorems 5-2.1 and 5-2.2 is the following theorem.

Theorem 5-2.3. If system (5-1.1) is stabilizable and detectable, it has a normal compensator of order $n-n_4$:

$$\begin{aligned} \dot{x}_c &= A_c x_c + B_c y \\ u &= F_c x_c + Fy \end{aligned} \tag{5-2.18}$$

In this theorem the conditions are the same as in the normal system case. Combining it with the necessity in Theorem 5-1.2 yields the following theorem.

Theorem 5-2.4. System (5-1.1) has a normal compensator (5-2.1) if and only if it is both stabilizable and detectable.

This theorem reveals an interesting phenomenon. The conditions for the existence of a singular compensator are the same as those for a normal one; the latter is more appropriate in system design for its convenience in realization. This is a fortunate thing for system design.

The design algorithm of normal compensators may be summarized as follows.

1. Take normalizability decomposition for a given system to obtain its normalizable subsystem (5-2.14) if the system is not normalizable.

2. Under the stabilizability and detectability assumption, choose matrices K_1, K_2, and G that satisfy

$$|\tilde{E}_{11}+\tilde{B}_1K_2| \neq 0, \qquad \sigma(\tilde{E}_{11},\tilde{A}_1-G\tilde{C}_1) \subset \mathbb{C}^-,$$
$$\sigma(\tilde{E}_{11}+\tilde{B}_1K_2,\tilde{A}_1+\tilde{B}_1K_1) \subset \mathbb{C}^-.$$

3. Compute the matrices

$$A_c = (\tilde{E}_{11}+\tilde{B}_1K_2)^{-1}(\tilde{A}_1-G\tilde{C}_1+\tilde{B}_1K_1)$$
$$B_c = (\tilde{E}_{11}+\tilde{B}_1K_2)^{-1}G$$
$$F_c = K_1 - K_2A_c$$
$$F = -K_2(\tilde{E}_{11}+\tilde{B}_1K_2)^{-1}G.$$

Then dynamic system (5-2.18) is a normal compensator for system (5-1.1).

From Corollary 5-2.1, and Theorems 5-2.2 and 5-2.3, we can further prove (Dai and Wang, 1987b) the following corollary.

Corollary 5-2.2. Assume that system (5-1.1) is stabilizable and detectable. If it is impulse controllable and impulse observable, it has a normal compensator (5-2.17) such that its closed-loop system has n_c+rankE stable finite poles.

Therefore, the closed-loop system has no infinite poles and no impulse terms exist in its state response.

5-3. Reduced-Order Normal Compensators

In control system design, the higher the order of a compensator, the more components needed for its realization, increasing investment and raising the cost, and the lower the reliability of the control system, decreasing its efficiency. Therefore, the principle in system design should be lowering the order of compensators as far as possible when allowed.

Now we will consider the existence and design methods of reduced-order compensators.

Theorem 5-3.1. Assume that the singular system

$$\begin{aligned} E\dot{x} &= Ax + Bu \\ y &= Cx \end{aligned} \tag{5-3.1}$$

is stabilizable and detectable. If it is dual normalizable and rank$C = r$, it has a normal compensator of order $n-r$ of the following form

$$\begin{aligned} \dot{x}_c &= A_c x_c + B_c y \\ u &= F_c x_c + Fy \end{aligned} \tag{5-3.2}$$

where $x_c \in \mathbb{R}^{n_c}$, $n_c = n-r$.

Proof. According to the stabilizability assumption, there exists a matrix $K \in \mathbb{R}^{m \times n}$ satisfying

$$\sigma(E, A+BK) \subset \mathbb{C}^-. \tag{5-3.3}$$

Subject to the assumption rank$C = r$, from the proof process of Theorem 4-2.1 we know that there exist nonsingular matrices P and Q such that

$$\begin{aligned} QEP &= \begin{bmatrix} E_{11} & 0 \\ E_{21} & I_{n-r} \end{bmatrix}, \quad QAP = \begin{bmatrix} A_{11} & A_{12} \\ A_{21} & A_{22} \end{bmatrix}, \quad QB = \begin{bmatrix} B_1 \\ B_2 \end{bmatrix}, \\ CP &= [I_r, 0] \end{aligned} \tag{5-3.4}$$

and that system (5-3.1) has a reduced-order normal state observer of order $n-r$:

$$\begin{aligned} \dot{x}_c &= \bar{A}_c x_c + \bar{B}_c u + \bar{G} y \\ w &= \bar{F}_c x_c + \bar{F} y \end{aligned}$$

such that $\lim_{t\to\infty}(w-x) = 0$, $\forall\ x(0)$, $x_c(0)$ in which

$$\begin{aligned} \bar{A}_c &= A_{22} - G_2 A_{12}, \qquad \bar{B}_c = B_2 - G_2 B_1 \\ \bar{G} &= A_{21} - G_2 A_{11} - (A_{22} - G_2 A_{12})(E_{21} - G_2 E_{11}) \\ \bar{F}_c &= P[0/I_{n-r}], \qquad \bar{F} = P[I_r/-E_{21}+G_2 E_{11}] \end{aligned} \tag{5-3.5}$$

and $G_2 \in \mathbb{R}^{(n-r)xr}$ is any matrix satisfying

$$\sigma(A_{22}-G_2A_{12}) \subset \mathbb{C}^- . \qquad (5\text{-}3.6)$$

Consider the control law

$$u = Kw = K\bar{F}_c x_c + K\bar{F}y$$

where K is defined in (5-3.3). We can prove that the following system

$$\begin{aligned} \dot{x}_c &= \bar{A}_c x_c + \bar{B}_c u + \bar{C}y \\ u &= Kw = K\bar{F}_c x_c + K\bar{F}y \end{aligned} \qquad (5\text{-}3.7)$$

is just a normal compensator for system (5-3.1).

In fact, substituting (5-3.7) into (5-3.1), we obtain the closed-loop system

$$\begin{bmatrix} E & 0 \\ 0 & I_{n-r} \end{bmatrix}\begin{bmatrix} \dot{x} \\ \dot{x}_c \end{bmatrix} = \begin{bmatrix} A+BK\bar{F}C & BK\bar{F}_c \\ \bar{B}_cK\bar{F}C+\bar{G}C & \bar{A}_c+\bar{B}_cK\bar{F}_c \end{bmatrix}\begin{bmatrix} x \\ x_c \end{bmatrix} . \qquad (5\text{-}3.8)$$

By denoting $KP = [K_1, K_2]$, paying attention to (5-3.3)-(5-3.6), and making elementary transformation, we know that the closed-loop pole set is

$$\sigma(\begin{bmatrix} E & 0 \\ 0 & I_{n-r} \end{bmatrix}, \begin{bmatrix} A+BK\bar{F}C & BK\bar{F}_c \\ \bar{B}_cK\bar{F}C+\bar{G}C & \bar{A}_c+\bar{B}_cK\bar{F}_c \end{bmatrix})$$

$$= \sigma(\begin{bmatrix} QEP & 0 \\ 0 & I_{n-r} \end{bmatrix}, \begin{bmatrix} QAP+QBK\bar{F}CP & QBK\bar{F}_c \\ \bar{B}_cK\bar{F}CP+\bar{G}CP & \bar{A}_c+\bar{B}_cK\bar{F}_c \end{bmatrix})$$

$$= \sigma(\begin{bmatrix} E_{11} & 0 & 0 \\ E_{21} & I & 0 \\ 0 & 0 & I \end{bmatrix}, \begin{bmatrix} A_{11}+B_1K_1-B_1K_2(E_{21}-G_2E_{11}) & A_{12} & B_1K_2 \\ A_{21}+B_2K_1-B_2K_2(E_{21}-G_2E_{11}) & A_{22} & B_2K_2 \\ \bar{G}+\bar{B}_cK_1-\bar{B}_cK_2(E_{21}-G_2E_{11}) & 0 & \bar{A}_c+\bar{B}_cK_2 \end{bmatrix})$$

$$= \sigma(\begin{bmatrix} E_{11} & 0 & 0 \\ E_{21} & I & 0 \\ 0 & E_{21}-G_2E_{11} & I \end{bmatrix}, \begin{bmatrix} A_{11}+B_1K_1 & A_{12} & B_1K_2 \\ A_{21}+B_2K_1 & A_{22} & B_2K_2 \\ A_{21}-G_2A_{11}+\bar{B}_cK_1 & 0 & \bar{A}_c+\bar{B}_cK_2 \end{bmatrix})$$

$$= \sigma(E,A+BK) \cup \sigma(A_{22}-G_2A_{12}) \subset \mathbb{C}^- .$$

Thus, the closed-loop system is stable, with the separation principle of pole sets.

Setting

$$A_c = \bar{A}_c + \bar{B}_cK\bar{F}_c, \qquad B_c = \bar{G} + \bar{B}_cK\bar{F},$$

$$F_c = K\bar{F}_c, \qquad F = K\bar{F},$$

system (5-3.7) is a normal compensator of order n-r. Q.E.D.

The process also proves the following corollary.

Corollary 5-3.1. If system (5-3.1) is R-controllable and observable, it always has a normal compensator (5-3.2) of order n-rankC whose pole set may be arbitrarily assigned.

Example 5-3.1. From Example 4-2.1, the singular system

$$\begin{bmatrix} 0 & 1 & 0 \\ 0 & 0 & 1 \\ 0 & 0 & 0 \end{bmatrix}\dot{x} = \begin{bmatrix} 1 & 0 & 0 \\ 0 & 1 & 0 \\ 0 & 0 & 1 \end{bmatrix}x + \begin{bmatrix} 0 & 0 \\ 1 & 0 \\ 0 & 1 \end{bmatrix}u$$
$$y = [1 \quad 0 \quad 0]x \tag{5-3.9}$$

has a normal observer of order n-r = 2 of the form

$$\dot{x}_c = \begin{bmatrix} 0 & -1 \\ 1 & -1 \end{bmatrix}x_c + \begin{bmatrix} 0 & -1 \\ 1 & -1 \end{bmatrix}u + \begin{bmatrix} 1 \\ 0 \end{bmatrix}y$$
$$w = \begin{bmatrix} 1 \\ 0 \\ 0 \end{bmatrix}y + \begin{bmatrix} 0 & 0 \\ 1 & 0 \\ 0 & 1 \end{bmatrix}x_c \tag{5-3.10}$$

such that $\lim_{t\to\infty}(x-w) = 0$, $\forall\ x(0),\ x_c(0)$. Let the feedback gain matrix be

$$K = \begin{bmatrix} 0 & 0 & 0 \\ 1 & 2 & 0 \end{bmatrix}.$$

We have $\sigma(E, A+BK) = \{-1, -1\} \subset \mathbb{C}^-$. By Theorem 5-3.1 we may construct its normal compensator:

$$\dot{x}_c = \begin{bmatrix} -2 & -1 \\ -1 & -1 \end{bmatrix}x_c + \begin{bmatrix} 0 \\ -1 \end{bmatrix}y$$
$$u = \begin{bmatrix} 0 & 0 \\ 2 & 0 \end{bmatrix}x_c + \begin{bmatrix} 0 \\ 1 \end{bmatrix}y$$

whose closed-loop system is

$$\begin{bmatrix} 0 & 1 & 0 & 0 & 0 \\ 0 & 0 & 1 & 0 & 0 \\ 0 & 0 & 0 & 0 & 0 \\ 0 & 0 & 0 & 1 & 0 \\ 0 & 0 & 0 & 0 & 1 \end{bmatrix}\begin{bmatrix} \dot{x} \\ \dot{x}_c \end{bmatrix} = \begin{bmatrix} 1 & 0 & 0 & 0 & 0 \\ 0 & 1 & 0 & 0 & 0 \\ 1 & 0 & 1 & 2 & 0 \\ 0 & 0 & 0 & -2 & -1 \\ -1 & 0 & 0 & -1 & -1 \end{bmatrix}\begin{bmatrix} x \\ x_c \end{bmatrix}$$

with the finite pole set $\{-1, -1, -\frac{1}{2}+\frac{3}{2}i, -\frac{1}{2}-\frac{3}{2}i\}$, which is the union of poles of state feedback and normal observer.

Theorem 5-3.2. Assume that system (5-3.1) is stabilizable and detectable. If it is impulse observable and impulse controllable, a normal compensator of order q = rankE may be designed:

$$\dot{x}_c = A_c x_c + B_c y$$
$$u = F_c x_c + Fy \tag{5-3.11}$$

where $x_c \in \mathbb{R}^{n_c}$, n_c = rankE, such that its closed-loop system has n_c+ rankE stable finite poles.

Thus the closed-loop system has no infinite poles.

Proof. By assumption, system (5-3.1) is impulse controllable and impulse observable. From Theorem 3-5.1, we know that a matrix $K_1 \in \mathbb{R}^{m\times r}$ may be chosen that satisfies

$$\deg(|sE - (A+BK_1C)|) = \text{rank}E. \tag{5-3.12}$$

Consider the feedback control

$$u = K_1y + v \tag{5-3.13}$$

where v is the new control input, and, when applied to (5-3.1), the closed-loop system is

$$\begin{aligned} E\dot{x} &= (A+BK_1C)x + Bv \\ y &= Cx. \end{aligned} \tag{5-3.14}$$

Accounting for equation (5-3.12), via suitable selection of transformation matrices Q and P system (5-3.14) is r.s.e. to

$$\begin{aligned} \dot{x}_1 &= A_1x_1 + B_1v \\ 0 &= x_2 + B_2v \\ y &= C_1x_1 + C_2x_2 \end{aligned} \tag{5-3.15}$$

where $x_1 \in \mathbb{R}^q$, $x_2 \in \mathbb{R}^{n-q}$, and

$$QEP = \text{diag}(I_q, 0), \qquad Q(A+BK_1C)P = \text{diag}(A_1, I_{n-q}),$$

$$QB = \begin{bmatrix} B_1 \\ B_2 \end{bmatrix}, \qquad CP = [C_1, C_2],$$

$$B_1 \in \mathbb{R}^{q\times m}, \quad B_2 \in \mathbb{R}^{(n-q)\times m}, \quad C_1 \in \mathbb{R}^{r\times q}, \quad C_2 \in \mathbb{R}^{r\times(n-q)}.$$

Furthermore, the stabilizability and detectability of system (5-3.1) shows that (A_1,B_1,C_1) is stabilizable and detectable. Thus, two matrices $K_2 \in \mathbb{R}^{m\times q}$, $G_2 \in \mathbb{R}^{q\times r}$ may be chosen such that $\sigma(A_1+B_1K_2) \subset \mathbb{C}^-$, and $\sigma(A_1-G_2C_1) \subset \mathbb{C}^-$. Now we will construct the observer for substate x_1 in (5-3.15):

$$\begin{aligned} \dot{x}_c &= (A_1-G_2C_1)x_c + (B_1+G_2C_2B_2)v + G_2y \\ w &= x_c \end{aligned} \tag{5-3.16}$$

where $x_c \in \mathbb{R}^q$, and $\lim_{t\to\infty}(w-x_1) = 0$, $\forall\, x(0), x_c(0)$.

Let the control be

$$v = K_2w = K_2x_c$$

From (5-3.13) and (5-3.16) we know the overall input satisfies:

$$\begin{aligned} \dot{x}_c &= (A_1-G_2C_1)x_c + (B_1+G_2C_2B_2)K_2x_c + G_2y \\ u &= K_1y + K_2x_c \end{aligned} \tag{5-3.17}$$

Applying it to the original system (5-3.1), the closed-loop system is

$$\begin{bmatrix} E & 0 \\ 0 & I \end{bmatrix}\begin{bmatrix} \dot{x} \\ \dot{x}_c \end{bmatrix} = \begin{bmatrix} A+BK_1C & BK_2 \\ G_2C & A_1-G_2C_1+B_1K_2+G_2C_2B_2K_2 \end{bmatrix}\begin{bmatrix} x \\ x_c \end{bmatrix}.$$

Its finite pole set is

$$\sigma(\begin{bmatrix} E & 0 \\ 0 & I \end{bmatrix}, \begin{bmatrix} A+BK_1C & BK_2 \\ G_2C & A_1-G_2C_1+B_1K_2+G_2C_2B_2K_2 \end{bmatrix})$$

$$= \sigma(A_1+B_1K_2) \cup \sigma(A_1-G_2C_1) \subset \mathbb{C}^-.$$

Therefore, (5-3.17) is a compensator for system (5-3.1). Its order is n_c=rankE and the closed-loop pole number is 2rankE = n_c+rankE, which is our conclusion. Q.E.D.

Compared with the similar result of Corollary 5-2.2, compensator (5-3.17) is of a lower order while they both finish the same design purpose; their closed-loop systems have no infinite poles.

Theorem 5-3.3. Singular system (5-3.1) has a normal compensator

$$\begin{aligned} \dot{x}_c &= A_cx_c + B_cy \\ u &= F_cx_c + Fy \end{aligned} \tag{5-3.18}$$

where $x_c \in \mathbb{R}^{n_c}$, such that its closed-loop system has no infinite poles, i.e.,

$$\deg(\left|\begin{bmatrix} E & 0 \\ 0 & I_{n_c} \end{bmatrix}s - \begin{bmatrix} A+BFC & BF_c \\ B_cC & A_c \end{bmatrix}\right|) = n_c + \text{rank}E \tag{5-3.19}$$

if and only if it is stabilizable, detectable, impulse controllable, and impulse observable.

Proof. Sufficiency is given by Theorem 5-3.2. The stabilizability and detectability parts of necessity have been proven by Theorem 5-2.4.

For the remaining part of necessity, we note that for matrices A_c, B_c, F_c, and F we have

$$\deg(\left|\begin{bmatrix} E & 0 \\ 0 & I_{n_c} \end{bmatrix}s - \begin{bmatrix} A+BFC & BF_c \\ B_cC & A_c \end{bmatrix}\right|) \leqslant n_c + \text{rank}E$$

and the equality holds if and only if $\deg(|sE-(A+BFC)|)$ = rankE, indicating the impulse controllability and impulse observability for system (5-3.1). Q.E.D.

Singular systems with no infinite poles, or, equivalently, with no impulse portion in their state response, are an important class of systems that possess the property of structural stability, or robustness of stability, which will be studied thoroughly in the next chapter.

Example 5-3.2. Consider the compensation problem for the simplified model of chemical process given in the introduction of Chapter 1:

$$\begin{aligned} \dot{x}_1 &= A_{11}x_1 + A_{12}x_2 + B_1u \\ 0 &= A_{21}x_1 + A_{22}x_2 + B_2u \\ y &= C_1x_1 + C_2x_2. \end{aligned} \tag{5-3.20}$$

Here $x_1 \in \mathbb{R}^{n_1}$, $x_2 \in \mathbb{R}^{n_2}$, $n_1 + n_2 = n$. System (5-3.20) is in the EF2 for singular system (5-3.1).

Assume that system (5-3.20) is stabilizable, detectable, impulse controllable, and impulse observable. Then $\mathrm{rank}[A_{22}/C_2] = \mathrm{rank}[A_{22}, B_2] = n_2$. Lemma 3-5.1 has proven that a matrix $K_2 \in \mathbb{R}^{m\times r}$ exists satisfying

$$|A_{22} + B_2K_2C_2| \neq 0. \tag{5-3.21}$$

For such a matrix K_2, we define the control

$$u = K_2y + v_1 \tag{5-3.22}$$

where v_1 is the new input. When applied to (5-3.20), it forms the closed-loop system

$$\begin{aligned} \dot{x}_1 &= (A_{11}+B_1K_2C_1)x_1 + (A_{12}+B_1K_2C_2)x_2 + B_1v_1 \\ 0 &= (A_{21}+B_2K_2C_1)x_1 + (A_{22}+B_2K_2C_2)x_2 + B_2v_1 \\ y &= C_1x_1 + C_2x_2. \end{aligned} \tag{5-3.23}$$

Noticing (5-3.21), $A_{22}+B_2K_2C_2$ is invertible. Defining the transformation matrices

$$Q = \begin{bmatrix} I_{n_1} & -(A_{12}+B_1K_2C_2)(A_{22}+B_2K_2C_2)^{-1} \\ 0 & I_{n_2} \end{bmatrix}$$

$$P = \begin{bmatrix} I_{n_1} & 0 \\ -(A_{22}+B_2K_2C_2)^{-1}(A_{21}+B_2K_2C_1) & (A_{22}+B_2K_2C_2)^{-1} \end{bmatrix}$$

then Q and P are nonsingular. Under the coordinate transformation

$$\begin{bmatrix} \tilde{x}_1 \\ \tilde{x}_2 \end{bmatrix} = P^{-1}\begin{bmatrix} x_1 \\ x_2 \end{bmatrix}, \qquad \tilde{x}_1 \in \mathbb{R}^{n_1}, \quad \tilde{x}_2 \in \mathbb{R}^{n_2}$$

system (5-3.23) is r.s.e. to

$$\begin{aligned} \dot{\tilde{x}}_1 &= A_o\tilde{x}_1 + B_ov_1 \\ 0 &= \tilde{x}_2 + B_2v_1 \\ y &= C_o\tilde{x}_1 + C_2\tilde{x}_2 \end{aligned}$$

where

$$A_o = A_{11} + B_1K_2C_1 - (A_{12}+B_1K_2C_2)(A_{22}+B_2K_2C_2)^{-1}(A_{21}+B_2K_2C_1)$$
$$B_o = B_1 - (A_{12}+B_1K_2C_2)(A_{22}+B_2K_2C_2)^{-1}B_2 \quad (5\text{-}3.24)$$
$$C_o = C_1 - C_2(A_{22}+B_2K_2C_2)^{-1}(A_{21}+B_2K_2C_1).$$

By assumption, system (5-3.20) is stabilizable and detectable. Then, (A_o,B_o,C_o) is stabilizable and detectable. Thus, suitable matrices K_1,G_2 may be selected to satisfy

$$\sigma(A_o+B_oK_1) \subset \mathbb{C}^-, \quad \sigma(A_o-G_2C_o) \subset \mathbb{C}^- \quad (5\text{-}3.25)$$

Now, following the proof process of Theorem 5-3.2, we can construct the normal compensator for system (5-3.20):

$$\begin{aligned} \dot{x}_c &= A_c x_c + B_c y \\ u &= F_c x_c + F y \end{aligned} \quad (5\text{-}3.26)$$

where $x_c \in \mathbb{R}^{n_1}$,

$$\begin{aligned} A_c &= A_o - G_2C_o + B_oK_1 + G_2C_2B_2K_1 \\ B_c &= G_2 \\ F_c &= K_1 \\ F &= K_2 \end{aligned} \quad (5\text{-}3.27)$$

and their closed-loop system has no infinite poles, or no impulse terms exist in their closed-loop state response.

In summary, the design algorithm of normal compensators for system (5-3.20) is as follows.

Under the sufficient conditions for the existence of normal compensators as in Theorem 5-3.3.

1. Choosing matrix K_2 so that (5-3.21) is fulfilled.

2. For such K_2, calculating A_o, B_o, C_o according to (5-3.24).

3. Choosing matrices K_1, G_2 satisfying (5-3.25) with expected pole sets $\sigma(A_o + B_oK_1)$ and $\sigma(A_o-G_2C_o)$.

4. Calculating coefficient matrices A_c, B_c, F_c, and F in accordance with (5-3.27).

5. Then (5-3.26) is the designed normal compensator with desired properties.

Compensators discussed so far have a common feature: Their output, or feedback control u(t), is a linear combination of measure output and the state of compensator. Now we will give one result for another type of compensators.

Theorem 5-3.4 (Dai and Wang, 1987b). Let system (5-3.1) be stabilizable, detectable, and impulse observable. Then it always has a compensator of order no greater than rankE of the following form:

$$\begin{aligned} \dot{x}_c &= A_c x_c + B_c u + Gy \\ w &= F_c x_c + Fy + Hu \\ u &= Kw \end{aligned} \qquad (5\text{-}3.28)$$

where $x_c \in \mathbb{R}^{n_c}$, $w \in \mathbb{R}^n$, $n_c \leqslant \text{rank}E$, and all matrices are constant, such that the closed-loop system has no infinite poles.

The compensator is a combination of P-state feedback control and the normal state observer in the form of (4-2.18).

The structure of compensator (5-3.28) is shown in Figure 5-3.1.

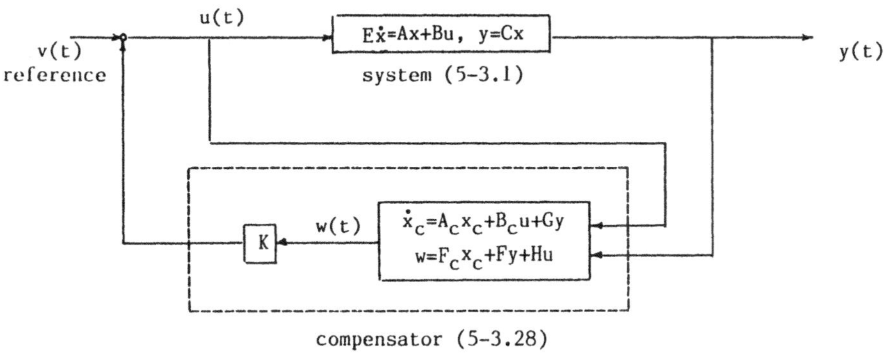

Figure 5-3.1. Dynamic Compensator (5-3.28)

Example 5-3.3. In the proof of Theorem 5-3.2, if the Luenberger reduced-order observer for system (5-3.15) is used instead of the full-order observer (5-3.16), and the same algorithm is processed, the final compensator is in the form of (5-3.28) instead of (5-3.11).

So far, we have discussed existence and design methods for various compensators. They have two common features. They are obtained based on state observers, and the separation principle holds in their design, enabling us to design the feedback and observer separately. Thus, it is convenient for use.

When $E = I_n$ is allowed, all the results become the well-known results for normal systems in linear system theory.

While all other compensators are in the form of dynamic output feedback as shown in Figure 5-2.1, the compensator in the form of (5-3.28) adopts a different version. Its output (which acts as the input for original system) visibly depends on input u(t), thus not possessing the form of direct dynamic feedback. Viewed from the point of transfer relationship, besides the dynamic output feedback compensation, it also includes a series compensator, which acts as a feedforward controller. This property may be seen clearly from Figure 5-3.2.

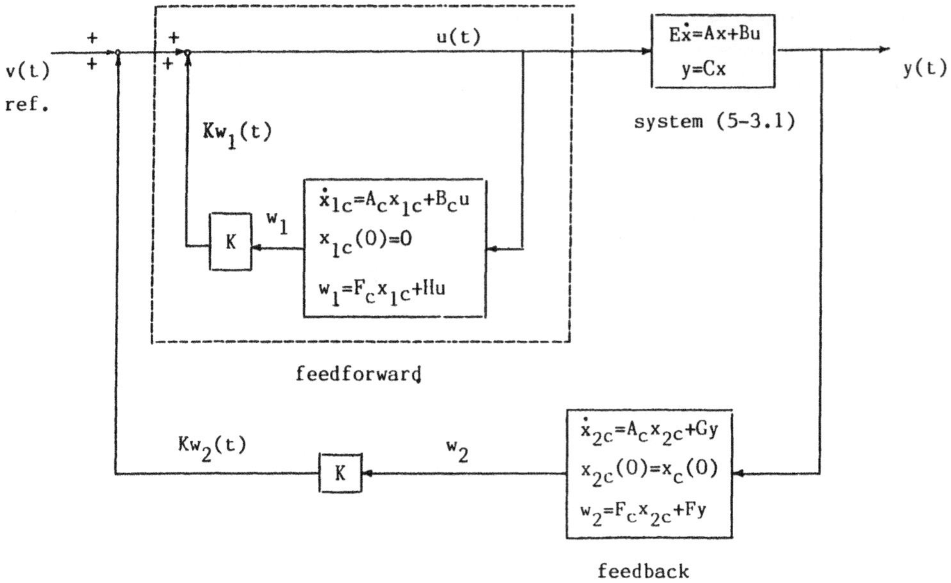

Figure 5-3.2. Compensator (5-3.28)

For a given system, the compensators that stabilize it are by no means unique. However, the more the requirements, such as lower order or normality, the less freedom in their design.

As proven in Section 2-5, for any regular system (5-3.1) there always exist two nonsingular matrices Q and P, such that under the coordinate transformation

$$P^{-1}x = [\tilde{x}_1/\tilde{x}_2/\tilde{x}_3/\tilde{x}_4], \qquad \tilde{x}_i \in \mathbb{R}^{\tilde{n}_i}, \qquad \sum_{i=1}^{4} \tilde{n}_i = n,$$

system (5-3.1) is r.s.e. to its canonical form:

$$\begin{bmatrix} E_{11} & 0 & E_{13} & 0 \\ E_{21} & E_{22} & E_{23} & E_{24} \\ 0 & 0 & E_{33} & 0 \\ 0 & 0 & E_{43} & E_{44} \end{bmatrix}\begin{bmatrix} \dot{\tilde{x}}_1 \\ \dot{\tilde{x}}_2 \\ \dot{\tilde{x}}_3 \\ \dot{\tilde{x}}_4 \end{bmatrix} = \begin{bmatrix} A_{11} & 0 & A_{13} & 0 \\ A_{21} & A_{22} & A_{23} & A_{24} \\ 0 & 0 & A_{33} & 0 \\ 0 & 0 & A_{43} & A_{44} \end{bmatrix}\begin{bmatrix} \tilde{x}_1 \\ \tilde{x}_2 \\ \tilde{x}_3 \\ \tilde{x}_4 \end{bmatrix} + \begin{bmatrix} \tilde{B}_1 \\ \tilde{B}_2 \\ 0 \\ 0 \end{bmatrix} u$$

$$y = [C_1, 0, C_3, 0][\tilde{x}_1/\tilde{x}_2/\tilde{x}_3/\tilde{x}_4] \tag{5-3.29}$$

where the subsystem

$$\begin{aligned} E_{11}\dot{\tilde{x}}_1 &= A_{11}\tilde{x}_1 + \tilde{B}_1 u \\ \tilde{y} &= C_1\tilde{x}_1 \end{aligned} \tag{5-3.30}$$

is both controllable and observable. If

$$\begin{aligned} E_c\dot{x}_c &= A_c x_c + B_c\tilde{y} \\ u &= F_c x_c + F\tilde{y} \end{aligned} \tag{5-3.31}$$

is its compensator. Here, E_c may be singular. Then the system determined by

$$\begin{aligned} E_c\dot{x}_c &= A_c x_c + B_c y \\ u &= F_c x_c + Fy \end{aligned} \tag{5-3.32}$$

is a compensator for system (5-3.1) since when applied to (5-3.1) the closed-loop system

$$\begin{bmatrix} E & 0 \\ 0 & E_c \end{bmatrix}\begin{bmatrix} \dot{x} \\ \dot{x}_c \end{bmatrix} = \begin{bmatrix} A+BFC & BF_c \\ B_cC & A_c \end{bmatrix}\begin{bmatrix} x \\ x_c \end{bmatrix} \tag{5-3.33}$$

has the finite pole set

$$\sigma\left(\begin{bmatrix} E & 0 \\ 0 & E_c \end{bmatrix}, \begin{bmatrix} A+BFC & BF_c \\ B_cC & A_c \end{bmatrix}\right)$$

$$= \sigma\left(\begin{bmatrix} E_{11} & 0 \\ 0 & E_c \end{bmatrix}, \begin{bmatrix} A_{11}+B_1FC_1 & B_1F_c \\ B_cC_1 & A_c \end{bmatrix}\right) \cup \sigma(E_{22},A_{22}) \cup \sigma(E_{33},A_{33}) \cup \sigma(E_{44},A_{44}),$$

which is obviously stable when the original system (5-3.1) is stabilizable and detectable.

Sparked by these thoughts, the general design algorithm of compensators (singular or normal) may be summarized as follows. Under the sufficient conditions of stabilizability and detectability for system (5-3.1).

1. Taking canonical decomposition to obtain (5-3.29) when it is not.

2. Choosing suitable feedback

$$u = K_1\tilde{x}_1 - K_2\dot{\tilde{x}}_1 \quad (\text{or } u = K_1\tilde{x}_1) \tag{5-3.34}$$

to stabilize subsystem (5-3.30).

3. Constructing for subsystem (5-3.30) the state observer (H may be zero matrix):

$$E_c\dot{x}_c = A_c x_c + B_c u + G\tilde{y}$$
$$w = F_c x_c + F\tilde{y} + Hu$$

such that $\lim_{t\to\infty}(w-x) = 0, \ \forall\ x(0),\ x_c(0)$.

4. The system

$$E_c\dot{x}_c = A_c x_c + B_c u + Gy$$
$$w = F_c x_c + Fy + Hu$$
$$u = K_1 w - K_2\dot{w}$$

is a compensator for system (5-3.1).

5-4. Singular Compensators in Output Regulation Systems

Now we will consider the compensation problem for singular systems with extra disturbance. In this section we will study the deterministic disturbance with models whose initial condition is unknown. In this problem, our compensators should have twofold properties: resistance of the disturbance and stability when disturbance disappears, the latter property is termed internal stability.

We would like to explain the disturbance. Assume that the disturbance $f(t)$ satisfies the singular equation

$$E_f\dot{f}(t) = A_f f(t) \tag{5-4.1}$$

where E_f and A_f are constant, and system (5-4.1) is regular. From the state representation, its fast substate is always zero once started, contributing nothing to the system. Thus, we will hereafter only consider the disturbance $f(t)$ that satisfies an ordinary differential equation (or the slow subsystem form):

$$\dot{f}(t) = Lf(t). \tag{5-4.2}$$

Consider the singular system with external disturbance $f(t)$:

$$
\begin{aligned}
E\dot{x}(t) &= Ax(t) + Bu(t) + Mf(t) \\
\dot{f}(t) &= Lf(t) \\
y(t) &= C_1x(t) + C_2f(t) \\
z(t) &= D_1x(t) + D_2f(t)
\end{aligned}
\tag{5-4.3}
$$

where $x(t) \in \mathbb{R}^n$ is state, $u(t) \in \mathbb{R}^m$ is the control input, $f(t) \in \mathbb{R}^p$ is the disturbance satisfying (5-4.2), $y(t) \in \mathbb{R}^r$ is the measure output and $z(t) \in \mathbb{R}^d$ is the vector to be regulated. E, A, B, M, L, C_1, C_2, D_1, and D_2 are constant matrices of appropriate dimensions. Since (asymptotically) stable disturbance input has no effects on asymptotically regulated output, we hereafter assume that $\sigma(L) \subset \bar{\mathbb{C}}^+$.

Let $u(t) \equiv 0$, $t \geqslant 0$. If system (5-4.3) is stable when $f(t) \equiv 0$, $t \geqslant 0$, i.e., $\lim_{t\to\infty} x(t) = 0$, $\forall\, x(0)$, system (5-4.3) will be termed internally stable; if

$$\lim_{t\to\infty} (z(t) - z_r(t)) = 0, \quad \forall\, x(0)$$

holds for reference signal $z_r(t)$, system (5-4.3) will be called output regulation. If $z_r(t)$ satisfies a singular differential equation, by using the state augmention method, the problem when $z_r(t) \neq 0$ may be changed into the problem when $z_r(t) = 0$. Thus we will assume that $z_r(t) \equiv 0$, $t \geqslant 0$.

<u>Example 5-4.1.</u> When $u(t) \equiv 0$, $t \geqslant 0$, singular system (4-5.8) with disturbance input becomes

$$
\begin{aligned}
\begin{bmatrix} 0 & 1 & 0 \\ 0 & 0 & 1 \\ 0 & 0 & 0 \end{bmatrix} \dot{x}(t) &= \begin{bmatrix} 1 & 0 & 0 \\ 0 & 1 & 0 \\ 0 & 0 & 1 \end{bmatrix} x(t) + \begin{bmatrix} 0 \\ 2 \\ 1 \end{bmatrix} f(t) \\
z(t) &= [1 \quad 0 \quad 0]x(t)
\end{aligned}
\tag{5-4.4}
$$

whose solution is

$$
\begin{aligned}
x(t) &= \begin{bmatrix} -2\dot{f}(t) - \ddot{f}(t) \\ -2f(t) - \dot{f}(t) \\ -f(t) \end{bmatrix} \\
z(t) &= -\,2\dot{f}(t) - \ddot{f}(t).
\end{aligned}
\tag{5-4.5}
$$

Let the disturbance $f(t)$ satisfies $\dot{f}(t) = 0$, i.e., a jump signal. From (5-4.5) we know that system (5-4.4) is not only internally stable but also output regulation.

If the disturbance $f(t)$ satisfies $\dot{f}(t) = f(t)$, also from (5-4.5), system (5-4.4) is no longer output regulation.

<u>Example 5-4.2.</u> When $f(t) \equiv 0$, the system

$$
\begin{aligned}
\begin{bmatrix} 1 & 0 & 0 \\ 0 & 1 & 0 \\ 0 & 0 & 0 \end{bmatrix} \dot{x} &= \begin{bmatrix} 1 & -1 & 0 \\ 0 & 0 & 2 \\ 0 & 0 & 1 \end{bmatrix} x + \begin{bmatrix} 1 \\ 0 \\ 1 \end{bmatrix} f \\
z &= [0 \quad 1 \quad 1]x + f
\end{aligned}
$$

is not stable. Thus it is not internally stable, or output regulation.

Consider the dynamic compensator

$$\begin{aligned} E_c\dot{x}_c(t) &= A_c x_c(t) + B_c y(t) \\ u(t) &= F_c x_c(t) + Fy(t) \end{aligned} \tag{5-4.6}$$

where $x_c(t) \in \mathbb{R}^{n_c}$ is its state; E_c, A_c, B_c, F_c, and F are constant matrices of appropriate dimensions. Systems (5-4.6) and (5-4.3) form the closed-loop system

$$\begin{aligned} \begin{bmatrix} E & 0 \\ 0 & E_c \end{bmatrix}\begin{bmatrix} \dot{x} \\ \dot{x}_c \end{bmatrix} &= \begin{bmatrix} A+BFC & BF_c \\ B_cC & A_c \end{bmatrix}\begin{bmatrix} x \\ x_c \end{bmatrix} + \begin{bmatrix} M+BFC_2 \\ B_cC_2 \end{bmatrix} f \\ \dot{f} &= Lf \\ y &= C_1x + C_2f \\ z &= D_1x + D_2f. \end{aligned} \tag{5-4.7}$$

Definition 5-4.1. Let (5-4.6) be a dynamic system. If

1. Closed-loop system (5-4.7) is internally stable, i.e.,

$$\sigma\left(\begin{bmatrix} E & 0 \\ 0 & E_c \end{bmatrix}, \begin{bmatrix} A+BFC & BF_c \\ B_cC & A_c \end{bmatrix}\right) \subset \mathbb{C}^- \tag{5-4.8}$$

2. System (5-4.7) is output regulation, i.e.,

$$\lim_{t\to\infty} z(t) = 0, \quad \forall\ x(0),\ x_c(0),\ f(0), \tag{5-4.9}$$

we will call (5-4.6) an output regulator for system (5-4.3), or regulator without confusion.

(a) If rank$E_c < n_c$, system (5-4.6) is termed a singular output regulator (SOR).

(b) Otherwise, rank$E_c = n_c$, $E_c = I_{n_c}$ is assumed without loss of generality, system (5-4.6) is termed a normal output regulator (NOR).

Only SORs are considered in this section.

Lemma 5-4.1. Assume that (E, A) is a regular pencil. Then the matrix equation

$$AV - EVL = M \tag{5-4.10}$$

has a unique solution for any matrix M if and only if

$$\sigma(E,A) \cap \sigma(L) = \emptyset. \tag{5-4.11}$$

Proof. Under the regularity assumption for (E,A), by Lemma 1-2.2 there exist nonsingular matrices Q and P such that

$$QEP = \text{diag}(I_{n_1}, N), \qquad QAP = \text{diag}(A_1, I_{n_2})$$

where $n_1 + n_2 = n$, $N \in \mathbb{R}^{n_2 \times n_2}$ is nilpotent. Left multiplying Q on both sides of (5-4.10), we obtain

$$\mathrm{diag}(A_1, I)P^{-1}V - \mathrm{diag}(I, N)P^{-1}VL = QM. \tag{5-4.12}$$

By denoting

$$P^{-1}V = [V_1/V_2], \qquad QM = [M_1/M_2],$$

equation (5-4.12) becomes

$$A_1V_1 - V_1L = M_1 \tag{5-4.13a}$$

$$V_2 - NV_2L = M_2. \tag{5-4.13b}$$

It may be verified that equation (5-4.13b) has a unique solution

$$V_2 = \sum_{i=0}^{h-1} N^i M_2 L^i \tag{5-4.14}$$

for any L and M_2, where h is the nilpotent index of N. Equation (5-4.13a) is a Sylvester equation, which has a unique solution for any matrix M_1 if and only if (Appendix B)

$$\sigma(A_1) \cap \sigma(L) = \emptyset.$$

Thus, from $\sigma(E,A) = \sigma(A_1)$, we know that (5-4.10) has a unique solution for any matrix M if and only if (5-4.11) holds. Q.E.D.

Lemma 5-4.2 (Elementary Lemma). Consider the singular system

$$\begin{aligned} E\dot{x} &= Ax + Mf \\ \dot{f} &= Lf \\ z &= D_1x + D_2f \end{aligned} \tag{5-4.15}$$

with $\sigma(E,A) \subset \mathbb{C}^-$. The property $\lim_{t\to\infty} z(t) = 0$, $\forall\ x(0), f(0)$ is true if and only if there exists a matrix V satisfying

$$\begin{aligned} AV - EVL &= M \\ D_1V &= D_2. \end{aligned} \tag{5-4.16}$$

Proof. Sufficiency: For any matrix $V \in \mathbb{R}^{n \times p}$ satisfying (5-4.16), we define $w = x + Vf$, which possesses the dynamic equation $E\dot{w} = Aw$, $z = D_1w$. Thus, by $\sigma(E,A) \subset \mathbb{C}^-$ we have

$$\lim_{t\to\infty} z(t) = \lim_{t\to\infty} D_1w(t) = 0, \ \forall\ x(0), f(0).$$

Necessity: Since we have assumed that $\sigma(E,A) \subset \mathbb{C}^-$ and $\sigma(L) \subset \bar{\mathbb{C}}^+$, it will be $\sigma(E,A) \cap \sigma(L) = \emptyset$. Lemma 5-4.2 shows that a unique matrix V exists that satisfies (5-4.10). Let $w = x+Vf$. We have

$$E\dot{w} = Aw$$
$$z = D_1 w + (D_2 - D_1 V)f.$$

Using $\sigma(E,A) \subset \mathbb{C}^-$ once more, $\lim_{t\to\infty} w(t) = 0$, $\forall$ $w(0)$. Therefore, $\lim_{t\to\infty} z(t) = 0$ is true for any f(0) (thus f(t)) if and only if

$$\lim_{t\to\infty} (D_2 - D_1 V)f(t) = 0, \quad \forall \ f(0),$$

together with $\sigma(L) \subset \bar{\mathbb{C}}^+$, implying $D_2 = D_1 V$. Q.E.D.

This lemma is needed in later discussions.

Example 5-4.3. In system (5-4.4),

$$E = \begin{bmatrix} 0 & 1 & 0 \\ 0 & 0 & 1 \\ 0 & 0 & 0 \end{bmatrix}, \qquad A = \begin{bmatrix} 1 & 0 & 0 \\ 0 & 1 & 0 \\ 0 & 0 & 1 \end{bmatrix}, \qquad M = \begin{bmatrix} 0 \\ 2 \\ 1 \end{bmatrix}.$$

Matrix $V = (I-E)^{-1}M = [3 \ \ 3 \ \ 1]^\tau$ is the unique solution of the Sylvester equation

$$AV - EVL = M.$$

However, $D_1 V = 3 \neq D_2 = 0$. Thus, this system is not output regulation according to Lemma 5-4.2 in which L = 1. This coincides with the result in Example 5-4.1.

A direct result of Lemma 5-4.2 is the following corollary.

Corollary 5-4.1. Assume that (5-4.6) is a dynamic compensator such that the closed-loop system (5-4.6) is internally stable. Then the closed-loop system is output regulation if and only if two matrices $V_1 \in \mathbb{R}^{n\times p}$, $V_2 \in \mathbb{R}^{n_c\times p}$ exist such that

$$\begin{aligned} &(A+BFC_1)V_1 + BF_cV_2 - EV_1L = M + BFC_2 \\ &B_cC_1V_1 + A_cV_2 - E_cV_2L = B_cC_2 \\ &D_1V_1 = D_2. \end{aligned} \tag{5-4.17}$$

Theorem 5-4.1. Consider the following singular system with disturbance input:

$$\begin{aligned} E\dot{x} &= Ax + Bu + Mf \\ \dot{f} &= Lf \\ y &= C_1 x \\ z &= D_2 x \end{aligned} \tag{5-4.18}$$

If

1. (E,A,B,C_1) is stabilizable.

2. rank[B M] = rankB.

3.

$$\text{rank}\begin{bmatrix} sE-A & -M \\ 0 & sI-L \\ C_1 & 0 \end{bmatrix} = n + p, \quad \forall\ s \in \bar{\mathbb{C}}^+,\ s \text{ finite},$$

system (5-4.18) has an SOR of the form (5-4.6).

Proof. By assumption 1 a matrix $K_1 \in \mathbb{R}^{m\times n}$ may be chosen to satisfy $\sigma(E,A+BK_1) \subset \mathbb{C}^-$; assumption 2 indicates the existence of a matrix K_2 satisfying

$$M = BK_2. \tag{5-4.19}$$

Apparently, the feedback control

$$u = K_1 x - K_2 f \tag{5-4.20}$$

when applied to (5-4.18), stabilizes its closed-loop system

$$E\dot{x} = (A+BK_1)x. \tag{5-4.21}$$

Thus $\lim_{t\to\infty} z = D_1 \lim_{t\to\infty} x = 0,\ \forall\ x(0)$.

Note that feedback control (5-4.20) cannot be realized directly. To realize it we consider the following composite system

$$\begin{aligned} \begin{bmatrix} E & 0 \\ 0 & I \end{bmatrix}\begin{bmatrix} \dot{x} \\ \dot{f} \end{bmatrix} &= \begin{bmatrix} A & M \\ 0 & L \end{bmatrix}\begin{bmatrix} x \\ f \end{bmatrix} + \begin{bmatrix} B \\ 0 \end{bmatrix}u \\ y &= [C_1 \quad 0][x/f]. \end{aligned} \tag{5-4.22}$$

It is detectable according to assumption 3, assuring the existence of its singular state observer:

$$\begin{aligned} \begin{bmatrix} E & 0 \\ 0 & I \end{bmatrix}\dot{x}_c &= \left(\begin{bmatrix} A & M \\ 0 & L \end{bmatrix} - G[C_1 \quad 0]\right)x_c + \begin{bmatrix} B \\ 0 \end{bmatrix}u + Gy \\ w &= x_c, \end{aligned} \tag{5-4.23}$$

where G satisfies

$$\sigma\left(\begin{bmatrix} E & 0 \\ 0 & I \end{bmatrix}, \begin{bmatrix} A & M \\ 0 & L \end{bmatrix} - G[C_1 \quad 0]\right) \subset \mathbb{C}^-, \tag{5-4.24}$$

such that $\lim_{t\to\infty}(w - [x/f]) = 0,\ \forall\ x(0),\ x_c(0),\ f(0)$.

Using the control $u = [K_1 \quad -K_2]x_c$ is lieu of (5-4.20), its substitution into (5-4.23) yields

$$\begin{aligned} E_c\dot{x}_c &= A_c x_c + B_c y \\ u &= F_c x_c + Fy \end{aligned} \tag{5-4.25}$$

where $x_c \in \mathbb{R}^{n+d}$, and coefficient matrices defined as

$$E_c = \begin{bmatrix} E & 0 \\ 0 & I \end{bmatrix}, \quad A_c = \begin{bmatrix} A & M \\ 0 & L \end{bmatrix} - G[C_1 \quad 0] + \begin{bmatrix} B \\ 0 \end{bmatrix}[K_1 \ -K_2] \tag{5-4.26}$$

$$B_c = G, \ F_c = [K_1 \quad -K_2], \quad F = 0.$$

Hence (5-4.25) is an SOR for system (5-4.18).

In fact, their closed-loop stsrem has finite pole set

$$\sigma(\begin{bmatrix} E & 0 \\ 0 & I \end{bmatrix}, \begin{bmatrix} A+BFC & BF_c \\ B_cC & A_c \end{bmatrix})$$

$$= \sigma(E, A+BK_1) \cup \sigma(\begin{bmatrix} E & 0 \\ 0 & I \end{bmatrix}, \begin{bmatrix} A & M \\ 0 & L \end{bmatrix} - G[C_1 \quad 0]) \subset \mathbb{C}^-$$

implying the closed-loop system is internally stable.

From (5-4.19) and (5-4.26), equation (5-4.17) is fulfilled with the matrices

$$V_1 = 0 \in \mathbb{R}^{n+d}, \qquad V_2 = [0/-I_d] \in \mathbb{R}^{(n+d)xp}.$$

Therefore, Corollary 5-4.1 shows (5-4.25) is an SOR for system (5-4.18). Q.E.D.

This theorem shows a sufficient condition for the existence and design method of SORs for systems with disturbance not directly involved in its measure and regulated outputs.

If $D_1 = I$, $z = x$. The regulated vector is the state whose regulator is called state regulator. The following Theorem 5-4.2 is a result concerning the state regulator.

<u>Theorem 5-4.2</u>. Consider the system

$$\begin{aligned} E\dot{x} &= Ax + Bu + Mf \\ \dot{f} &= Lf \\ y &= C_1x \\ z &= x. \end{aligned} \tag{5-4.27}$$

If rankM = p, system (5-4.27) has a singular state regulator (5-4.6) if and only if

1. (E,A,B,C) is stabilizable.

2. rank[B M] = rankB.

3. $$(\begin{bmatrix} E & 0 \\ 0 & I \end{bmatrix}, \begin{bmatrix} A & M \\ 0 & L \end{bmatrix}, [C_1 \quad 0])$$

is detectable.

Its proof is omitted here (see Dai and Wang, 1987d).

The theorem gives the sufficient and necessary condition of the singular state regulators only for systems with rankM = p. The case is much complicated for general systems, or even for system (5-4.18).

As seen in Theorems 5-4.1 and 5-4.2, the detectability of the composite system (5-4.22) is given as a sufficient condition, which indicates that the disturbance should be (asymptotically) observed in order to be rejected by applying a dynamic compensator (regulator). This is in accordance with our common knowledge, otherwise, by no means may we mention the problem of disturbance rejection without knowledge of the disturbance itself.

The output regulator is a special dynamic feedback mechanism that achieves a particular goal, resisting the influence of disturbance on the regulated vector.

Example 5-4.4. Consider the singular system

$$\begin{bmatrix} 1 & 0 & 0 \\ 0 & 1 & 1 \\ 0 & 0 & 0 \end{bmatrix}\dot{x} = \begin{bmatrix} 1 & -1 & 0 \\ 0 & 1 & -1 \\ 1 & -2 & 0 \end{bmatrix}x + \begin{bmatrix} 0 & 0 \\ 0 & 1 \\ 1 & 0 \end{bmatrix}u + \begin{bmatrix} 0 \\ 1 \\ -1 \end{bmatrix}f$$

$$y = [1 \quad 0 \quad 0]x \qquad (5\text{-}4.28)$$

$$z = x$$

with the jump disturbance input f(t) satisfying

$$f = 0. \qquad (5\text{-}4.29)$$

Now consider the design of its state regulator. In system (5-4.28)

$$E = \begin{bmatrix} 1 & 0 & 0 \\ 0 & 1 & 1 \\ 0 & 0 & 0 \end{bmatrix}, \quad A = \begin{bmatrix} 1 & -1 & 0 \\ 0 & 1 & -1 \\ 1 & -2 & 0 \end{bmatrix}, \quad B = \begin{bmatrix} 0 & 0 \\ 0 & 1 \\ 1 & 0 \end{bmatrix}, \quad M = \begin{bmatrix} 0 \\ 1 \\ -1 \end{bmatrix},$$

$$C_1 = [1 \quad 0 \quad 0], \quad C_2 = 0, \quad D_1 = I_3, \quad D_2 = 0, \quad L = 0.$$

Since rank[sE-A, B] = 3 = n, $\forall$ s $\in \bar{\mathbb{C}}^+$, s finite, system (E,A,B,C) is stabilizable.

Furthermore, rank[B M] = 2 = rankB and

$$\text{rank}\begin{bmatrix} sE-A & -M \\ 0 & sI-L \\ C_1 & 0 \end{bmatrix} = 4 = n + p, \quad \forall\ s \in \bar{\mathbb{C}}^+, \text{ s finite},$$

i.e.,

$$\left(\begin{bmatrix} E & 0 \\ 0 & I \end{bmatrix}, \begin{bmatrix} A & M \\ 0 & L \end{bmatrix}, [C_1 \quad 0]\right)$$

is detectable.

From Theorem 5-4.2, we are sure of the existence of a singular state regulator. Le Let

$$K_1 = \begin{bmatrix} 2 & 0 & 0 \\ 0 & 0 & 0 \end{bmatrix}.$$

We have

$$\sigma(E, A+BK_1) = \{-1, -0.5\} \subset \mathbb{C}^-. \qquad (5\text{-}4.30)$$

If we choose the feedback control

$$u = \begin{bmatrix} 2 & 0 & 0 \\ 0 & 0 & 0 \end{bmatrix} x + \begin{bmatrix} 1 \\ -1 \end{bmatrix} f \tag{5-4.31}$$

when applied to system (5-4.28), its closed-loop system

$$E\dot{x} = (A+BK_1)x, \qquad z = x$$

is stable according to (5-4.30). That is, $\lim_{t\to\infty} z(t) = 0$ for any $x(0)$.

To realize the feedback control (5-4.31), we consider the composite system

$$\begin{bmatrix} 1 & 0 & 0 & 0 \\ 0 & 1 & 1 & 0 \\ 0 & 0 & 0 & 0 \\ 0 & 0 & 0 & 1 \end{bmatrix} \begin{bmatrix} \dot{x} \\ \dot{f} \end{bmatrix} = \begin{bmatrix} 1 & -1 & 0 & 0 \\ 0 & 1 & -1 & 1 \\ 1 & -2 & 0 & -1 \\ 0 & 0 & 0 & 0 \end{bmatrix} \begin{bmatrix} x \\ f \end{bmatrix} + \begin{bmatrix} 0 & 0 \\ 0 & 1 \\ 1 & 0 \\ 0 & 0 \end{bmatrix} u$$

$$y = [1 \;\; 0 \;\; 0 \;\; 0][x/f], \tag{5-4.32}$$

which corresponds to system (5-4.23), and is detectable as testified earlier. In fact, matrix $G = [0 \;\; 0 \; -5 \;\; 2]^{\tau}$ fills the equation

$$\sigma\left(\begin{vmatrix} E & 0 \\ 0 & I \end{vmatrix}, \begin{vmatrix} A & M \\ 0 & L \end{vmatrix} + G[C_1 \;\; 0]\right) = \{-1, -1\} \subset \mathbb{C}^-.$$

Therefore, system

$$\begin{bmatrix} 1 & 0 & 0 & 0 \\ 0 & 1 & 1 & 0 \\ 0 & 0 & 0 & 0 \\ 0 & 0 & 0 & 1 \end{bmatrix} \dot{x}_c = \begin{bmatrix} 1 & -1 & 0 & 0 \\ 0 & 1 & -1 & 1 \\ 6 & -2 & 0 & -1 \\ -2 & 0 & 0 & 0 \end{bmatrix} x_c + \begin{bmatrix} 0 & 0 \\ 0 & 1 \\ 1 & 0 \\ 0 & 0 \end{bmatrix} u + \begin{bmatrix} 0 \\ 0 \\ -5 \\ 2 \end{bmatrix} y$$

$$w = x_c$$

is a singular observer for system (5-4.32).

Let the control input

$$u = \begin{bmatrix} 2 & 0 & 0 & 1 \\ 0 & 0 & 0 & -1 \end{bmatrix} x_c$$

be used to substitute (5-4.31). Then the singular state regulator is

$$\begin{bmatrix} 1 & 0 & 0 & 0 \\ 0 & 1 & 1 & 0 \\ 0 & 0 & 0 & 0 \\ 0 & 0 & 0 & 1 \end{bmatrix} \dot{x}_c = \begin{bmatrix} 1 & -1 & 0 & 0 \\ 0 & 1 & -1 & 0 \\ 8 & -2 & 0 & 0 \\ -2 & 0 & 0 & 0 \end{bmatrix} x_c + \begin{bmatrix} 0 \\ 0 \\ -5 \\ 2 \end{bmatrix} y$$

$$u = \begin{bmatrix} 2 & 0 & 0 & 1 \\ 0 & 0 & 0 & -1 \end{bmatrix} x_c \tag{5-4.33}$$

according to Theorem 5-4.1. For any constant disturbance $f(t)$, system (5-4.33) makes its closed-loop system internally stable and output regulation.

5-5. Normal Compensators in Output Regulation Systems

Compared with singular compensators, normal ones are more convenient to physically realize and more insensitive to external nois inputs.

For a given singular system with disturbance input

$$\begin{aligned} E\dot{x} &= Ax + Bu + Mf \\ \dot{f} &= Lf \\ y &= C_1x + C_2f \\ z &= D_1x + D_2f \end{aligned} \tag{5-5.1}$$

where the items are determined by (5-4.3), its normal output regulator (NOR) is an output regulator in the form of

$$\begin{aligned} \dot{x}_c &= A_cx_c + B_cy \\ u &= F_cx_c + Fy \end{aligned} \tag{5-5.2}$$

where $x_c \in \mathbb{R}^{n_c}$, A_c, B_c, F_c, and F are constant matrices of appropriate dimensions.

System (5-5.1) and compensator (5-5.2) form the closed-loop system

$$\begin{aligned} \begin{bmatrix} E & 0 \\ 0 & I \end{bmatrix}\begin{bmatrix} \dot{x} \\ \dot{x}_c \end{bmatrix} &= \begin{bmatrix} A+BFC & BF_c \\ B_cC & A_c \end{bmatrix}\begin{bmatrix} x \\ x_c \end{bmatrix} + \begin{bmatrix} M+BFC_2 \\ B_cC_2 \end{bmatrix} f \\ \dot{f} &= Lf \\ y &= C_1x + C_2f \\ z &= D_1x + D_2f. \end{aligned} \tag{5-5.3}$$

As a special case of $E_c = I$ whenever it is allowed in Corollary 5-4.1, we have Corollary 5-5.1.

<u>Corollary 5-5.1</u>. Consider system (5-5.1). Let closed-loop system (5-5.3) be internally stable. Then closed-loop system (5-5.3) is output regulation if and only if two matrices $V_1 \in \mathbb{R}^{nxp}$, $V_2 \in \mathbb{R}^{n_cxp}$ exist such that

$$\begin{aligned} &(A+BFC_1)V_1 + BF_cV_2 - EV_1L = M + BFC_2 \\ &B_cC_1V_1 + A_cV_2 - V_2L = B_cC_2 \\ &D_1V_1 - D_2. \end{aligned} \tag{5-5.4}$$

The key difference in the design of normal regulators from that of singular regulators is that either a normal observer is used to observe the state and the disturbance, or derivative state feedback is used in the design process. In general, normal regulators have a more complicated form than singular regulators.

Theorem 5-5.1. Consider the singular system with disturbance

$$\begin{aligned} E\dot{x} &= Ax + Bu + Mf \\ \dot{f} &= Lf \\ y &= C_1 x \\ z &= D_1 x \end{aligned} \tag{5-5.5}$$

in which disturbance is not involved visibly in measure and regulated outputs. If

1. (E,A,B,C_1) is stabilizable.
2. $\text{rank}[B \;\; M] = \text{rank}B$.
3. $$\left(\begin{bmatrix} E & 0 \\ 0 & I \end{bmatrix}, \begin{bmatrix} A & M \\ 0 & L \end{bmatrix}, [C_1 \;\; 0]\right)$$

is detectable, system (5-5.5) always has an NOR in the form of (5-5.2).

Proof. The proof is constructive. As seen in normalizability decomposition (5-2.12), there exist two nonsingular matrices Q and P such that

$$QEP = \begin{bmatrix} E_{11} & E_{12} \\ 0 & N_{22} \end{bmatrix}, \quad QAP = \begin{bmatrix} A_1 & 0 \\ 0 & I \end{bmatrix}, \quad QB = \begin{bmatrix} B_1 \\ 0 \end{bmatrix} \tag{5-5.6}$$

where E_{11}, $A_1 \in \mathbb{R}^{n_1 \times n_1}$, $N_{22} \in \mathbb{R}^{n_2 \times n_2}$, $n_1 + n_2 = n$, N_{22} is nilpotent, and (E_{11}, A_1, B_1) is normalizable. Let

$$QM = \begin{bmatrix} M_1 \\ M_2 \end{bmatrix}, \qquad M_1 \in \mathbb{R}^{n_1 \times p}, \quad M_2 \in \mathbb{R}^{n_2 \times p}.$$

Then from assumption 2 we have $M_2 = 0$. Therefore, under the coordinate transformation,

$$x = P\begin{bmatrix} x_1 \\ x_2 \end{bmatrix}, \quad x_1 \in \mathbb{R}^{n_1}, \quad x_2 \in \mathbb{R}^{n_2},$$

system (5-5.5) is r.s.e. to

$$\begin{aligned} \begin{bmatrix} E_{11} & E_{12} \\ 0 & N_{22} \end{bmatrix}\begin{bmatrix} \dot{x}_1 \\ \dot{x}_2 \end{bmatrix} &= \begin{bmatrix} A_1 & 0 \\ 0 & I \end{bmatrix}\begin{bmatrix} x_1 \\ x_2 \end{bmatrix} + \begin{bmatrix} B_1 \\ 0 \end{bmatrix} u + \begin{bmatrix} M_1 \\ 0 \end{bmatrix} f \\ \dot{f} &= Lf \\ y &= C_{11}x_1 + C_{12}x_2 \\ z &= D_{11}x_1 + D_{12}x_2 \end{aligned} \tag{5-5.7}$$

where

$$C_1P = [C_{11}, C_{12}], \qquad D_1P = [D_{11}, D_{12}]. \tag{5-5.8}$$

It amy be verified that the subsystem

$$\begin{aligned} E_{11}\dot{x}_1 &= A_1x_1 + B_1u + M_1f \\ \dot{f} &= Lf \\ \bar{y} &= C_{11}x_1 \\ \bar{z} &= D_{11}x_1 \end{aligned} \tag{5-5.9}$$

possesses the following properties.

1. $(E_{11}, A_1, B_1, C_{11})$ is stabilizable and normalizable.

2. $\mathrm{rank}[B_1, M_1] = \mathrm{rank}B_1$.

3. $$\left(\begin{bmatrix} E_{11} & 0 \\ 0 & I \end{bmatrix}, \begin{bmatrix} A_1 & M_1 \\ 0 & L \end{bmatrix}, [C_{11}, 0]\right)$$

is detectable.

From assumption 1, there exist matrices $K_{21} \in \mathbb{R}^{m\times n_1}$ and $K_{11} \in \mathbb{R}^{m\times n_1}$ satisfying

$$\mathrm{rank}(E_{11}+B_1K_{21}) = n_1, \qquad \sigma(E_{11}+B_1K_{21}, A_1+B_1K_{11}) \subset \mathbb{C}^-. \tag{5-5.10}$$

Assumption 2 indicates a matrix $K \in \mathbb{R}^{n_1\times p}$ may be chosen to satisfy

$$M_1 = B_1K. \tag{5-5.11}$$

Clearly, the control law

$$u = K_{11}x_1 - K_{21}\dot{x}_1 - Kf \tag{5-5.12}$$

such that $\lim_{t\to\infty} z(t) = D_1 \lim_{t\to\infty} x_1(t) = 0, \ \forall \ x_1(0)$.

To realize the control (5-5.12), consider the system

$$\begin{aligned} \begin{bmatrix} E_{11} & 0 \\ 0 & I \end{bmatrix}\begin{bmatrix} \dot{x}_1 \\ \dot{f} \end{bmatrix} &= \begin{bmatrix} A_1 & M_1 \\ 0 & L \end{bmatrix}\begin{bmatrix} x_1 \\ f \end{bmatrix} + \begin{bmatrix} B_1 \\ 0 \end{bmatrix} u \\ \bar{y} &= [C_{11}, 0][x_1/f] \end{aligned} \tag{5-5.13}$$

By assumption 3, this system is detectable. Thus there exists a matrix $G \in \mathbb{R}^{r\times(n_1+p)}$ such that

$$\sigma\left(\begin{bmatrix} E_{11} & 0 \\ 0 & I_p \end{bmatrix}, \begin{bmatrix} A_1 & M_1 \\ 0 & L \end{bmatrix} - G[C_{11}, 0]\right) \subset \mathbb{C}^-. \tag{5-5.14}$$

Therefore, the system

$$\begin{bmatrix} E_{11} & 0 \\ 0 & I_p \end{bmatrix} \dot{x}_c = \left(\begin{bmatrix} A_1 & M_1 \\ 0 & L \end{bmatrix} - G[C_{11}, 0] \right) x_c + \begin{bmatrix} B_1 \\ 0 \end{bmatrix} u + G\bar{y} \tag{5-5.15}$$

$$w = x_c$$

is a singular observer for system (5-5.13), where $x_c \in \mathbb{R}^{n_1+p}$.

Using x_c in lieu of $[x/f]$ in (5-5.12), we obtain the control

$$u = [K_{11}, -K]x_c - [K_{21}, 0]\dot{x}_c,$$

which, when applied to (5-5.15), results in the system

$$\dot{x}_c = A_c x_c + B_c \bar{y}$$
$$u = F_c x_c + F\bar{y}$$

where

$$A_c = \begin{bmatrix} (E_{11}+B_1K_{21})^{-1}(A_1-G_1C_{11}+B_1K_{11}) & 0 \\ -G_2C_{11} & L \end{bmatrix}$$

$$B_c = \begin{bmatrix} (E_{11}+B_1K_{21})^{-1}G_1 \\ G_2 \end{bmatrix}, \qquad G = \begin{bmatrix} G_1 \\ G_2 \end{bmatrix}, \tag{5-5.16}$$

$$F_c = [K_{11}-K_{21}(E_{11}+B_1K_{21})^{-1}(A_1-G_1C_{11}+B_1K_{11}), \ -K]$$

$$F = -K_{21}(E_{11}+B_1K_{21})^{-1}.$$

It can be proven that the compensator

$$\dot{x}_c = A_c x_c + B_c y$$
$$u = F_c x_c + Fy \tag{5-5.17}$$

is an NOR for system (5-5.5) (Dai and Wang, 1987d). Q.E.D.

Example 5-4.4. Consider system (5-4.28)-(5-4.29). This system is normalizable, and the matrix

$$K_{21} = \begin{bmatrix} 0 & 0 & 1 \\ 0 & 0 & 0 \end{bmatrix}$$

satisfies $\text{rank}(E+BK_{21}) = 3 = n$. If we let

$$K_{11} = \begin{bmatrix} 0 & 0 & 0 \\ 9 & -3 & 0 \end{bmatrix}$$

it will be $\sigma(E+BK_{21}, A+BK_{11}) = \{-1, -1, -1\} \subset \mathbb{C}^-$. Following the proof process of Theorem 5-5.1, we choose the control as

$$u = \begin{bmatrix} 0 & 0 & 0 & 1 \\ 9 & -3 & 0 & -1 \end{bmatrix} x_c - \begin{bmatrix} 0 & 0 & 1 & 0 \\ 0 & 0 & 0 & 0 \end{bmatrix} \dot{x}_c \tag{5-5.18}$$

where x_c is determined by

$$\begin{bmatrix} 1 & 0 & 0 & 0 \\ 0 & 1 & 1 & 0 \\ 0 & 0 & 0 & 0 \\ 0 & 0 & 0 & 1 \end{bmatrix} \dot{x}_c = \begin{bmatrix} 1 & -1 & 0 & 0 \\ 0 & 1 & -1 & 1 \\ 6 & -2 & 0 & -1 \\ -2 & 0 & 0 & 0 \end{bmatrix} x_c + \begin{bmatrix} 0 & 0 \\ 0 & 1 \\ 1 & 0 \\ 0 & 0 \end{bmatrix} u + \begin{bmatrix} 0 \\ 0 \\ -5 \\ 2 \end{bmatrix} y, \qquad (5\text{-}5.19)$$

which is a singular observer for system (5-4.28)-(5-4.29).

According to the results of Theorem 5-5.1, from (5-5.18) and (5-5.19) we obtain the NOR

$$\dot{x}_c = \begin{bmatrix} 1 & 0 & 0 & 0 \\ 15 & -4 & -1 & 0 \\ 6 & -2 & 0 & 0 \\ -2 & 0 & 0 & 0 \end{bmatrix} x_c + \begin{bmatrix} 0 \\ -5 \\ -5 \\ 2 \end{bmatrix} y$$

$$u = \begin{bmatrix} -6 & 2 & 0 & 1 \\ 9 & -3 & 0 & -1 \end{bmatrix} x_c + \begin{bmatrix} 5 \\ 0 \end{bmatrix} y,$$

which makes its closed-loop system both internally stable and output regulation for any constant disturbance $f(t)$.

Comparing the conditions of Theorems 5-4.1 and 5-5.1 we see that they are the same. The normal regulator may be designed if the system has a singular one, the former is certainly of application significance.

As indicated from the proof process of Theorem 5-4.1, if a normal observer in the form of (it exists under certain conditions by Chapter 4):

$$\dot{x}_c = A_c x_c + B_c u + Gy$$
$$u = F_c x_c + Fy$$

is used in lieu of (5-4.23) to observe the state of (5-4.22), instead of Theorem 4-2.1 the proof process will yield the following theorem.

Theorem 5-5.2 (Dai and Wang, 1987d). Let (E,A,B,C_1) be normalizable, assumptions 1-3 in Theorem 5-5.1 be satisfied, and $\text{rank}C_1 = r$. Then system (5-5.5) has an NOR of order $n+p-r$ of the form

$$\dot{x}_c = \bar{A}_c x_c + \bar{B}_c y$$
$$u = \bar{F}_c x_c + \bar{F} y.$$

Its proof is omitted here.

As for the normal state regulator, we have the following theorem.

Theorem 5-5.3. Assume that $D_1 = I_n$, $\text{rank}M = p$. System (5-5.5) with disturbance input $f(t)$ has a normal state regulator

$$\dot{x}_c = A_c x_c + B_c y$$
$$u = F_c x_c + Fy$$

if and only if assumptions 1-3 in Theorem 5-5.1 are satisfied.

Proof. Its proof is given by combining Theorem 5-5.1 with Theorem 5-4.2. Q.E.D.

It is worth pointing out that the output regulators (singular or normal) in either the last section or this section consist of a model of external disturbance, i.e., $\sigma(E_c, A_c) = \sigma(L) \cup \sigma(\bar{A})$, or $\sigma(A_c) = \sigma(L) \cup \sigma(\bar{A})$. Therefore, the compensators themselves are not stable. This is a property that must exist in order to achieve output regulation.

5-6. Notes and References

Singular compensators were first given in Wang (1984).

In this chapter, compensators are discussed parallel to linear system theory. For details of some results whose proofs are omitted here, Dai and Wang (1987b,d) would be useful.

CHAPTER 6

STRUCTURALLY STABLE COMPENSATION IN SINGULAR SYSTEMS

Mathematical models are often used to describe systems of interests in system analysis and synthesis. However, choosing the appropriate models is a somewhat difficult task even though it is important in control theory. Currently, there are two main approaches in obtaining the model: from the physical relationships among variables; and the identification method. The model itself, on the other hand, is only an approximation of the system, since any practical system is too complicated to be accurately described. Furthermore, influenced by many external factors such as environments or executive components, the coefficients for a system are by no means fixed; deviations exist in them. In this sense, coefficients in a system always have uncertainties. Here, we will call them perturbations, defined as the deviation from nominal points. In some cases where perturbation is serious, crucial problem may occur, as shown in the following example.

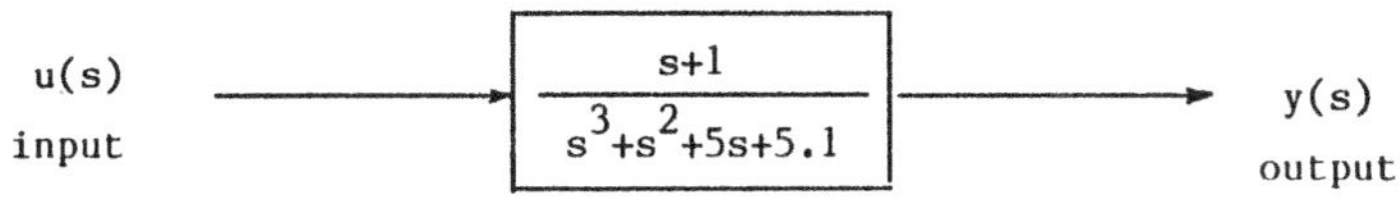

Figure 6.1

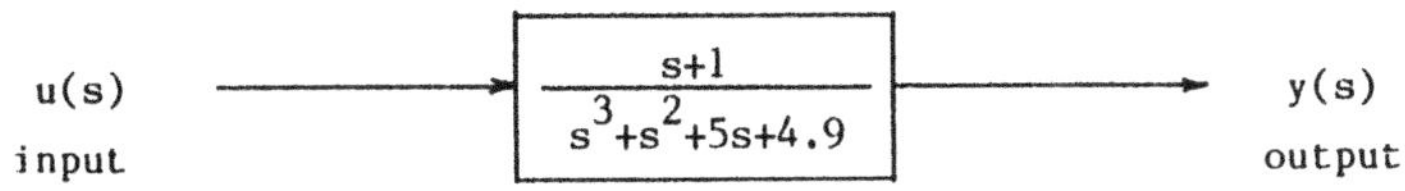

Figure 6.2

The system described by the input-output relation shown in Figure 6.1 is unstable according to the Routh criterion (Fortmann and Hitz, 1977). However, if errors exist in modeling and the system is mismodeled as shown in Figure 6.2. The system is stable.

Note that only a minor error 0.2 occurs between these two models to make a completely opposite conclusion on the stability of the system.

In this chapter, we will study the structural stability for singular systems with perturbations (or uncertainties, deviations, errors in other terms) in coefficient matrices, as well as their compensation strategies.

6-1. Structural Stability in Homogenous Singular Systems

It is wellknown that if the homogeneous normal system $\dot{x}(t) = Ax(t)$ where $x(t) \in \mathbb{R}^n$, $A \in \mathbb{R}^{n\times n}$ is stable, i.e., $\sigma(A) \subset \mathbb{C}^-$, a positive constant $\delta > 0$ exists such that

$$\sigma(A+\delta A) \subset \mathbb{C}^-$$

for arbitrary deviation δA in coefficient matrix A provided δA is within $\|\delta A\| < \delta$. Here $\|A\|$ is any consistent matrix norm. This property is termed structural stability here.

Thus a stable homogeneous system is structurally stable.

But this statement is unfortunately not applicable to the singular system

$$E\dot{x}(t) = Ax(t) \tag{6-1.1}$$

where $x(t) \in \mathbb{R}^n$, E, $A \in \mathbb{R}^{n\times n}$ are constant matrices. It is assumed that system (6-1.1) is regular and $q \triangleq \text{rank}E < n$. The stability of system (6-1.1) is sensitive to parameter perturbations, and thus makes the structural stability more complicated than the case in normal systems.

Example 6-1.1. Consider the singular system

$$\begin{bmatrix} 1 & 0 & 0 \\ 0 & 1 & 0 \\ 0 & 0 & 0 \end{bmatrix} \dot{x} = \begin{bmatrix} -1 & -1 & 0 \\ 0 & 0 & 1 \\ 0 & 1 & 0 \end{bmatrix} x. \tag{6-1.2}$$

It is stable with finite pole set $\sigma(E,A) = \{-1\} \subset \mathbb{C}^-$. When a perturbation

$$\delta A = \begin{bmatrix} 0 & 0 & 0 \\ 0 & 0 & 0 \\ 0 & 0 & -\varepsilon \end{bmatrix}, \qquad \varepsilon > 0 \text{ is a scalar,}$$

is present, we have

$$\sigma(E,A+\delta A) = \{ -1, \ \tfrac{1}{\varepsilon} \}.$$

Thus, the perturbed system $E\dot{x}(t) = (A+\delta A)x(t)$ has an unstable finite pole $\frac{1}{\varepsilon}$ (thus it is unstable) no matter how small ε (thus δA) is.

In system (6-1.1), E and A are called the structural parameters for the system. The deviation δA caused by any reason is termed perturbation in the parameter matrix A.

Definition 6-1.1. Data point $P\{E,A\}$ for system (6-1.1) is defined as a vector formed by listing the elements of E, A in arbitrary order.

For example, listing the elements in coefficient matrices in system (6-1.2) in the row order

$$P_1\{E,A\} = [1\ 0\ 0\ 0\ 1\ 0\ 0\ 0\ 0\ -1\ -1\ 0\ 0\ 0\ 1\ 0\ 1\ 0]^{\tau}$$

or in the column order

$$P_2\{E,A\} = [1\ 0\ 0\ 0\ 1\ 0\ 0\ 0\ 0\ -1\ 0\ 0\ -1\ 0\ 1\ 0\ 1\ 0]^{\tau},$$

both $P_1\{E,A\}$ and $P_2\{E,A\}$ are data points for system (6-1.2).

Data point may be either a column or a row vector. Here it is restricted to a column vector.

For any vector $a \in \mathbb{R}^n$, $\delta > 0$, the set

$$U_\delta(a) = \{ x \mid \|x-a\| < \delta ,\ x \in \mathbb{R}^n \}$$

is termed a neighborhood at a with radius δ, here $\|x\|$ is the consistent vector norm.

Definition 6-1.2. Assume that system (6-1.1) is stable, $\sigma(E,A) \subset \mathbb{C}^-$. If there exists a certain neighborhood $U_\delta(P\{E,A\})$ at data point $P\{E,A\}$ such that the stability of system (6-1.1) is preserved in the presence of arbitrary perturbations of structural parameters in this neighborhood, we will term the system (6-1.1) structurally stable at data point $P\{E,A\}$.

By definition, a structurally stable system is stable. But the inverse statement is not true, as shown in Example 6-1.1.

Theorem 6-1.1. Let system (6-1.1) be stable. It is structurally stable at data point $P\{E\}$ if and only if

$$\text{rank}E = n.$$

Proof. If rankE = n, system (6-1.1) becomes $\dot{x} = E^{-1}Ax$. Noticing that E is nonsingular in a certain neighborhood at $P\{E\}$, the previous system is structurally stable. This gives the sufficiency. Next we will prove the necessity.

Assume that system (6-1.1) is structurally stable at data point $P\{E\}$. If rankE $\neq$ n is false, rankE < n. Note that $\sigma(E,A) \subset \mathbb{C}^-$ (implying regularity). There must exist two nonsingular matrices Q and P such that system (6-1.1) is r.s.e. to EF1:

$$\dot{x}_1 = A_1 x_1$$
$$N\dot{x}_2 = x_2$$

where $x_1 \in \mathbb{R}^{n_1}$, $x_2 \in \mathbb{R}^{n_2}$, $n_1 + n_2 = n$, and

$$QEP = \mathrm{diag}(I_{n_1}, N), \qquad QAP = \mathrm{diag}(A_1, I).$$

Without loss of generality, we assume that N is in the form

$$N = \begin{bmatrix} 0 & 1 & & & \\ & 0 & 1 & & \\ & & \ddots & \ddots & \\ & & & \ddots & 1 \\ & & & & 0 \end{bmatrix} \in \mathbb{R}^{n_2 \times n_2}.$$

Let the perturbation be

$$\delta E = Q^{-1}\begin{bmatrix} 0_{n_1 \times n_1} & 0 & 0 \\ 0 & 0_{(n_2-1)\times 1} & 0 \\ 0 & \varepsilon & 0_{1\times(n_2-1)} \end{bmatrix} P^{-1} \in \mathbb{R}^{n\times n}, \qquad \delta A = 0,$$

where $\varepsilon > 0$ is a sufficiently small scalar. Then direct calculation gives the characteristic polynomial of perturbed system:

$$|s(E+\delta E) - A| = |Q^{-1}||P^{-1}||sI-A_1|(-1)^{n_2-1}(\varepsilon s^{n_2} - 1)$$

indicating that the perturbed system has an unstable pole $Re(s) > 0$ no matter how small $\varepsilon > 0$ is. Therefore, system (6-1.1) is not structurally stable, which conflicts with our assumption. Thus, rankE = n. Q.E.D.

This theorem shows that a singular system could not be structurally stable unless it becomes a normal one. Hence, painstaking care must be done in the determination of matrix E in system modeling so as to avoid false conclusion on stability to happen.

Theorem 6-1.2. Assume that system (6-1.1) is stable. It is structurally stable at data point P{A} if and only if

$$\deg(|sE-A|) = \mathrm{rank}E, \tag{6-1.3}$$

or in other words, the system has no infinite poles.

Proof. Sufficiency: If $\sigma(E,A) \subset \mathbb{C}^-$ and (6-1.3) holds, there must exist two nonsingular matrices $\tilde{Q}$ and $\tilde{P}$ such that

$$\tilde{Q}E\tilde{P} = \mathrm{diag}(I_q, 0), \qquad \tilde{Q}A\tilde{P} = \mathrm{diag}(A_1, I), \quad q = \mathrm{rank}E.$$

Denote

$$\delta A = \tilde{Q}^{-1}\begin{bmatrix} \delta A_{11} & \delta A_{12} \\ \delta A_{21} & \delta A_{22} \end{bmatrix}\tilde{P}^{-1}.$$

From the inequality

$$\| \delta A_{22} \| < \| [\delta A_{21}, \ \delta A_{22}] \| < \| \tilde{Q}\delta A\tilde{P} \| ,$$

we know that $\| \delta A_{22}\|$ is sufficiently small when $\| \delta A\|$ is. $(I+\delta A_{22})$ is nonsingular. Thus we have

$$|sE-A-\delta A| = |\tilde{Q}^{-1}||\tilde{P}^{-1}|\begin{bmatrix} sI_q-A_1-\delta A_{11} & -\delta A_{12} \\ -\delta A_{21} & -(I+\delta A_{22}) \end{bmatrix}$$

$$= |\tilde{Q}^{-1}||\tilde{P}^{-1}||-(I+\delta A_{22})||sI_q-A_1-\delta A_{11}-\delta A_{12}(I+\delta A_{22})^{-1}\delta A_{21}| .$$

Under the assumption $\sigma(A_1) = \sigma(E,A)\subset \mathbb{C}^-$, it is obvious that when δA is sufficiently small the perturbation

$$\delta A_{11} + \delta A_{12}(I+\delta A_{22})^{-1}\delta A_{21}$$

may be so small that

$$\sigma(A_1 + \delta A_{11}+ \delta A_{12}(I+\delta A_{22})^{-1} \delta A_{21}) \subset \mathbb{C}^-,$$

i.e., $\sigma(E,A+\delta A)\subset \mathbb{C}^-$, indicating that system (6-1.1) is structurally stable at data point $P\{A\}$.

Necessity: Let system (6-1.1) be structurally stable at $P\{A\}$, with $\sigma(E,A)\subset \mathbb{C}^-$. Since system (6-1.1) is regular, there exist nonsingular matrices Q and P such that it is r.s.e. to EF1:

$$\dot{x}_1 = A_1x_1, \qquad N\dot{x}_2 = x_2$$

where $x_1 \in \mathbb{R}^{n_1}$, $x_2 \in \mathbb{R}^{n_2}$, $n_1 + n_2 = n$, $N \in \mathbb{R}^{n_2 \times n_2}$ nilpotent. Without loss of generality, we assume

$$N = \begin{bmatrix} 0 & 1 & & & \\ & 0 & 1 & & \\ & & \ddots & \ddots & \\ & & & \ddots & 1 \\ & & & & 0 \end{bmatrix} \in \mathbb{R}^{n_2 \times n_2}$$

and the perturbation term δA

$$\delta A = Q^{-1}\begin{bmatrix} 0_{n_1\times n_1} & 0 \\ 0 & 0_{(n_2-1)\times n_2} \\ 0 & [\gamma, 0, \dots, 0] \end{bmatrix}P^{-1}.$$

Here $\gamma < 0$ is sufficiently small. In this case, we consider the polynomial

$$|Q||P||sE-(A+\delta A)| = |sI-A_1|\begin{bmatrix} -1 & s & & & \\ & -1 & s & & \\ & & \ddots & \ddots & \\ & & & \ddots & s \\ -\gamma & 0 & & & -1 \end{bmatrix}$$

$$= |sI-A_1|(-1)^{n_2}(\gamma s^{n_2-1}+1). \qquad (6\text{-}1.4)$$

Under the assumption of structural stability for system (6-1.1) at P{A}, $\sigma(E,A) \subset \mathbb{C}^-$ when $|\gamma| > 0$ is sufficiently small. (6-1.4) is a stable polynomial. Thus it must be $n_2 = 1$, or equivalently, N = 0, which is equivalent to deg(|sE-A|) = rankE. Q.E.D.

A simple but inaccurate explanation of this theorem is that system (6-1.1) is structurally stable at data point P{A} if and only if no impulse terms exist in its state response. This sparking result implies that when impulse terms are involved in the state response, the singular system is more sensitive to parameter variations.

Example 6-1.2. For system (6-1.2), we have deg(|sE-A|) = 1 < 2 = rankE < n = 3. By Theorem 6-1.2, this system is not structurally stable at either P{A} or P{E,A}, which coincides with the results of Example 6-1.1.

Example 6-1.3. Consider the system

$$E\dot{x} = Ax \qquad (6\text{-}1.5)$$

in which $x \in \mathbb{R}^3$ and

$$E = \begin{bmatrix} 1 & 0 & 0 \\ 0 & 0 & 0 \\ 0 & 0 & 0 \end{bmatrix}, \qquad A = \begin{bmatrix} -1 & 0 & 0 \\ 0 & 1 & 1 \\ 0 & 1 & 0 \end{bmatrix}.$$

For this system we have deg(|sE-A|) = deg(s+1) = 1 = rankE. Thus, system (6-1.5) is structurally stable at P{A} according to Theorem 6-1.2.

In a real system, the parameters often don't hold an equivalent position. Some parameters (often nonzero ones) may be uncertain, the perturbation occurs, while some are completely accurate, especially the zero elements. Furthermore, some elements may have deviations on only one direction, for example, with the increase of operating time of a mechanism, the temperature of its components always tend to increase instead of decrease. In this sense, perturbation in a system often subjects to constraints that may be generally described as

$$g(\delta E, \delta A) = 0. \qquad (6\text{-}1.6)$$

We define

$$U_1 = \{ P\{E+\delta E, A+\delta A\} \mid g(\delta E, \delta A) = 0 \} \qquad (6\text{-}1.7)$$

as a constrained neighborhood at data point P{E,A}, on which the structural stability has the following theorem.

Theorem 6-1.3 (Dai and Wang, 1987c). Let system (6-1.1) be stable and q_1 = deg(|sE-A|). If

$$U_1 = \{ P\{E+\delta E, A+\delta A\} \mid \deg(|s(E+\delta E) - (A+\delta A)|) - q_1 = 0 \}$$

for any sufficiently small $[\delta E, \delta A]$ and $P\{E+\delta E, A+\delta A\} \in U_1$, we have

$$\sigma(E+\delta E, A+\delta A) \subset \mathbb{C}^-.$$

In this case (6-1.1) is called constrained structurally stable at data point $P\{E,A\}$ with constraints U_1.

The condition given in this theorem is only sufficient. The perturbed system (6-1.1) may also be stable for perturbation revoking this condition as shown in the following Example 6-1.5.

Example 6-1.4. For system (6-1.2), $\deg(|sE-A|) = 1 < \mathrm{rank}E = 2$. Thus it is not structurally stable at $P\{A\}$ by Theorem 6-1.2. However, let the perturbation be in the form of

$$\delta E = 0, \qquad \delta A = \begin{bmatrix} \delta a_{11} & \delta a_{12} & \delta a_{13} \\ \delta a_{21} & \delta a_{22} & \delta a_{23} \\ \delta a_{31} & \delta a_{32} & 0 \end{bmatrix}. \tag{6-1.8}$$

When the perturbation δA is sufficiently small, $|\delta a_{32}| < 1$, $|\delta a_{23}| < 1$ and the equation

$$\deg(|s(E+\delta E) - (A+\delta A)|) = \deg\left(\left|\begin{bmatrix} s+1-\delta a_{11} & 1-\delta a_{12} & -\delta a_{13} \\ -\delta a_{21} & s-\delta a_{22} & -1-\delta a_{23} \\ -\delta a_{31} & -1-\delta a_{32} & 0 \end{bmatrix}\right|\right)$$

$$= 1 = \deg(|sE-A|)$$

holds. Therefore, by Theorem 6-1.3 the perturbed system $(E+\delta E)\dot{x} = (A+\delta A)x$ is stable for such perturbation provided it is sufficiently small.

Although the perturbation is constrained in this example, its wide form enables it to represent most of the real cases.

Example 6-1.5. Consider the system (6-1.2) with perturbation

$$\delta E = 0, \qquad \delta A = \begin{bmatrix} \delta a_{11} & \delta a_{12} & \delta a_{13} \\ 0 & \delta a_{22} & \delta a_{23} \\ 0 & \delta a_{32} & \delta a_{33} \end{bmatrix}, \qquad \delta a_{33} > 0. \tag{6-1.9}$$

Its characteristic polynomial

$$|s(E+\delta E)-(A+\delta A)|$$

$$= (s+1-\delta a_{11})(-1)(\delta a_{33}s+1+\delta a_{23}+\delta a_{32}+\delta a_{23}\delta a_{32}-\delta a_{22}\delta a_{33}).$$

together with $\delta a_{33} > 0$, clearly shows $\sigma(E+\delta E), A+\delta A) \subset \mathbb{C}^-$ provided δA is sufficiently small to satisfy

$$|\delta a_{11}| < 1, \qquad |\delta a_{23}+\delta a_{32}+\delta a_{23}\delta a_{32}-\delta a_{22}\delta a_{33}| < 1.$$

This is a single-direction perturbation problem since $\delta a_{33} > 0$ is constrained. This example shows that a perturbed system may be stable even though Theorem 6-1.3 is not applicable here.

Theorem 6-1.3 is inconvenient for use for its complicated form. Of interests often is its special form such as these in Theorems 6-1.1 and 6-1.2. A further result is the following corollary.

Corollary 6-1.1. Let system (6-1.1) be stable, rankE = deg(|sE-A|), and

$$U_2 = \{ P\{E+\delta E, A+\delta A\} \mid \text{rank}(E+\delta E) = \text{rank}E \}$$

Then system (6-1.1) is structurally stable on U_2.

The combination of this corollary with Theorem 6-1.2 shows that system (6-1.1) is structurally stable if the slow-fast subsystem structure is preserved in the presence of perturbation. This property is in accordance with common sense.

The physical sense of Theorem 6-1.3 lies in that the stability of a stable singular system is preserved if the perturbation doesn't create a new pole (or the number of poles remains the same), which is often unstable.

6-2. Compensation Function on Structural Stability via Dynamic Feedback

As discussed in the last chapter, under certain conditions we may design dynamic compensators for singular systems. But we must be aware of the fact that the design procedure is applicable only to the case of accurate coefficient matrices. Since uncertainties are unavoidable to exist in any real system, the compensators should not only finish the stabilization achievement but also guarantee the closed-loop systems to be structurally stable, making them applicable in real control system design.

For simplicity and practical significance, in this chapter only normal compensators are discussed.

Consider the singular system

$$\begin{aligned} E\dot{x}(t) &= Ax(t) + Bu(t) \\ y(t) &= Cx(t) \end{aligned} \tag{6-2.1}$$

where $x(t) \in \mathbb{R}^n$, $y(t) \in \mathbb{R}^r$, $u(t) \in \mathbb{R}^m$ are its state, measure output, and control input, respectively; E, $A \in \mathbb{R}^{n\times n}$, $B \in \mathbb{R}^{n\times m}$, $C \in \mathbb{R}^{r\times n}$ are constant matrices. The system is assumed to be regular with singular E, and the normal dynamic compensator:

$$\begin{aligned} \dot{x}_c(t) &= A_c x_c(t) + B_c y(t) \\ u(t) &= F_c x_c(t) + Fy(t) \end{aligned} \tag{6-2.2}$$

where $x_c(t) \in \mathbb{R}^{n_c}$, A_c, B_c, F_c, and F are constant matrices of appropriate dimensions.

System (6-2.1) and compensator (6-2.2) form the closed-loop system

$$\begin{bmatrix} E & 0 \\ 0 & I \end{bmatrix}\begin{bmatrix} \dot{x} \\ \dot{x}_c \end{bmatrix} = \begin{bmatrix} A+BFC & BF_c \\ B_cC & A_c \end{bmatrix}\begin{bmatrix} x \\ x_c \end{bmatrix}. \tag{6-2.3}$$

Definition 6-2.1. Consider system (6-2.1) and its compensator (6-2.2). The closed-loop system (6-2.3) is assumed to be stable, i.e.,

$$\sigma\left(\begin{bmatrix} E & 0 \\ 0 & I \end{bmatrix}, \begin{bmatrix} A+BFC & BF_c \\ B_cC & A_c \end{bmatrix}\right) \subset \mathbb{C}^-. \tag{6-2.4}$$

The compensator (6-2.2) will be termed a structurally stable normal compensator (SSNC) at data point P{E,A,B,C} if the closed-loop system is structurally stable at P{E,A,B,C}; It will be termed a constrained SSNC at P{E,A} with constraints U_1 if (6-2.3) is constrained structurally stable at P{E,A} with constraints U_1.

Theorem 6-2.1. Compensator (6-2.2) is an SSNC at data point P{A,B,C} for system (6-2.1) if and only if the closed-loop system (6-2.3) is stable and

$$\deg\left(\left|\begin{bmatrix} E & 0 \\ 0 & I \end{bmatrix}s - \begin{bmatrix} A+BFC & BF_c \\ B_cC & A_c \end{bmatrix}\right|\right) = n_c + \text{rank}E.$$

Proof. Sufficiency is shown in Theorem 6-1.2. Now we will prove necessity. For any regular system (6-2.1), there exist two nonsingular matrices Q and P such that

$$QEP = \text{diag}(I_{n_1}, N), \qquad QAP = \text{diag}(A_1, I_{n_2})$$

where $n_1 + n_2 = n$, $A_1 \in \mathbb{R}^{n_1 \times n_1}$, $N \in \mathbb{R}^{n_2 \times n_2}$ is nilpotent. For the sake of simplicity in writing, and without loss of generality, we assume

$$N = \begin{bmatrix} 0 & 1 & & & \\ & 0 & 1 & & \\ & & \ddots & \ddots & \\ & & & \ddots & 1 \\ & & & & 0 \end{bmatrix} \in \mathbb{R}^{n_2 \times n_2}.$$

Let the perturbation be

$$\delta A = Q^{-1}\text{diag}(0, \ \delta A_{22})P^{-1} \in \mathbb{R}^{n \times n}, \qquad \delta A_{22} = [\varepsilon, 0, \ldots, 0] \in \mathbb{R}^{1 \times n_2}.$$

Then in the presence of such perturbation, the perturbed closed-loop ssystem has the characteristic polynomial

$$\Delta(s) \triangleq \left|\begin{bmatrix} E & 0 \\ 0 & I \end{bmatrix}s - \begin{bmatrix} A+\delta A+BFC & BF_c \\ B_cC & A_c \end{bmatrix}\right|$$

$$= |Q^{-1}P^{-1}| \begin{vmatrix} sI-(A_1+B_1FC_1) & -B_1FC_2 & -B_1F_c \\ -B_2FC_1 & sN-(I+\delta A_{22}+B_2FC_2) & -B_2F_c \\ -B_cC_1 & -B_cC_2 & sI-A_c \end{vmatrix}. \qquad (6\text{-}2.5)$$

Let

$$B_2FC_2 = \begin{bmatrix} b_{11} & b_{12} & \cdots & b_{1n_2} \\ b_{21} & b_{22} & \cdots & b_{2n_2} \\ & \cdots & \cdots & \\ b_{n_21} & b_{n_22} & \cdots & b_{n_2n_2} \end{bmatrix}.$$

Next we consider the term of the highest order (HOT), which is

$$\text{HOT}(\Delta(s))$$

$$= |Q^{-1}P^{-1}| \, \text{HOT}(|sI - (A_1+B_1FC_1)|) \cdot \text{HOT}(|sN-(I+\delta A_{22}+B_2FC_2)|)\text{HOT}(|sI-A_c|)$$

$$= |Q^{-1}P^{-1}| \, \text{HOT}(|sN-(I+\delta A_{22}+B_2FC_2)|)s^{n_1+n_c}. \qquad (6\text{-}2.6)$$

Since

$$\text{HOT}(|sN - (I+\delta A_{22}+B_2FC_2)|)$$

$$= \text{HOT}\left(\left|\begin{bmatrix} -1 & s & & \\ & -1 & s & \\ & & \ddots & s \\ -\varepsilon & 0 & 0 \cdots & -1 \end{bmatrix} - \begin{bmatrix} b_{11} & b_{12} & \cdots & b_{1n_2} \\ \cdots & \cdots & \cdots & \\ b_{n_21} & b_{n_22} & \cdots & b_{n_2n_2} \end{bmatrix}\right|\right)$$

$$= (-\varepsilon - b_{n_21})s^{n_2-1}$$

from (6-2.6) we know the term of the highest order in $\Delta(s)$ is

$$(-\varepsilon - b_{n_21})s^{n+n_c-1}. \qquad (6\text{-}2.7)$$

On the other hand, it is assumed that the closed-loop system is stable, i.e., (6-2.4) holds. Therefore

$$\alpha = \left| - \begin{bmatrix} A+\delta A+BFC & BF_c \\ B_cC & A_c \end{bmatrix} \right|$$

is either a positive or negative scalar for all sufficiently small perturbations ε (thus, δA). Without loss of generality we assume $\alpha > 0$. Combining it with (6-2.7) we have

$$\Delta(s) = (-\varepsilon - b_{n_21})s^{n+n_c-1} + \dots + \alpha . \qquad (6\text{-}2.8)$$

Under the assumption of structural stability for the closed-loop system at $P\{A,B,C\}$, $\Delta(s)$ is a stable polynomial for any sufficiently small ε, either positive or negative. Thus, by noticing $\alpha > 0$, we know

$$-b_{n_{21}} > 0.$$

In this case we have

$$\deg\left(\left|\begin{bmatrix} E & 0 \\ 0 & I \end{bmatrix} s - \begin{bmatrix} A+BFC & BF_c \\ B_cC & A_c \end{bmatrix}\right|\right) = n_c + \text{rank}E,$$

which is the condition in the theorem. Q.E.D.

Example 6-2.1. According to Theorem 6-2.1, it may be testified that the compensators given in Examples 5-2.1, 5-3.1, and 5-3.2 are all SSNCs at P{A,B,C} for systems (5-1.8), (5-3.9), and (5-3.20), respectively.

Combining Theorem 6-2.1 with Theorem 5-3.2, we have the following theorem.

Theorem 6-2.2. System (6-2.1) has an SSNC (6-2.2) at data point P{A,B,C} if and only if it is stabilizable, detectable, impulse controllable, and impulse observable.

Therefore, the concepts of impulse controllability and impulse observability have a close relationship with structural stability. This theorem states only the existence of SSNCs in the form of (6-2.2). The design method may follow that in Chapter 5.

The combination of Theorems 6-1.2 and 6-2.1 yields the following corollary.

Corollary 6-2.1. If (6-2.2) is an SSNC at data point P{A,B,C} for system (6-2.1), the closed-loop system (6-2.3) is definitely structurally stable at data point $P\{A,B,C,A_c,B_c,F_c,F\}$.

In engineering, the compensators may be realized either using hardwares such as electric components, control meters, etc., or on a computer; In the former case, deviations occur due to factors, such as aging of components, admissible error in manufacturing, and environmental influence; in the latter case, uncertainties exist because of the restriction on word length. Therefore, perturbation always exist in the realization of compensators. What is stated in Corollary 6-2.1 is that some sufficiently small perturbations are allowed without revoking stability. This endows the SSNC with the practical significance.

Similar to the proof process of Theorem 6-2.1, it is easy to prove the following theorem.

Theorem 6-2.3. The normal compensator (6-2.2) could not be an SSNC at P{E} for system (6-2.1).

This fact is unfavorable for system design. This nevertheless weakens the value of usage of compensator (6-2.2).

To have a further understanding of this point, we examine the singular system in the EF2:

$$\dot{x}_1 = A_{11}x_1 + A_{12}x_2 + B_1u \tag{6-2.9a}$$

$$0 = A_{21}x_1 + A_{22}x_2 + B_2u \tag{6-2.9b}$$

$$y = C_1x_1 + C_2x_2.$$

In this EF2, dynamic equation (6-2.9a) may be sometimes viewed as a composite system formed by subsystems; meanwhile, the algebraic equation (6-2.9b) characterizes the interconnection among these subsystems. In the system modeling, provided we have a fair examination of system structure so as not to mistake the dynamic equation for the algebraic one, the matrix will suffers no perturbation; in case if happens, the perturbation should be the constrained perturbation with the constraints of fixed rank

$$\text{rank}(E+\delta E) = \text{rank}E.$$

Just as the order of a system should be determined in advance in the modeling of normal systems, in the analysis of singular systems, we should have the ability to separate the dynamic part from the algebraic part. This is a basic task in system analysis. In this sense, the closed-loop system (6-2.3) is structurally stable by Corollary 6-2.1.

Concerning the constrained perturbation, we have the following theorem.

Theorem 6-2.4. Let closed-loop system (6-2.3) be stable and

$$n_c + q_1 \triangleq \deg\left(\left|\begin{bmatrix} E & 0 \\ 0 & I \end{bmatrix} s - \begin{bmatrix} A+BFC & BF_c \\ B_cC & A_c \end{bmatrix}\right|\right).$$

Then (6-2.2) is a constrained SSNC at data point $P\{E,A,B,C\}$ with constraints U_1:

$$U_1 = \{P\{E+\delta E, A+\delta A, B+\delta B, C+\delta C\} \mid n_c + q_1 =$$

$$= \deg\left(\left|\begin{bmatrix} E+\delta E & 0 \\ 0 & I \end{bmatrix} s - \begin{bmatrix} A+\delta A+(B+\delta B)F(C+\delta C) & (B+\delta B)F_c \\ B_c(C+\delta C) & A_c \end{bmatrix}\right|\right)\}.$$

6-3. Structurally Stable Normal Compensators

As shown in Theorem 6-2.3, normal compensators of the form of (6-2.2) cannot be structurally stable for system (6-2.1) at data point P{E,A,B,C}. To construct an SSNC for a singular system at all structural parameters P{E,A,B,C}, we must consider compensators of other forms. If the derivative of y(t) is available, we may consider compensators of the following form

$$\begin{aligned} \dot{x}_c(t) &= A_c x_c(t) + B_c y(t) + H_c \dot{y}(t) \\ u(t) &= F_c x_c(t) + Fy(t) + H\dot{y}(t) \end{aligned} \tag{6-3.1}$$

where $x_c \in \mathbb{R}^{n_c}$, A_c, B_c, H_c, F_c, F, and H are constant matrices of appropriate dimensions.

First, if we denote $\bar{x}_c(t) = x_c(t) - H_c y(t)$, compensator (6-3.1) may be written as

$$\begin{aligned} \dot{\bar{x}}_c(t) &= A_c \bar{x}_c(t) + (B_c + A_c H_c) y(t) \\ u(t) &= F_c \bar{x}_c(t) + (F + F_c A_c H_c) y(t) + H\dot{y}(t). \end{aligned} \tag{6-3.2}$$

The dynamic equation doesn't involve the differential term y(t). Therefore, without loss of generality, we may assume that $H_c = 0$ in (6-3.1), so as to become

$$\begin{aligned} \dot{x}_c(t) &= A_c x_c(t) + B_c y(t) \\ u(t) &= F_c x_c(t) + Fy(t) + H\dot{y}(t). \end{aligned} \tag{6-3.3}$$

<u>Theorem 6-3.1</u>. Compensator (6-3.2) is an SSNC for system (6-3.1) at data point P{E,A,B,C} if and only if the following two conditions are fulfilled:

(i). System (6-2.1) is stabilizable and detectable.

(ii). The system is normalizable and dual normalizable.

<u>Proof</u>. Necessity: The compensator (6-3.2) and system (6-2.1) form the closed-loop system

$$\begin{bmatrix} E-BHC & 0 \\ 0 & I \end{bmatrix} \begin{bmatrix} \dot{x} \\ \dot{x}_c \end{bmatrix} = \begin{bmatrix} A+BFC & BF_c \\ B_c C & A_c \end{bmatrix} \begin{bmatrix} x \\ x_c \end{bmatrix}. \tag{6-3.3}$$

By Theorem 6-2.3, the closed-loop system (6-3.3) is structurally stable at data point P{E} (thus at P{E-BHC}) if and only if rank diag(E-BHC, I_{n_c}) = n + n_c or rank(E-BHC) = n. Thus system (6-2.1) is normalizable and dual normalizable.

(i) may be easily obtained from the stability of (6-3.3).

Sufficiency: Sufficiency is proven in a constructive way. Let (i).hold. A matrix $H \in \mathbb{R}^{m \times r}$ may be chosen so that (Lemma 3-5.1)

$$|E-BHC| \neq 0. \tag{6-3.4}$$

Choosing feedback control

$$u = H\dot{y} + v \tag{6-3.5}$$

where $v(t)$ is the new input, feedback control (6-3.5) and system (6-2.1) form the closed-loop system

$$(E-BHC)\dot{x} = Ax + Bv$$
$$y = Cx.$$

Noticing (6-3.4), this system becomes

$$\begin{aligned} \dot{x} &= \bar{A}x + \bar{B}v \\ y &= Cx \end{aligned} \tag{6-3.6}$$

where

$$\bar{A} = (E-BHC)^{-1}A, \qquad \bar{B} = (E-BHC)^{-1}B. \tag{6-3.7}$$

Under the assumption (i), normal system (6-3.6) is both stabilizable and detectable. Therefore a matrix K_1 exists satisfying

$$\sigma(\bar{A}+\bar{B}K_1) \subset \mathbb{C}^-. \tag{6-3.8}$$

Thus, the feedback control

$$v = K_1x \tag{6-3.9}$$

when applied to (6-3.6) stabilizes its closed-loop system. To physically realize this control law, we now consider constructing an observer for system (6-3.6). Since (6-3.6) is detectable, there exists a matrix $G \in \mathbb{R}^{nxr}$ so that

$$\sigma(\bar{A}-GC) \subset \mathbb{C}^-.$$

Thus, the system

$$\begin{aligned} \dot{x}_c &= (\bar{A}-GC)x_c + \bar{B}u + Gy \\ w &= x_c \end{aligned} \tag{6-3.10}$$

is a state observer for system (6-3.6) such that

$$\lim_{t\to\infty} (x - w) = 0, \qquad \forall\ x(0),\ x_c(0).$$

By using the control $v = K_1x_c$ in lieu of (6-3.9), and from the preceding equation and feedback control (6-3.5) and system (6-3.10), we have

$$\begin{aligned} \dot{x}_c &= (\bar{A}-GC+\bar{B}K_1)x_c + Gy \\ u &= K_1x_c + H\dot{y}, \end{aligned} \tag{6-3.11}$$

which, when applied to system (6-2.1), forms the closed-loop system

$$\begin{bmatrix} E-BHC & 0 \\ 0 & I \end{bmatrix}\begin{bmatrix} \dot{x} \\ \dot{x}_c \end{bmatrix} = \begin{bmatrix} A & BK_1 \\ GC & \bar{A}-GC+\bar{B}K_1 \end{bmatrix}\begin{bmatrix} x \\ x_c \end{bmatrix}.$$

That is

$$\begin{bmatrix} \dot{x} \\ \dot{x}_c \end{bmatrix} = \begin{bmatrix} \bar{A} & \bar{B}K_1 \\ GC & \bar{A}-GC+\bar{B}K_1 \end{bmatrix}\begin{bmatrix} x \\ x_c \end{bmatrix}, \tag{6-3.12}$$

which is stable (thus structurally stable) with finite pole set $\sigma(\bar{A}+\bar{B}K_1) \cup \sigma(\bar{A}-GC) \subset \mathbb{C}^-$. Hence, (6-3.11) is in the form of (6-3.2) and an SSNC for ssytem (6-2.1) at data point $P\{E,A,B,C\}$. Q.E.D.

The constructive proof of this theorem provides a design method of SSNCs of the form of (6-3.2).

Example 6-4.1. It is easy to testify that the singular system

$$\begin{bmatrix} 0 & 1 & 0 \\ 0 & 0 & 1 \\ 0 & 0 & 0 \end{bmatrix}\dot{x} = \begin{bmatrix} 1 & 0 & 0 \\ 0 & 1 & 0 \\ 0 & 0 & 1 \end{bmatrix}x + \begin{bmatrix} 0 & 0 \\ 1 & 0 \\ 0 & 1 \end{bmatrix}u$$
$$y = [1 \quad 0 \quad 0]x \tag{6-3.13}$$

is stabilizable, detectable, normalizable, and dual normalizable.

Choosing $H = [0 \quad 1]^{\tau}$, the feedback $u = H\dot{y} + v$ and system (6-3.13) form the closed-loop system

$$\dot{x} = \begin{bmatrix} 0 & 0 & -1 \\ 1 & 0 & 0 \\ 0 & 1 & 0 \end{bmatrix}x + \begin{bmatrix} 0 & -1 \\ 0 & 0 \\ 1 & 0 \end{bmatrix}v = \bar{A}x + \bar{B}v$$
$$y = [1 \quad 0 \quad 0]x = Cx. \tag{6-3.14}$$

Let

$$K_1 = \begin{bmatrix} 0 & 0 & 0 \\ 3 & 3 & 0 \end{bmatrix}, \qquad G = [3 \quad 0 \quad -3]^{\tau}.$$

It will be $\sigma(\bar{A}+\bar{B}K_1) = \{-1, -1, -1\}$, $\sigma(\bar{A}-GC) = \{-1, -1, -1\}$. Now we construct for system (6-3.14) the state observer

$$\dot{x}_c = \begin{bmatrix} -3 & 0 & -1 \\ 1 & 0 & 0 \\ 3 & 1 & 0 \end{bmatrix}x_c + \begin{bmatrix} 0 & -1 \\ 0 & 0 \\ 1 & 0 \end{bmatrix}v + \begin{bmatrix} 3 \\ 0 \\ -3 \end{bmatrix}y$$

and the control input $v = K_1x_c$ to form the system

$$\dot{x}_c = \begin{bmatrix} -6 & -3 & -1 \\ 1 & 0 & 0 \\ 3 & 1 & 0 \end{bmatrix}x_c + \begin{bmatrix} 3 \\ 0 \\ -3 \end{bmatrix}y$$

$$u = \begin{bmatrix} 0 & 0 & 0 \\ 3 & 3 & 0 \end{bmatrix}x_c + \begin{bmatrix} 0 \\ 1 \end{bmatrix}\dot{y}$$

which is an SSNC for system (6-3.13) at all structural parameters by Theorem 6-3.1.

It has been shown in Theorem 6-2.3 that for a general system (6-2.1), the compensators of the form of (6-2.2) is at most a constrained SSNC at data point P{E,A} , especially referring to E. However, the compensators constructed by using both measure and derivative may assure the structural stability of the closed-loop system at all structural parameters. Thus, to achieve the structural stability property, measuring mechanism must be added to obtain both output and derivative. This is the cost for ensuring the structural stability of the closed-loop system at all structural parameters.

It is widely noted that the differentials always amplify high frequency noises. In engineering, derivatives of a signal are often avoided for this point. Therefore, in system design, compensators of the form (6-3.2) should be avoided as far as possible, especially the case where noises exist in measure. Nevertheless, in some cases, $\dot{y}(t)$ may be available. For example, when y(t) represents the position, $\dot{y}(t)$ is the speed that may be obtained by a speed meter.

As a matter of fact, system (6-3.2) is a simple generalization of PID control in linear control system theory.

The compensation principle of (6-3.2) is shown in Figure 6-3.1.

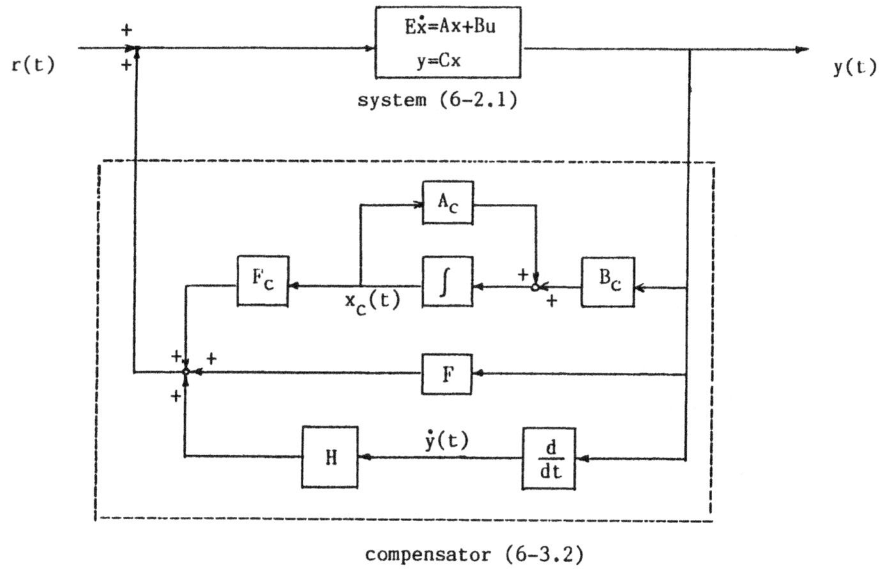

Figure 6-3.1. Structurally Stable Normal Compensator (6-3.2)

As seen in the proving process of Theorem 6-3.1, if the sufficient conditions are satisfied and rank$C = r$, instead of constructing a full-order state observer (6-3.10) we may construct a Luenberger observer for system (6-3.6) of order $n-r$:

$$\begin{aligned} \dot{x}_c &= (\bar{A}_{22}-\bar{G}_2\bar{A}_{12})x_c + [\bar{A}_{21}+\bar{G}_2\bar{A}_{11}+(\bar{A}_{22}-\bar{G}_2\bar{A}_{12})\bar{G}_2]y + (\bar{B}_2-\bar{G}_2\bar{B}_1)v \\ w &= \bar{P}^{-1}[0/I]x_c + \bar{P}^{-1}[I_r/\bar{G}_2]y \end{aligned} \qquad (6\text{-}3.15)$$

where $x_c \in \mathbb{R}^{n-r}$, $w \in \mathbb{R}^n$, and $\bar{P} = [C/\bar{C}]$, $\bar{C}$ is any $(n-r)\times n$ matrix satisfying rank$\bar{P} = n$,

$$\bar{P}A\bar{P}^{-1} = \begin{bmatrix} \bar{A}_{11} & \bar{A}_{12} \\ \bar{A}_{21} & \bar{A}_{22} \end{bmatrix}, \qquad \bar{P}B = \begin{bmatrix} \bar{B}_1 \\ \bar{B}_2 \end{bmatrix}, \qquad C\bar{P}^{-1} = [I_r \quad 0] \qquad (6\text{-}3.16)$$

$\bar{G}_2$ satisfies $\sigma(\bar{A}_{22}-\bar{G}_2\bar{A}_{12}) \subset \mathbb{C}^-$, such that

$$\lim_{t\to\infty} (x - w) = 0, \ \forall\ x(0),\ x_c(0).$$

Similarly, here we also use the control $v = K_1 w$ in lieu of (6-3.9). From (6-3.16), we have

$$\begin{aligned} \dot{x}_c &= A_c x_c + B_c y \\ v &= F_c x_c + Fy \end{aligned} \qquad (6\text{-}3.17)$$

where

$$\begin{aligned} A_c &= \bar{A}_{22} - \bar{G}_2\bar{A}_{12} + (\bar{B}_2-\bar{G}_2\bar{B}_1)K_1\bar{P}^{-1}[0/I_{n-r}] \\ B_c &= \bar{A}_{21} - \bar{G}_2\bar{A}_{11} + (\bar{A}_{22}-\bar{G}_2\bar{A}_{12})\bar{G}_2 + (\bar{B}_2-\bar{G}_2\bar{B}_1)K_1\bar{P}^{-1}[I_r/\bar{G}_2] \\ F_c &= K_1\bar{P}^{-1}[0/I_{n-r}] \\ F &= K_1\bar{P}^{-1}[I_r/\bar{G}_2]. \end{aligned}$$

It is easy to testify that (6-3.17) and (6-3.6) form the normal closed-loop system

$$\begin{bmatrix} \dot{x} \\ \dot{x}_c \end{bmatrix} = \begin{bmatrix} \bar{A}+\bar{B}FC & \bar{B}F_c \\ B_cC & A_c \end{bmatrix}\begin{bmatrix} x \\ x_c \end{bmatrix},$$

which is not only stable but also structurally stable at data point $P\{E,A,B,C\}$.

Therefore, the system

$$\begin{aligned} \dot{x}_c &= A_c x_c + B_c y \\ u &= F_c x_c + Fy + H\dot{y} \end{aligned} \qquad (6\text{-}3.18)$$

proven in such a way is an SSNC for system (6-2.1) at all structural parameters.

Summing up the preceding discussion, we have proven the following theorem.

<u>Theorem 6-3.2</u>. Assume that system (6-2.1) is stabilizable, detectable, norma-

lizable, and dual normalizable, rankC = r. Then it has a reduced-order SSNC (6-3.18) of order n-r at data point P{E,A,B,C}.

Example 6-3.2. Consider system (6-3.13). If the following Luenberger observer is used instead of the full-order observer in Example 6-3.1 for system (6-3.14):

$$\dot{x}_c = \begin{bmatrix} 0 & -1 \\ 1 & -1 \end{bmatrix} x_c + \begin{bmatrix} 2 \\ 0 \end{bmatrix} y + \begin{bmatrix} 0 & -1 \\ 1 & -1 \end{bmatrix} v$$

$$w = \begin{bmatrix} 0 & 0 \\ 1 & 0 \\ 0 & 1 \end{bmatrix} x_c + \begin{bmatrix} 1 \\ -1 \\ -1 \end{bmatrix} y$$

and choose the control input $v = K_1 w$ we obtain the compensator

$$\dot{x}_c = \begin{bmatrix} -3 & -1 \\ -2 & -1 \end{bmatrix} x_c + \begin{bmatrix} 2 \\ 0 \end{bmatrix} y$$

$$u = \begin{bmatrix} 0 & 0 \\ 3 & 0 \end{bmatrix} x_c + \begin{bmatrix} 0 \\ 1 \end{bmatrix} \dot{y},$$

which is in the form of (6-3.2). According to Theorem 6-3.2, it is an SSNC for system (6-3.13) at data point P{E,A,B,C}, of the lower order of 2.

6-4. Structurally Stable Normal Compensators in Output Regulation Systems

For a system with disturbance input, its compensators stabilizing the closed-loop system should have twofold properties: the stability, especially structural stability, and the external disturbance resistibility (output regulation). Thus, in this section we will study the compensators in regulation systems that possess the previously mentioned properties.

Consider the following system with disturbance input:

$$\begin{aligned} E\dot{x} &= Ax + Bu + Mf \\ \dot{f} &= Lf \\ y &= C_1 x + C_2 f \\ z &= D_1 x + D_2 f \end{aligned} \tag{6-4.1}$$

where $x \in \mathbb{R}^n$ is its state, $u \in \mathbb{R}^m$ is its control input, $y \in \mathbb{R}^r$ is its measure output, $z \in \mathbb{R}^d$ is the vector to be regulated, and $f \in \mathbb{R}^p$ is external disturbance. E, A, B, M, L, C_1, C_2, D_1, and D_2 are constant matrices of appropriate dimensions. It is assumed that (E,A) is regular and $\sigma(L) \subset \bar{\mathbb{C}}^+$.

Let the compensator be of the form:

$$\dot{x}_c = A_c x_c + B_c y$$
$$u = F_c x_c + Fy \tag{6-4.2}$$

where $x_c \in \mathbb{R}^{n_c}$, A_c, B_c, F_c, and F are constant matrices of appropriate dimensions. Assume that it is a normal output regulator for system (6-4.1). Therefore, its closed-loop system is internally stable and output regulation.

Definition 6-4.1. Consider system (6-4.1) and its output regulator (6-4.2). If the internal stability and output regulation are preserved in the presence of perturbations in a certain neighborhood at data point P{E,A,B,C}, we will call (6-4.2) a structurally stable normal output regulator (SSNOR) for system (6-4.1) at data point P{E,A,B,C}.

Thus, according to the definition, an SSNOR has the regulation property to resist both disturbance input and perturbation in parameters.

Definition 6-4.2. We say that z is readable from y if there exists a matrix $S \in \mathbb{R}^{dxr}$ such that z = Sy, or $D_1 = SC_1$ and $D_2 = SC_2$. Without loss of generality, we assume that $S = [0 \quad I_d]$, or in other words, z is part of y.

Definition 6-4.3. Consider systems (6-4.1) and (6-4.2).

1. If there exists an $n_c x n_c$ nonsingular matrix T such that

$$T^{-1}A_cT = \begin{bmatrix} A_{c_1} & A_{c_3} \\ 0 & A_{c_2} \end{bmatrix} \tag{6-4.3}$$

where $A_{c_1} \in \mathbb{R}^{n_{c1} x n_{c1}}$, $A_{c_2} \in \mathbb{R}^{n_{c2} x n_{c2}}$, $A_{c_3} \in \mathbb{R}^{n_{c1} x n_{c2}}$, $n_{c_1} + n_{c_2} = n_c$, and A_{c_2} has at least d invariant factors that can be divided by the minimal polynomial of L, we will say that (6-4.2) incorporates an internal model of L.

2. Let (6-4.2) incorporate an internal model of L. If

$$\text{rank}\begin{bmatrix} sI-A_c \\ F_c \end{bmatrix} = n_c, \qquad \forall \quad s \in \sigma(L),$$

we will say that the model is observable from u.

3. Let (6-4.2) incorporate an internal model of L. If

$$T^{-1}B_c = \begin{bmatrix} \overset{m-d}{B_{c_1}} & \overset{d}{B_{c_3}} \\ 0 & B_{c_2} \end{bmatrix} \begin{matrix} \} \; n_{c_1} \\ \} \; n_{c_2} \end{matrix} \tag{6-4.4}$$

and

$$\text{rank}[sI-A_{c_2}, B_{c_2}] = n_{c_2}, \; \forall \; s \in \sigma(L). \tag{6-4.5}$$

we will say that the model is controllable about z.

Remark 1. System (6-4.2) is always unstable if it incorporates an internal model of L since $\sigma(L) \subset \bar{C}^+$.

For instance, let the invariants of L be 1, (s-1), (s-1), and the invariants of A_c are s+1, (s-1), ... , (s-1) (of number d). Then A_c incorporates an internal model of L.

Proofs of the results in this section is somewhat lengthy and tedious, so we will only introduce fundamental results without any proofs that may be found in Dai and Wang (1987d).

Theorem 6-4.1. Consider system (6-4.1) and its normal compensator (6-4.2). If their closed-loop system is internally stable, it must be

1. (E,A,B,C_1) is stabilizable and detectable.
2. (A_c,B_c,F_c) is stabilizable and detectable.

On one hand, there exist perturbations in their parameters; on the other hand, to be an SSNOR for system (6-4.1), the compensator (6-4.2) should remain some special structure even in the presence of perturbations. For this reason we stipulate that

1. The readability of z from y is preserved in the presence of perturbation.

2. The internal model is fixed, and its controllability about z is preserved in the presence of perturbation, or more precisely, perturbation in B_c only happens in the matrices B_{c_1}, B_{c_2} and B_{c_3}.

These two conditions are the elementary assumptions of perturbation.

Theorem 6-4.2. Let (6-4.2) be an SSNOR for system (6-4.1) at data point $P\{M,B_c\}$. Then

1. z is readable from y.
2. System (6-4.2) incorporates an internal model of L. This model is observable from u, and controllable about z.

Theorem 6-4.3. Subject to the assumptions 1 and 2 in Theorem 6-4.2, (6-4.2) is an output regulator for system (6-4.1) provided their closed-loop system is internally stable.

Stability is a basic requirement of a real system. This theorem shows that the internal model plays an important role in the output regulation systems. The next result further examines this problem.

Theorem 6-4.4. System (6-4.2) is an SSNOR for system (6-4.1) at data point $P\{A,B,M,C_1,C_2,D_1,D_2,B_c,F_c,F\}$ if and only if

1. (E,A,B,C_1) is stabilizable and detectable.

2. z is readable from y.

3. The closed-loop system is internally stable.

4. System (6-4.2) incorporates an internal model of L. This model is observable from u and controllable about z.

5.
$$\deg\left(\left| \begin{bmatrix} E & 0 \\ 0 & I_{n_c} \end{bmatrix} s - \begin{bmatrix} A+BFC & BF_c \\ B_cC & A_c \end{bmatrix} \right| \right) = n_c + \operatorname{rank}E. \qquad (6\text{-}4.6)$$

We would like to make some comments on this theorem.

A. As seen in Theorem 6-4.1, condition 1 may be deduced from condition 3. To emphasize this point, we rule out as one condition.

B. For the sake of simplicity in explanation, perturbation on A_c are not mentioned in this theorem. However, this theorem is applicable to the case where perturbation may occur in A_c without A_{c_2}, which incorporates an internal model of L. Thus, while some errors may be allowed in the design of compensator (6-4.2), the matrices A_{c_2} and B_{c_2} must be designed precisely. This must be considered in the selection of materials in the physical realization of (6-4.2).

C. Impulse controllability and impulse observability are necessary conditions of (6-4.6).

D. Theorem 6-4.4 may be false if the elementary assumption on perturbation doesn't hold.

Example 6-5.1. Consider the singular system

$$\begin{bmatrix} 1 & 0 & 0 \\ 0 & 1 & 0 \\ 0 & 0 & 0 \end{bmatrix} \dot{x} = \begin{bmatrix} -1 & 1 & 0 \\ 1 & 0 & -1 \\ 0 & 0 & -1 \end{bmatrix} x + \begin{bmatrix} 1 & 0 \\ 0 & 0 \\ 0 & 1 \end{bmatrix} u + \begin{bmatrix} 1 \\ -1 \\ 2 \end{bmatrix} f$$

$$\dot{f} = f \qquad (6\text{-}4.7)$$

$$y = \begin{bmatrix} 1 & 0 & 0 \\ 2 & 3 & -4 \end{bmatrix} x + \begin{bmatrix} 0 \\ 1 \end{bmatrix} f$$

$$z = [2 \quad 3 \quad -4]x + f$$

and its compensator

$$\dot{x}_c = \begin{bmatrix} -1 & 0 \\ 0 & 1 \end{bmatrix} x_c + \begin{bmatrix} 0 & 0 \\ 0 & 1 \end{bmatrix} y$$
$$u = x_c \qquad (6\text{-}4.8)$$

with the closed-loop system

$$\begin{bmatrix} 1 & 0 & 0 & 0 & 0 \\ 0 & 1 & 0 & 0 & 0 \\ 0 & 0 & 0 & 0 & 0 \\ 0 & 0 & 0 & 1 & 0 \\ 0 & 0 & 0 & 0 & 1 \end{bmatrix} \begin{bmatrix} \dot{x} \\ \\ \dot{x}_c \end{bmatrix} = \begin{bmatrix} -1 & 1 & 0 & 1 & 0 \\ 1 & 0 & -1 & 0 & 0 \\ 0 & 0 & -1 & 0 & 1 \\ 0 & 0 & 0 & -1 & 0 \\ 2 & 3 & -4 & 0 & 1 \end{bmatrix} \begin{bmatrix} x \\ \\ x_c \end{bmatrix} + \begin{bmatrix} 1 \\ -1 \\ 2 \\ 0 \\ 1 \end{bmatrix} f \qquad (6\text{-}4.9a)$$

$$z = [2 \quad 3 \ -4]x + f. \tag{6-4.9b}$$

In this case, L = 1, z = [0,1]y, z is readable from y. System (6-4.8) incorporates an internal model of L. This model is observable from u and controllable about z. The closed-loop system (6-4.9) is internally stable and (6-4.6) holds. Thus, by Theorem 6-4.4, system (6-4.8) is an SSNOR for system (6-4.7) at data point $F\{A,B,C_1,C_2,D_1,D_2,M,B_c,F_c,F\}$.

Denote (6-4.9) by

$$\begin{aligned} \bar{E}\dot{\bar{x}} &= \bar{A}\bar{x} + \bar{M}f \\ z &= Cx + f. \end{aligned} \tag{6-4.10}$$

Then (6-4.10) with coefficient matrices defined in (6-4.9) is an SSNOR for system (6-4.7). However, the output regulation may be not achievable if the elementary assumptions on perturbation are revoked. Let us have a look at the following cases.

1. In this case, the elementary assumption 1 is not satisfied.

Let the perturbation be $\delta D_1 = 0$; $\delta C_1 = 0$; $\delta C_2 = 0$; $\delta D_2 = \varepsilon$. From (6-4.7)-(6-4.9) we know

$$z' = [2 \quad 3 \ -4]x + (1+\varepsilon)f = z + \varepsilon f.$$

Here z' denotes z with perturbation δD_2. Since $\lim_{t\to\infty} z(t) = 0,\ \forall\ x_c(0), x(0)$, we have

$$\lim_{t\to\infty} z' = \lim_{t\to\infty} (\varepsilon f) = \text{doesn't exist}, \quad \forall\ x(0),\ x_c(0).$$

Thus the perturbed closed-loop system is not output regulation, no matter how small $\varepsilon \neq 0$ is.

2. In this case, elementary assumption 2 doesn't hold.

Let the perturbation in the compensator (6-4.8) be

$$\delta B_c = \begin{bmatrix} 0 & 0 \\ \varepsilon & 0 \end{bmatrix}, \quad \delta F_c = 0, \quad \delta A_c = 0, \quad \delta F = 0.$$

The perturbed closed-loop system is

$$\begin{aligned} \bar{E}\dot{\bar{x}} &= \bar{A}'\bar{x} + \bar{M}f \\ z &= [2 \quad 3 \ -4]x + f \end{aligned} \tag{6-4.11}$$

where

$$\bar{A}' = \begin{bmatrix} -1 & 1 & 0 & 1 & 0 \\ 1 & 0 & -1 & 0 & 0 \\ 0 & 0 & -1 & 0 & 1 \\ 00 & 0 & 0 & -1 & 0 \\ 2+\varepsilon & 3 & -4 & 0 & 1 \end{bmatrix}.$$

Paying attention to $\sigma(\bar{E},\bar{A}) \subset \mathbb{C}^-$ and $\deg(|s\bar{E}-\bar{A}|) = \text{rank}\bar{E}$, from Theorem 6-1.2 we know

$\sigma(\bar{E},\bar{A}') \subset \mathbb{C}^-$ provided $|\varepsilon|$ is sufficiently small. Thus the matrix equation $\bar{A}'V - \bar{E}VL = \bar{M}$ has a unique solution

$$V = (\bar{A}'-\bar{E})^{-1}\bar{M} = \frac{1}{12+\varepsilon}[-2+\varepsilon,\ 8+\varepsilon,\ 2-\varepsilon,\ 0,\ 28+2\varepsilon].$$

By denoting $e(t) = \bar{x} + Vf = [x/x_c] + Vf$ it may be verified from (6-4.11) that $e(t)$ and $z(t)$, respectively, satisfy

$$\bar{E}\dot{e}(t) = \bar{A}'e(t) \tag{6-4.12}$$

and

$$z(t) = [2 \quad 3\ -4 \quad 0 \quad 0]e(t) + \frac{-9\varepsilon}{12}f. \tag{6-4.13}$$

Therefore

$$\lim_{t\to\infty} e(t) = 0, \quad \forall\ x(0),\ x_c(0)$$

according to (6-4.12) and $\sigma(\bar{E},\bar{A}') \subset \mathbb{C}^-$. Combining it with (6-4.13) we have

$$\lim_{t\to\infty} z(t) = \lim_{t\to\infty} (-9\varepsilon/12)f = \text{doesn't exist}, \quad \forall\ x(0),\ x_c(0),\ \varepsilon \neq 0.$$

The perturbed closed-loop system is not output regulation.

Theorem 6-4.5. System (6-4.2) is an SSNOR for system (6-4.1) at data point $P\{E,M,B_c\}$ if and only if

1. (E,A,B,C_1) is stabilizable and detectable.
2. The closed-loop system is internally stable.
3. z is readable from y.
4. System (6-4.2) incorporates an internal model of L. This model is observable from u and controllable about z.
5. E is nonsingular.

From condition 5, we see once more the important fact that the singularity of E again destroys the structural stability of the closed-loop system. This again shows us that the comprehension of parameter E is of the most importance in system modeling.

Combining Theorems 6-2.4, 6-4.2, and 6-4.3, we have the following corollary.

Corollary 6-4.1. Assume that conditions 1-4 in Theorem 6-4.5 hold. Then, system (6-4.2) is a constrained SSNOR for system (6-4.1) at data point $P\{\bar{E},\bar{A}\}$ with constraints

$$\deg(|s\,\mathrm{diag}(\bar{E}+\delta\bar{E},\ I) - (\bar{A}+\delta\bar{A})|) = \deg(|s\bar{E}-\bar{A}|)$$

where

$$\bar{E} = \begin{bmatrix} E & 0 \\ 0 & I_{n_c} \end{bmatrix}, \qquad \bar{A} = \begin{bmatrix} A+BFC & BF_c \\ B_cC & A_c \end{bmatrix}$$

and $\delta\bar{A}$ denotes the perturbation in $\bar{A}$ caused by perturbations in A, B, C_1, C_2, D_1, D_2, M, B_c, F_c, and F.

Particularly, this corollary is applicable to the case with constraints on δE:

$$\text{rank}(E+\delta E) = \text{rank}E$$

and

$$\deg(|s\bar{E}-\bar{A}|) = n_c + \text{rank}E.$$

Theorems 6-4.3 - 6-4.5 show that to be an SSNOR for system (6-4.1) in the presence of perturbation, system (6-4.2) must incorporate an internal model of L. When z is readable from y, this model should be observable from u and controllable about z. Then, system (6-4.2) is an output regulator provided the closed-loop system is internally stable. This is usually termed the internal model principle.

So far, we have discussed the general structure of SSNORs. Now we explore their existence and design methods.

<u>Theorem 6-4.6</u>. Consider system (6-4.1). If z is readable from y, rankD_1 = d, the system has an SSNOR of the form (6-4.2) at data point P{A,B,M} if and only if

(1). (E,A,B,C_1) is stabilizable and detectable.

(2). (E,A,B,C_1) is impulse controllable and impulse observable.

(3).

$$\text{rank}\begin{bmatrix} sE-A & B_1 \\ D_1 & 0 \end{bmatrix} = n + d, \quad \forall\ s \in \sigma(L).$$

<u>Proof</u>. Necessity: (1) and (2) are easy to obtain from Theorem 6-4.4 and remark C. Next we prove (3).

Let (6-4.2) be an SSNOR for system (6-4.1) at data point P{A,B,M}. Then, Corollary 5-5.1 shows that there exist two matrices $V_1 \in \mathbb{R}^{n\times p}$, $V_2 \in \mathbb{R}^{n_c\times p}$ satisfying

$$\begin{aligned} &(A+BFC_1)V_1 + BF_cV_2 - EV_1L = M + BFC_2 \\ &D_1V_1 = D_2 \end{aligned} \qquad (6\text{-}4.14)$$

and two more matrices $V_1 + \delta V_1 \in \mathbb{R}^{n\times p}$, $V_2 + \delta V_2 \in \mathbb{R}^{n_c\times p}$ satisfying

$$\begin{aligned} &(A+BFC_1)(V_1+\delta V_1) + BF_c(V_2+\delta V_2) - E(V_1+\delta V_1)L = M+\delta M+BFC_2 \\ &D_1(V_1 + \delta V_1) = D_2. \end{aligned} \qquad (6\text{-}4.15)$$

Subtracting these corresponding equations in (6-4.14) and (6-4.15) yields:

$$(A+BFC_1)\delta V_1 - E\delta V_1 L + BF_c\delta V_2 = \delta M$$
$$D_1\delta V_1 = 0. \tag{6-4.16}$$

Let $\lambda \in \sigma(L)$ be any eigenvalue of L and η be the eigenvector belonging to λ. Then $L\eta = \lambda\eta$ Right multiplying both sides of (6-4.16) we have

$$[(A+BFC_1) - \lambda E]\delta V_1\eta + BF_c\delta V_2\eta = \delta M\eta$$
$$D_1\delta V_1\eta = 0 \tag{6-4.17}$$

For the arbitrariness of δM, we may choose δM_i, i = 1, 2, ..., n, so that $\delta M_1\eta$, $\delta M_2\eta$, ... , $\delta M_n\eta$ are independent vectors. Corresponding to each δM_i, there exist δV_1^i, δV_2^i, i = 1, 2, ... , n, satisfying (6-4.17). Defining $\alpha_i = \delta V_1^i\eta$, $\beta_i = \delta V_2^i\eta$, $\gamma_i = \delta M_i\eta$, i = 1, 2, ... , n, from (6-4.17) we obtain

$$\begin{bmatrix} A+BFC_1-\lambda E & BF_c \\ D_1 & 0 \end{bmatrix}\begin{bmatrix} \alpha_1 & \alpha_2 & \cdots & \alpha_n \\ \beta_1 & \beta_2 & \cdots & \beta_n \end{bmatrix} = \begin{bmatrix} \gamma_1 & \gamma_2 & \cdots & \gamma_n \\ 0 & 0 & \cdots & 0 \end{bmatrix}. \tag{6-4.18}$$

Equation (6-4.18) has solutions α_i, β_i, i = 1, 2, ... , n, if and only if

$$\text{rank}\begin{bmatrix} A+BFC_1-\lambda E & BF_c \\ D_1 & 0 \end{bmatrix} = \text{rank}\begin{bmatrix} A+BFC_1-\lambda E & BF_c & \gamma_1, & \gamma_2, & \ldots, & \gamma_n \\ D_1 & 0 & 0 & 0 & \ldots & 0 \end{bmatrix}$$

$$= n + d, \ \forall \ \lambda \in \sigma(L).$$

On the other hand, the preceding equation may be rewritten as

$$\text{rank}\begin{bmatrix} A+BFC_1-\lambda E & BF_c \\ D_1 & 0 \end{bmatrix} = \text{rank}\begin{bmatrix} -I & 0 \\ 0 & I \end{bmatrix}\begin{bmatrix} \lambda E-A & B \\ D_1 & 0 \end{bmatrix}\begin{bmatrix} I & 0 \\ -FC_1 & -F_c \end{bmatrix}$$

$$= n + d, \ \forall \ \lambda \in \sigma(L).$$

Therefore

$$\text{rank}\begin{bmatrix} sE-A & B \\ D_1 & 0 \end{bmatrix} = n + d, \ \forall \ s \in \sigma(L),$$

which is (3).

Sufficiency: Now we will constructively prove the sufficiency:

a. Let $g(s) = s^\mu + \ldots + a_1 s + a_0$ be the minimal polynomial of L. Constructing the matrix

$$J = \begin{bmatrix} 0 & 1 & & \\ & 0 & 1 & \\ & & \ddots & \\ & & & 1 \\ -a_{\mu-1} & -a_{\mu-2} & \cdots & -a_0 \end{bmatrix}$$

then $|sI-J| = g(s)$.

b. Constructing the system

$$\dot{x}_{c_2} = A_{c_2}x_{c_2} + B_{c_2}z \tag{6-4.19}$$

where $x_{c_2} \in \mathbb{R}^{n_{c2}}$, $n_{c_2} = d\mu$,

$$A_{c_2} = \mathrm{diag}(\overbrace{J, J, \ldots J}^{d}) \in \mathbb{R}^{n_{c2} \times n_{c2}}$$

$$B_{c_2} = \mathrm{diag}(\overbrace{e_1, e_1, \ldots, e_1}^{d}) \in \mathbb{R}^{n_{c2} \times d}, \quad e_1 = [0 \ 0 \ \ldots \ 0 \ 1]^\tau \in \mathbb{R}^{\mu}$$

it is obvious that (A_{c_2}, B_{c_2}) is controllable, and A_{c_2} has d invariants that may be divided by g(s).

c. When $f(t) \equiv 0$, systems (6-4.1) and (6-4.19) form the composite system

$$\begin{bmatrix} E & 0 \\ 0 & I_{n_{c2}} \end{bmatrix}\begin{bmatrix} \dot{x} \\ \dot{x}_{c2} \end{bmatrix} = \begin{bmatrix} A & 0 \\ B_{c_2}D_1 & A_{c2} \end{bmatrix}\begin{bmatrix} x \\ x_{c2} \end{bmatrix} + \begin{bmatrix} B \\ 0 \end{bmatrix}u$$

$$\begin{bmatrix} y \\ x_{c_2} \end{bmatrix} = \begin{bmatrix} C_1 & 0 \\ 0 & I \end{bmatrix}\begin{bmatrix} x \\ x_{c_2} \end{bmatrix}. \tag{6-4.20}$$

Subject to (1) and (3), we have

$$\mathrm{rank}\begin{bmatrix} sE-A & 0 & B \\ -B_{c_2}D_1 & sI-A_{c_2} & 0 \end{bmatrix}$$

$$= \mathrm{rank}\begin{bmatrix} I & 0 & 0 \\ 0 & -B_{c2} & sI-A_{c2} \end{bmatrix}\begin{bmatrix} sE-A & B & 0 \\ D_1 & 0 & 0 \\ 0 & 0 & I \end{bmatrix}\begin{bmatrix} I & 0 & 0 \\ 0 & 0 & I \\ 0 & I & 0 \end{bmatrix}$$

$$= \mathrm{rank}\begin{bmatrix} I & 0 & 0 \\ 0 & B_{c2} & sI-A_{c_2} \end{bmatrix} = n + n_{c_2}, \quad \forall\ s \in \sigma(L).$$

Thus, by assumption (1), we know

$$\left(\begin{bmatrix} E & 0 \\ 0 & I \end{bmatrix}, \begin{bmatrix} A & 0 \\ B_{c_2}D_1 & A_{c_2} \end{bmatrix}, \begin{bmatrix} B \\ 0 \end{bmatrix}\right)$$

is stabilizable, so is (6-4.20). Noticing (1) and (2), it is easy to verify that system (6-4.20) is detectable, impulse controllable, and impulse observable. Hence, according to Theorem 5-3.2 we know that the system has a normal compensator

$$\dot{x}_{c_1} = A_{c_1}x_{c_1} + [A_{c_3},\ B_{c_1}][x_{c_2}/y]$$
$$u = F_{c_1}x_{c_1} + [F_{c_2},\ F][x_{c_2}/y] \qquad (6\text{-}4.21)$$

where $x_{c_1} \in \mathbb{R}^{n_{c_1}}$, whose closed-loop system has $n+n_{c_1}+n_{c_2}$ stable finite poles.

For the matrices determined as earlier, we construct the system

$$\dot{x}_c = A_c x_c + B_c y$$
$$u = F_c x_c + Fy \qquad (6\text{-}4.22)$$

where $x_c \in \mathbb{R}^{n_c}$, $n_c = n_{c_1}+n_{c_2}$, and

$$x_c = \begin{bmatrix} x_{c_1} \\ x_{c_2} \end{bmatrix}, \qquad A_c = \begin{bmatrix} A_{c_1} & A_{c_3} \\ 0 & A_{c_2} \end{bmatrix}, \qquad B_c = \begin{bmatrix} B_{c_1} \\ [0\ \ B_{c_2}] \end{bmatrix},$$

$$F_c = [F_{c_1},\ F_{c_2}], \qquad S = [0\ \ I_d].$$

It may be testified from Theorem 6-4.4 that system (6-4.22) is an SSNOR for system (6-4.1) at data point $P\{A,B,M\}$. Q.E.D.

The constructive proof also provides us with a design method for SSNORs. It is worth pointing out that the SSNOR designed by this method is of order $n_{c_1} + n_{c_2}$, which is generally high.

<u>Example 6-4.1</u>. Consider the system

$$E\dot{x} = Ax + Bu + Mf$$
$$\dot{f} = Lf$$
$$y = C_1x + C_2f \qquad (6\text{-}4.23)$$
$$z = D_1x + D_2f$$

with coefficient matrices

$$E = \begin{bmatrix} 1 & 0 & 0 \\ 0 & 1 & 0 \\ 0 & 0 & 0 \end{bmatrix}, \quad A = \begin{bmatrix} -1 & 1 & 0 \\ 1 & 0 & -1 \\ 0 & 0 & -1 \end{bmatrix}, \quad B = \begin{bmatrix} 1 & 0 \\ 0 & 0 \\ 0 & 1 \end{bmatrix}, \quad M = \begin{bmatrix} 1 \\ -1 \\ 2 \end{bmatrix},$$

$$C_1 = \begin{bmatrix} 1 & 0 & 0 \\ 2 & 3 & -4 \end{bmatrix}, \quad C_2 = \begin{bmatrix} 0 \\ 1 \end{bmatrix}, \quad L = 1, \quad D_1 = [2\ \ 3\ \ -4], \quad D_2 = 1.$$

Following the procedure in Theorem 6-4.6, we design its SSNOR.

a. $L = 1$, $g(s) = s-1$, thus $J = [1]$.

b. Since $d = \mu = 1$, we construct

$$A_{c_2} = [1], \qquad B_{c_2} = [1];$$

thus (6-4.20) becomes

$$\begin{bmatrix} 1 & 0 & 0 & 0 \\ 0 & 1 & 0 & 0 \\ 0 & 0 & 0 & 0 \\ 0 & 0 & 0 & 1 \end{bmatrix} \begin{bmatrix} \dot{x} \\ \dot{x}_{c2} \end{bmatrix} = \begin{bmatrix} -1 & 1 & 0 & 0 \\ 1 & 0 & -1 & 0 \\ 0 & 0 & -1 & 0 \\ 2 & 3 & -4 & 1 \end{bmatrix} \begin{bmatrix} x \\ x_{c2} \end{bmatrix} + \begin{bmatrix} 1 & 0 \\ 0 & 0 \\ 0 & 1 \\ 0 & 0 \end{bmatrix} u$$

$$\begin{bmatrix} y \\ x_{c_2} \end{bmatrix} = \begin{bmatrix} 1 & 0 & 0 & 0 \\ 2 & 3 & -4 & 0 \\ 0 & 0 & 0 & 1 \end{bmatrix} \begin{bmatrix} x \\ x_{c_2} \end{bmatrix} . \tag{6-4.24}$$

c. Direct computation verifies that system (6-4.24) is stabilizable, detectable, impulse controllable and impulse observable. Following the design procedure in the last chapter (Section 5-3), we design its compensator

$$\dot{x}_{c_1} = -0.75x_{c_1} + [0.5 \quad 0.25 \quad 0] \begin{bmatrix} y \\ x_{c_2} \end{bmatrix}$$

$$u = \begin{bmatrix} -2.4 \\ 0 \end{bmatrix} x_{c_1} + \begin{bmatrix} -4 & -6.2 & 0 \\ 0 & 0 & 0 \end{bmatrix} \begin{bmatrix} y \\ x_{c_2} \end{bmatrix}$$

with pole set of closed-loop system $\{-1, -1, -2, -0.75\} \subset \mathbb{C}^-$.

d. Hence, the system

$$\dot{x}_c = \begin{bmatrix} -0.75 & 0 \\ 0 & 1 \end{bmatrix} x_c + \begin{bmatrix} 0.5 & 0.25 \\ 0 & 1 \end{bmatrix} y$$

$$u = \begin{bmatrix} -2.4 & 0 \\ 0 & 0 \end{bmatrix} x_c + \begin{bmatrix} -4 & -6.2 \\ 0 & 0 \end{bmatrix} y$$

is just an SSNOR for system (6-4.23) at data point $P\{A,B,M\}$.

For general singular systems, matrix E is singular, from Theorem 6-4.5 we know that (6-4.2) cannot be an SSNOR for system (6-4.1) at data point $P\{E,B_c,M\}$. If $\dot{y}(t)$ is available for our use, we may consider the compensator of the form of

$$\begin{aligned} \dot{x}_c &= A_c x_c + B_c y \\ u &= F_c x_c + Fy + H\dot{y} \end{aligned} \tag{6-4.25}$$

where $x_c \in \mathbb{R}^{n_c}$ and all matrices are constant of appropriate dimensions. The following are two results concerning such compensators.

<u>Theorem 6-4.7</u>. Subject to the elementary assumption on perturbation, system (6-4.25) is an SSNOR for system (6-4.1) at data point $P\{E,A,B,C_1,C_2,D_1,D_2,M,B_c,F_c,F\}$ if and only if

1. (E,A,B,C_1) is stabilizable and detectable.
2. z is readable from y.
3. The closed-loop system is internally stable.

4. System (6-4.25) incorporates an internal model of L. This model is observable from u and controllable about z.

5. $\mathrm{rank}(E-BHC_1) = n$.

Normalizability and dual normalizability are necessary for the tenability of condition 5.

Theorem 6-4.8. Assume that the regulated vector z is readable from measure y, and $\mathrm{rank}D_1 = d$. Then (6-4.1) has an SSNOR of the form of (6-4.25) at data point $P\{E,A,B,M\}$ if and only if

1. (E,A,B,C_1) is stabilizable and detectable.
2. (E,A,B,C_1) is normalizable and dual normalizable.
3. $$\mathrm{rank}\begin{bmatrix} sE-A & B \\ D_1 & 0 \end{bmatrix} = n + d, \quad \forall\ s \in \sigma(L).$$

Remarks on Theorems 6-4.6 and 6-4.8 follow.

Remark 1. Subject to the sufficient condition, the compensator (6-4.25) may be designed the same as for (6-4.2), which is given in Theorem 6-4.6.

Remark 2. The same rank condition

$$\mathrm{rank}\begin{bmatrix} sE-A & B \\ D_1 & 0 \end{bmatrix} = n + d, \quad \forall\ s \in \sigma(L) \tag{6-4.26}$$

is involved in the theorems. Since $B \in \mathbb{R}^{n\times m}$, $D_1 \in \mathbb{R}^{d\times n}$, if the preceding equation holds, $m \geqslant d$, that is, the inputs must not be less than the regulated vector.

Remark 3. When disturbance disappears ($f(t) \equiv 0$), the transition zeros from u to the regulated output z satisfy

$$\mathrm{rank}\begin{bmatrix} sE-A & B \\ D_1 & 0 \end{bmatrix} < n + \min(d,m), \quad \forall\ s \in \mathbb{C},$$

together with (6-4.26), implying that the eigenvalues of L (the poles of disturbance) are not such transition zeros.

6-5. Notes and References

This chapter is based on Dai and Wang (1987c,d), in which another constrained structurally stable compensation problem was considered, in which case the perturbation in matrix E is constrained to rank($E+\delta E$) < n, in thinking that the perturbation should not revoke the singularity of singular systems. The process is the same as used here with a slight modification.

Singular compensators were also studied in Wang (1984).

Materials on structurally stable compensation problem in normal output regulation systems may be found in Wonham (1974).

CHAPTER 7

SYSTEM ANALYSIS VIA TRANSFER MATRIX

As pointed out in Chapter 2, the transfer matrix characterizes the input and output relationship of a system. Therefore, we may base our analysis and synthesis on the transfer matrix. This method is termed system analysis via transfer matrix here. Three main problems are studied in this chapter: transfer matrix structure under state feedback, input function observers, and decoupling control.

7-1. Transfer Matrix in Singular Systems

We continue our discussion from Section 2-6, studying further properties of the transfer matrix.

Consider the singular system

$$\begin{aligned} E\dot{x} &= Ax + Bu \\ y &= Cx \end{aligned} \tag{7-1.1}$$

where $x \in \mathbb{R}^n$, $u \in \mathbb{R}^m$, $y \in \mathbb{R}^r$ are its state, control input, and measure output, respectively. E, $A \in \mathbb{R}^{n\times n}$, $B \in \mathbb{R}^{n\times m}$, and $C \in \mathbb{R}^{r\times n}$ are constant matrices. It is assumed that system (7-1.1) is regular and rank$E < n$.

System (7-1.1) has transfer matrix G(s):

$$G(s) = C(sE-A)^{-1}B. \tag{7-1.2}$$

First, we have the following theorem.

<u>Theorem 7-1.1</u>. Let G(s) be the transfer matrix of system (7-1.1). Then

$$G(s) = C(sE-A)^{-1}B = \tau\hat{C}(\tau I-\hat{E})^{-1}\hat{B} \triangleq \hat{G}(\tau) \tag{7-1.3}$$

where

$$\hat{E} = (\alpha E+A)^{-1}E, \qquad \hat{B} = (\alpha E+A)^{-1}B, \qquad \hat{C} = -C$$
$$\tau = \frac{1}{s+\alpha}, \quad \alpha \quad \text{satisfies } |\alpha E+A| \neq 0. \tag{7-1.4}$$

Proof. It is easy to obtain (7-1.3) from

$$(sE-A)^{-1} = ((s+\alpha)E-\alpha E-A)^{-1} = (\tau^{-1}\hat{E}-I)^{-1}(\alpha E+A)^{-1} = -\tau(\tau I-\hat{E})^{-1}(\alpha E+A)^{-1}.$$

Q.E.D.

The transformation

$$\tau = \frac{1}{s+\alpha}$$

is called the time-scale transformation for singular systems. It implies a relationship between infinite and finite.

The simple Theorem 7-1.1 shows an important fact: Via time-scale transformation, the transfer matrix G(s) of any regular singular system may be changed into a strictly proper matrix in τ(stands for the transfer matrix of a certain normal system) multiplying τ. Thus, from a mathematical point of view, the singular system differs from the normal system only by a factor τ. Therefore, the singular system may sometimes be studied by using the abundant results in linear system theory.

Further, noticing

$$\hat{G}(\tau) = \hat{C}\hat{B} + \hat{C}(\tau I-\hat{E})^{-1}\hat{B},$$

$\hat{G}(\tau)$ is the transfer matrix in τ of a normal system with input involved visibly in measure output. The preceding equation is also an expression of transfer matrix under the EF3.

<u>Example 7-1.1</u>. For the singular system described by the transfer function

$$G(s) = \frac{s+1}{s^2+s+1} + s^2 + 1,$$

if we take the time-scale transformation $\tau = \frac{1}{s}$

$$\hat{G}(\tau) = 2 - \frac{1}{\tau^2+\tau+1} + \frac{1}{\tau^2} .$$

Let G(s) be rational and square. Then its determination |G(s)| is a rational function. If $|G(s)| \neq 0$, G(s) will be termed nonsingular or invertible.

For example, let

$$G(s) = \begin{bmatrix} \frac{1}{s+2} & \frac{s^2+1}{s+2} \\ \frac{1}{s+1} & \frac{s^2}{s+1} \end{bmatrix}.$$

From $|G(s)| = -\frac{1}{(s+1)(s+2)} \neq 0$, G(s) is nonsingular. But the rational matrix H(s)

$$H(s) = \begin{bmatrix} \frac{s+1}{s+2} & \frac{s^2+s+1}{s+2} \\ 1 & \frac{s^2+s+1}{s+1} \end{bmatrix}$$

has $|H(s)| \equiv 0$. Thus, H(s) is not nonsingular or singular.

Let G(s) be nonsingular. Rational matrix H(s) = adj(G(s))/ G(s) satisfies

$$G(s)H(s) = H(s)G(s) = I. \tag{7-1.5}$$

H(s) is called the inverse of G(s) and is denoted by $H(s) = G^{-1}(s)$.

The rank of G(s) is defined as the highest order of nonsingular subdeterminations in G(s), and is denoted by rankG(s).

For instance, in the preceding two examples, rankG(s) = 2 and rankH(s) = 1.

For the general rectangular matrix G(s) (not necessarily be square), we define

Definition 7-1.1. Let $G(s) \in \mathbb{R}^{r\times m}$ be rational. If there exists a rational matrix $H_L(s)$ ($H_R(s)$) such that

$$H_L(s)G(s) = I_m \quad (G(s)H_R(s) = I_r), \tag{7-1.6}$$

G(s) will be called left (right) invertible.

$H_L(s)$ ($H_R(s)$) are termed the left (right) inverses of G(s).

According to (7-1.6), if G(s) is left (right) invertible, $r \geqslant m$ ($r \leqslant m$), or the measures must be not less (more) than the inputs.

Obviously, $H_L(s) = H_R(s) = G^{-1}(s)$ when G(s) is nonsingular. The transfer function of single-input and single-output system is always nonsingular if it is not identically zero.

For G(s) in Example 7-1.1,

$$G_L(s) = G_R(s) = \frac{s^2+s+1}{s^4+s^3+2s^2+2s+1} .$$

Theorem 7-1.2. The rational matrix $G(s) \in \mathbb{R}^{r\times m}$ is left (right) invertible if and only if rankG(s) = m (rankG(s) = r).

Proof. Necessity is obvious. Now we will prove the sufficiency.

Let rankG(s) = m. Then there exists a rational matrix $T(s) \in \mathbb{R}^{m\times r}$ such that T(s)G(s) is square and nonsingular. Choosing

$$H_L(s) = (T(s)G(s))^{-1}T(s)$$

we have

$$H_L(s)G(s) = (T(s)G(s))^{-1}T(s)G(s) = I_m.$$

Thus, G(s) is left invertible. Right invertibility may be proved in a similar way. Q.E.D.

Example 7-1.2. The rank of

$$G(s) = \begin{bmatrix} \frac{1}{s+1} & \frac{1}{s+2} & 1 \\ \frac{s}{s+1} & \frac{s+1}{s+2} & 2 \end{bmatrix}$$

is 2. Thus it is right invertible. Let

$$T(s) = \begin{bmatrix} 1 & 0 \\ 0 & 1 \\ 0 & 0 \end{bmatrix}.$$

Then

$$G(s)T(s) = \begin{bmatrix} \frac{1}{s+1} & \frac{1}{s+2} \\ \frac{s}{s+1} & \frac{s+1}{s+2} \end{bmatrix}$$

and

$$H_R(s) = T(s)(G(s)T(s))^{-1} = \begin{bmatrix} (s+1)^2 & -(s+1) \\ -s(s+2) & s+2 \\ 0 & 0 \end{bmatrix}.$$

Let T(s) be chosen as

$$T_1(s) = \begin{bmatrix} 1 & 0 \\ 0 & 0 \\ 0 & 1 \end{bmatrix}.$$

Then

$$G(s)T_1(s) = \begin{bmatrix} \frac{1}{s+1} & 1 \\ \frac{s}{s+1} & 2 \end{bmatrix}$$

and

$$\bar{H}_R(s) = T_1(s)(G(s)T_1(s))^{-1} = \begin{bmatrix} \frac{-2(s+1)}{s-2} & \frac{s+1}{s-2} \\ 0 & 0 \\ \frac{s}{s-2} & \frac{-1}{s-2} \end{bmatrix}.$$

Thus, the right or left inverse is not unique.

Since the transfer matrix for a singular system is determined by its controllable and observable subsystem, and under state feedback control either uncontrollable or unobservable subsystems contribute nothing to transfer matrix. Thus, we assume in this chapter that system (7-1.1) is controllable and observable unless specified.

7-2. State Feedback and Transfer Matrix: Single-Input Single-Output Systems

While state feedback control may change the poles of the closed-loop system, its transfer matrix is changed at the same time. Now we will consider the influence of state feedback control on transfer matrix.

First we consider the single-input single-output system

$$\begin{aligned} E\dot{x} &= Ax + bu \\ y &= cx \end{aligned} \tag{7-2.1}$$

where $x \in \mathbb{R}^n$, $b, c^{\tau} \in \mathbb{R}^n$, $E, A \in \mathbb{R}^{n\times n}$ are constant matrices. It is assumed that rank$E$ $< n$ and system (7-2.1) is regular. System (7-2.1) has the transfer function of

$$G(s) = c(sE-A)^{-1}b = \frac{b(s)}{a(s)} \tag{7-2.2}$$

where

$$a(s) = s^{l_1} + \sum_{i=0}^{l_1-1} a_i s^i, \qquad b(s) = \sum_{i=0}^{k} b_i s^i, \qquad b_k \neq 0,$$

$a(s)$ and $b(s)$ are coprime, i.e., the greatest common divisor of $a(s)$ and $b(s)$ is 1.

If $k < l_1$, transfer function (7-2.2) is a strictly proper function, thus it is the transfer function of a certain normal system which doesn't attempt to study it. Thus we assume $k \geqslant l_1$.

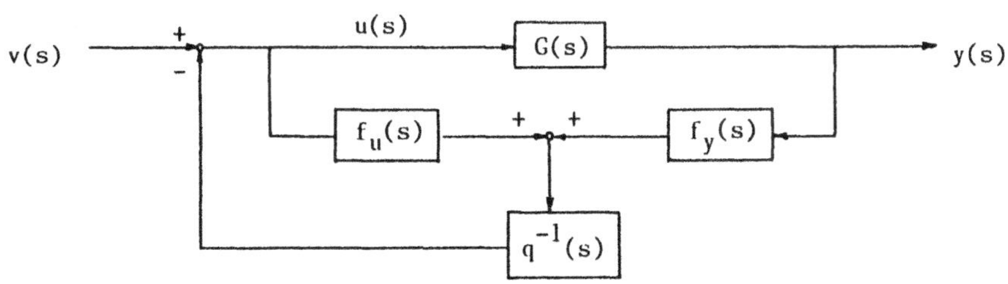

Figure 7-2.1. Dynamic Compensation of (7-2.1)

To have an audio-visual understanding of analysis in time domain, we will look at its compensation of frequency domain. Consider the compensator described by Figure 7-2.1. Direct computation shows the transfer relationship from v(s) to y(s) is

$$y(s) = \frac{B(s)q(s)}{q(s)a(s) + (b(s)f_y(s)+a(s)f_u(s))} v(s). \tag{7-2.3}$$

From assumption, a(s) and b(s) are coprime, there exist two polynomials $\alpha(s)$ and $\beta(s)$ so that $\alpha(s)a(s) + \beta(s)b(s) = 1$ (Kailath, 1980). For any polynomial f(s) with order not greater than k, we choose q(s) to be any stable polynomial with order k. Multiplying the preceding equation by q(s)f(s), we obtain

$$\alpha(s)a(s)q(s)f(s) + \beta(s)b(s)q(s)f(s) = f(s)q(s). \tag{7-2.4}$$

For the polynomials $\alpha(s)q(s)f(s)$ and b(s), by division algorithm for polynomial there exist $\gamma(s)$ and r(s) such that

$$\alpha(s)q(s)f(s) = b(s)\gamma(s) + r(s) \tag{7-2.5}$$

and r(s) has an order less than that of b(s). Substituting the preceding equation into (7-2.4) yields

$$b(s)f_y(s) = f(s)q(s) - a(s)r(s) \tag{7-2.6}$$

where $f_y(s) = \beta(s)q(s)f(s) + a(s)\gamma(s)$. Since b(s) is of order k, the polynomial on the right side has order not greater than 2k. Thus $f_y(s)$ has order not greater than k. Let $f_u(s) = r(s)$. $f_u(s)$, $f_y(s)$, and q(s) satisfy

1. $a(s)f_u(s) + b(s)f_y(s) = q(s)f(s)$.
2. $q^{-1}(s)f_u(s)$ and $q^{-1}(s)f_y(s)$ are proper.
3. q(s) is stable.

For such $f_u(s)$, $f_y(s)$ and q(s), transfer function (7-2.3) becomes

$$y(s) = \frac{b(s)q(s)}{(a(s)-f(s))q(s)} v(s) = \frac{b(s)}{a(s)-f(s)} v(s). \tag{7-2.7}$$

Particularly, let f(s) = a(s) - b(s). Then the order of f(s) is not greater than k and (7-2.7) becomes

$$y(s) = v(s). \tag{7-2.8}$$

The special feature of system (7-2.8) is that the measure output is precisely the same as the new input v(s), which will be termed a pure prediction system (PPS).

From the constructing process of $f_u(s)$ and $f_y(s)$ we note that the key point lies in design PPS (7-2.8) is that b(s) has an order not less than that of a(s), i.e., system (7-2.1) is a "true" singular system. Here we use the term "true" to imply that its controllable and observable subsystem is a singular one.

Now we will study the possibility of the preceding design: In the process from (7-2.7) to (7-2.8) we see that zero-pole cancellation happens, which is not a serious problem if b(s) is stable. However, if it is not the case, b(s) is unstable. The cancellation is impossible provided perturbation happens in b(s) to cause the closed-loop system to be unstable. Thus the design is applicable to the minimal phrase system.

Example 7-2.1. Consider the system y(s) = G(s)u(s) with

$$G(s) = \frac{1+s}{1+s} = 1.$$

Suppose that perturbation exists in G(s) to become $G_\varepsilon(s)$:

$$G_\varepsilon(s) = \frac{1+(1+\varepsilon)s}{1+s} = G(s) + \frac{\varepsilon s}{1+s}.$$

In this case zero-pole cancellation is impossible. But if we denote $y_\varepsilon(s) = G_\varepsilon(s)u(s)$, $e_y(s) = y(s) - y_\varepsilon(s)$, it will be $e_y(s) = -(\varepsilon s/(1+s))u(s)$.

Let u be a unit jump signal, $u(s) = \frac{1}{s}$. Then $e_y(s) = -\varepsilon/(1+s)$. Thus

$$e_y(t) = -\varepsilon e^{-t}, \qquad t \geqslant 0,$$

with the response locus shown in Figure 7-2.2, from which we know $|e_y(t)| \leqslant \varepsilon$, $t \geqslant 0$. Therefore, in the presence of perturbation, the output response differs only a little from the nominal one provided ε is small enough.

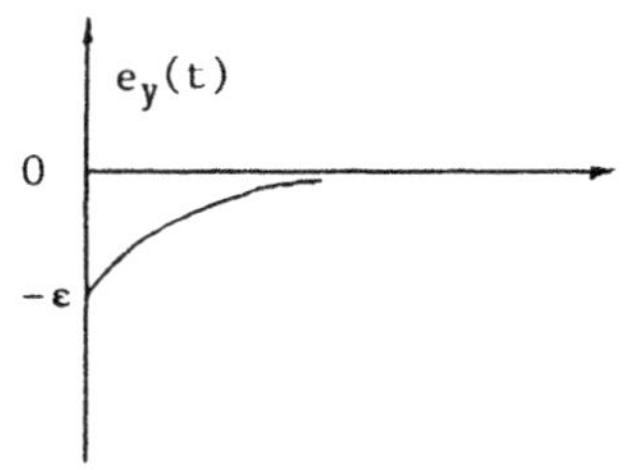

Figure 7-2.2

Example 7-2.2. For the system with unstable zero-pole cancellation

$$y(s) = G(s)u(s), \qquad G(s) = \frac{1-s}{1-s} = 1.$$

When perturbation exists in G(s) to become $G_\varepsilon(s)$:

$$G_\varepsilon(s) = \frac{1-(1+\varepsilon)s}{1-s} = G(s) - \frac{\varepsilon s}{1-s}.$$

Let $y_\varepsilon(s) = G_\varepsilon(s)u(s)$ be the perturbed output response, and $e_y(t) = y(t) - y_\varepsilon(t)$ be the response deviation caused by perturbation. Then when $u(t)$ is a unit jump signal, $u(s) = \frac{1}{s}$, we have $e_y(s) = y(s) - y_\varepsilon(s) = \varepsilon/(1-s)$. Hence

$$e_y(t) = -\varepsilon e^t, \qquad t \geqslant 0,$$

with the response locus shown in Figure 7-2.3, showing that $\lim_{t\to\infty} |e_y(t)| = \infty$ no matter how small $\varepsilon \neq 0$ is. This implies that unstable zero-pole cancellation is impossible. This point should be paid special attention in the design of PPS.

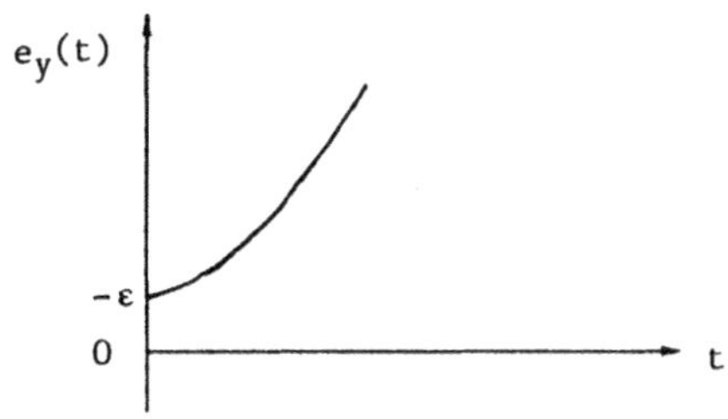

Figure 7-2.3

The preceding discussion shows that by using a dynamic compensator, the closed-loop system (7-2.2) may be a PPS. Note that state feedback control is usually realized via compensators. Thus we discuss this problem for state feedback control. We separately study this problem for P-, and P-D state feedbacks.

7-2.1. Pure proportional (P-) state feedback

Consider the P-state feedback:

$$u = Kx + v, \qquad K = [k_1, k_2, \ldots, k_n] \tag{7-2.9}$$

where $v \in \mathbb{R}^m$ is the new input. Feedback control (7-2.9) and system (7-2.1) form the closed-loop system

$$\begin{aligned} E\dot{x} &= (A+bK)x + bv \\ y &= cx. \end{aligned} \tag{7-2.10}$$

Under the assumption of regularity, there exist two nonsingular matrices Q and P such that system (7-2.1) is r.s.e. to

$$\mathrm{diag}(I_{n_1}, N)\dot{\bar{x}} = \mathrm{diag}(A_1, I_{n_2})\bar{x} + [b_1/b_2]u \tag{7-2.11a}$$

$$y = [c_1,\ c_2]\bar{x} \tag{7-2.11b}$$

where $\bar{x} = P^{-1}x$, $n_1 + n_2 = n$, $A_1 \in \mathbb{R}^{n_1 \times n_1}$, $N \in \mathbb{R}^{n_2 \times n_2}$ is nilpotent, and

$$QEP = \mathrm{diag}(I_{n_1},\ N), \qquad QAP = \mathrm{diag}(A_1,\ I_{n_2}),$$

$$Qb = [b_1/b_2], \qquad cP = [c_1,\ c_2].$$

Subject to the controllability assumption for system (7-2.1), (A_1,b_1) and (N,b_2) are controllable pairs. Without loss of generality, they are assumed to be in the controllability canonical form

$$A_1 = \begin{bmatrix} 0 & & & & * \\ 1 & 0 & & & * \\ & 1 & \ddots & & \vdots \\ & & & 0 & * \\ & & & 1 & * \end{bmatrix} \in \mathbb{R}^{n_1 \times n_1}, \qquad b_1 = \begin{bmatrix} 1 \\ 0 \\ \vdots \\ 0 \end{bmatrix} \in \mathbb{R}^{n_1}$$

$$N = \begin{bmatrix} 0 & & & & \\ 1 & 0 & & & \\ & 1 & \ddots & & \\ & & & 0 & \\ & & & 1 & 0 \end{bmatrix} \in \mathbb{R}^{n_2 \times n_2}, \qquad b_2 = \begin{bmatrix} 1 \\ 0 \\ \vdots \\ 0 \end{bmatrix} \in \mathbb{R}^{n_2}$$

where "*" represents the possibly nonzero elements. Let

$$K = \bar{K}P^{-1}, \qquad \bar{K} = [\overbrace{0,\ \ldots\ ,\ 0,\ 1}^{n_1},\ \overbrace{1,\ 0,\ \ldots\ ,\ 0}^{n_2}]. \tag{7-2.12}$$

Then the characteristic polynomial of closed-loop system (7-2.10) is

$$|sE - (A+bK)| = |sQEP - Q(A+bK)P|$$

$$= \left| \begin{bmatrix} sI-A_1 & 0 \\ 0 & sN-I \end{bmatrix} - \begin{bmatrix} b_1 \\ b_2 \end{bmatrix} \bar{K} \right| = \text{constant}.$$

Thus, there exist two nonsingular matrices $\bar{Q}$ and $\bar{P}$ such that (7-2.10) is r.s.e. to

$$\tilde{N}\dot{\tilde{x}} = \tilde{x} + b_o v$$
$$y = c_o\tilde{x} \tag{7-2.13}$$

where $\tilde{x} = \bar{P}^{-1}x$, and

$$\tilde{N} = \bar{Q}E\bar{P} = \begin{bmatrix} 0 & 1 & & \\ & 0 & 1 & \\ & & \ddots & 1 \\ & & & 0 \end{bmatrix} \qquad \bar{Q}b = b_o = \begin{bmatrix} 0 \\ \vdots \\ 0 \\ 1 \end{bmatrix}$$

$$\bar{Q}(A+bK)\bar{P} = I, \qquad c_o = c\bar{P} = [c_1^o,\ c_2^o,\ \ldots\ ,\ c_n^o].$$

Note that only a fast subsystem exists in (7-2.13). For any feedback $v = K_o\tilde{x} + \gamma$,

$K_o = [k_1^o, k_2^o, \dots, k_n^o]$, γ is the new input (or reference signal), the closed-loop system formed by v and (7-2.13) is

$$N\dot{\tilde{x}} = (I+b_oK_o)\tilde{x} + b_o\gamma$$
$$y = c_o\tilde{x}$$

with input output relationship $y(s) = c_o(sN - (I+b_oK_o)^{-1}b_o\gamma(s)$.

Denoting $\tau = \frac{1}{s}$, $\bar{k}_i^o = k_i^o\tau, i = 1, 2, \dots, n$, we have

$$y(s) = - c_o\tau[\tau I - (N-b_oK_o\tau)]^{-1}b_o\gamma(s)$$

$$= - \frac{c_1^o + c_2^o\tau + \dots + c_n^o\tau^{n-1}}{\bar{k}_1^o + \bar{k}_2^o\tau + \dots + \bar{k}_n^o\tau^{n-1} + \tau^n}\gamma(s)$$

$$= - \frac{c_1^o s^{n-1} + c_2^o s^{n-2} + \dots + c_n^o}{k_1^o s^{n-1} + k_2^o s^{n-2} + \dots + (1+k_n^o)}\gamma(s). \qquad (7\text{-}2.14)$$

From (7-2.14) we see that the P-state feedback cannot change the numerator of the transfer function unless zero-pole cancellation occurs. But its denominator may take any polynomial of order not greater than n-1.

Particularly, if we choose $K_o = [-c_1^o, -c_2^o, \dots, -c_{n-1}^o, -c_n^o-1]$, transfer function (7-2.14) becomes

$$y(s) = \gamma(s), \qquad (7\text{-}2.15)$$

which is a PPS. This corresponds with the result of dynamic compensation.

Example 7-2.3. Consider the two-order system

$$\begin{bmatrix} 1 & 0 \\ 0 & 0 \end{bmatrix}\dot{x} = \begin{bmatrix} 1 & 2 \\ 0 & 1 \end{bmatrix}x + \begin{bmatrix} 1 \\ 1 \end{bmatrix}u$$

$$y = [1 \quad 0]x.$$

For any state feedback in the form of $u = Kx + v = [k_1, k_2]x + v$, its closed-loop system has the following input-output transfer relationship:

$$y(s) = [1 \quad 0](s\begin{bmatrix} 1 & 0 \\ 0 & 0 \end{bmatrix} - \begin{bmatrix} 1 & 2 \\ 0 & 1 \end{bmatrix} - \begin{bmatrix} 1 \\ 1 \end{bmatrix}[k_1 \quad k_2])^{-1}\begin{bmatrix} 1 \\ 1 \end{bmatrix}v(s)$$

$$= \frac{1}{-(1+k_2)s + (1-k_1-k_2)}v(s).$$

If we select $k_1 = -1$, $k_2 = -1$, it will be $y(s) = v(s)$. Noticing that in this case, the closed-loop system has only infinite poles, this design is not applicable in practice.

Example 7-2.4. The following system

$$\begin{bmatrix} 1 & 0 \\ 0 & 0 \end{bmatrix}\dot{x} = \begin{bmatrix} 1 & 0 \\ 2 & 1 \end{bmatrix}x + \begin{bmatrix} 1 \\ 1 \end{bmatrix}u$$

$$y = [0 \quad 1]x$$

and the state feedback $u = [k_1 \quad k_2]x + v$ form the closed-loop system

$$\begin{bmatrix} 1 & 0 \\ 0 & 0 \end{bmatrix}\dot{x} = \begin{bmatrix} 1+k_1 & k_2 \\ 2+k_1 & 1+k_2 \end{bmatrix}x + \begin{bmatrix} 1 \\ 1 \end{bmatrix}v$$

$$y = [0 \quad 1]x, \tag{7-2.16}$$

which has the input-output relationship

$$y(s) = \frac{s+1}{-(1+k_2)s + 1+k_1-k_2} v(s).$$

Choosing $k_1 = k_2 = -2$, we have $y(s) = (s+1)/(s+1)v(s) = v(s)$, which is also a PPS. This system has no infinite poles. And stable zero-pole cancellation occurs here.

7-2.2. P-D state feedback

Now we consider the P-D state feedback:

$$u = K_1x + K_2\dot{x} + r(t). \tag{7-2.17}$$

According to the previous discussion, a P-state feedback $u = Kx + v(t)$ may be selected such that (7-2.10) is r.s.e. to (7-2.13). For any feedback (7-2.17), we denote

$$\hat{K}_1 = K_1 - K = [k_1^1, k_2^1, \dots, k_n^1]\bar{P}^{-1}, \quad K_2 = [k_1^2, k_2^2, \dots, k_n^2]\bar{P}^{-1}$$

$$v = \hat{K}_1x + K_2\dot{x} + r(t).$$

Then the closed-loop system formed (7-2.1) and (7-2.17) is

$$(E-bK_2)\dot{x} = (A+bK_1)x + br(t)$$

$$y = cx, \tag{7-2.18}$$

which has the transfer relationship

$$y(s) = c(s(E-bK_2)-(A+bK_1))^{-1}br(s) = \frac{c_1^0s^{n-1}+c_2^0s^{n-2}+\dots+c_n^0}{k_1^2s^n+(k_2^2-k_1^1)s^{n-1}+\dots+(-1-k_n^1)} r(s) \tag{7-2.19}$$

showing that as in the case of P-state feedback, under any P-D state feedback in the form of (7-2.17) the closed-loop system (7-2.18) has a fixed numerator unless zero-pole cancellation occurs. However, its denominator may take any polynomial of order not greater than n.

7-3. State Feedback and Transfer Matrix: Multiinput Multioutput Systems

The results on transfer matrix structure under state feedback for single-input single-output systems, although simple, are applicable to multiinput multioutput systems:

$$\begin{aligned} E\dot{x} &= Ax + Bu \\ y &= Cx \end{aligned} \tag{7-3.1}$$

where $x \in \mathbb{R}^n$, $u \in \mathbb{R}^m$, $y \in \mathbb{R}^r$ are its state, control input, and output, respectively; and $E, A \in \mathbb{R}^{n\times n}$, $B \in \mathbb{R}^{n\times m}$, $C \in \mathbb{R}^{r\times n}$ are constant matrices. System (7-3.1) is assumed to be regular and $\mathrm{rank}E < n$.

Consider the P-D state feedback control

$$u = K_1x + K_2\dot{x} + Fv \tag{7-3.2}$$

where $K_1, K_2 \in \mathbb{R}^{m\times n}$, $F \in \mathbb{R}^{m\times m}$ are constant matrices, F is nonsingular, and v(t) is the new input. Under the control (7-3.2), system (7-3.1) has the closed-loop system

$$\begin{aligned} (E-BK_2)\dot{x} &= (A+BK_1)x + BFv \\ y &= Cx \end{aligned} \tag{7-3.3}$$

with the transfer matrix

$$G_F(s) \triangleq C(s(E-BK_2) - (A+BK_1))^{-1}BF, \tag{7-3.4}$$

whose structure is dependent of the selection of K_1, K_2, F.

Subject to the controllability and observability assumptions for system (7-3.1), for any α satisfying $|\alpha E+A| \neq 0$, $(\hat{E},\hat{B},C)$ is controllable and observable (Section 2-4), where

$$\hat{E} = (\alpha E+A)^{-1}E, \qquad \hat{B} = (\alpha E+A)^{-1}B. \tag{7-3.5}$$

System (7-3.1) is r.s.e. to its EF3:

$$\begin{aligned} \hat{E}\dot{x} &= (I-\alpha\hat{E})x + \hat{B}u \\ y &= Cx \end{aligned} \tag{7-3.6}$$

with the closed-loop transfer matrix

$$G_F(s) = C((s+\alpha)\hat{E} - s\hat{B}K_2 - (I+\hat{B}K_1))^{-1}\hat{B}F = C((s+\alpha)(\hat{E}-\hat{B}K_2)-(I+\hat{B}(K_1-\alpha K_2)))^{-1}\hat{B}F. \tag{7-3.7}$$

We will start from (7-3.7) to study the transfer matrix structure under P-D state feedback control.

Theorem 7-3.1 (Dai, 1986b). Let system (7-3.1) be controllable and observable, rankB = m. Then for any two α's, $\alpha_1 \neq \alpha_2$, satisfying $|\alpha E+A| \neq 0$, $(\hat{E}_1,\hat{B}_1)$ and $(\hat{E}_2,\hat{B}_2)$ have the same controllability indices. Here, $\hat{E}_i = (\alpha_i E+A)^{-1}E$, $\hat{B}_i = (\alpha_i E+A)^{-1}B$, i = 1, 2.

This theorem shows that the controllability indices of $(\hat{E},\hat{B})$ are independent of the selection of α provided α satisfies $|\alpha E+A| \neq 0$. We call these numbers the controllability indices for system (7-3.1) and denote them by

$$\sigma_1^c \leqslant \sigma_2^c \leqslant \cdots \leqslant \sigma_m^c. \tag{7-3.8}$$

Particularly, when A is nonsingular, α may be chosen as zero. The controllability indices of $(\hat{E},\hat{B})$ are those of $(A^{-1}E,A^{-1}B)$.

Similarly, if rankC = r, we may prove that the observability indices of $(\hat{E},C)$ are also independent of α satisfying $|\alpha E+A| \neq 0$. We define them as the observability indices for system (7-3.1) and denote them by

$$\sigma_1^o \leqslant \sigma_2^o \leqslant \cdots \leqslant \sigma_r^o.$$

Example 7-3.1. Consider the system

$$\begin{bmatrix} 1 & 0 & 0 \\ -1 & 1 & 0 \\ 0 & 0 & 0 \end{bmatrix}\dot{x} = \begin{bmatrix} 1 & 0 & 1 \\ -2 & -1 & 0 \\ 0 & 1 & 0 \end{bmatrix}x + \begin{bmatrix} 1 & 0 \\ 0 & 0 \\ 0 & 1 \end{bmatrix}u$$
$$y = [0 \quad 0 \quad 1]x. \tag{7-3.9}$$

In this system, matrix A is nonsingular. Thus, the controllability indices for this system are that of $(A^{-1}E,A^{-1}B)$.

Since

$$E = A^{-1}E = \begin{bmatrix} 0.5 & -0.5 & 0 \\ 0 & 0 & 0 \\ 0.5 & 0.5 & 0 \end{bmatrix}, \qquad B = A^{-1}B = \begin{bmatrix} 0 & 0.5 \\ 0 & 1 \\ 1 & 0.5 \end{bmatrix},$$

which has the controllability indices 1 and 2, system (7-3.9) has controllability indices 1 and 2.

It may be testified that this system is observable. Hence, $(\hat{E},C)$ is observable with observability index 3, which is that for system (7-3.9).

Let system (7-3.1) be controllable, rankB = m, and $\sigma_1^c \leqslant \sigma_2^c \leqslant \cdots \leqslant \sigma_m^c$ be the controllability indices for system (7-3.1), which are also of $(\hat{E},\hat{B})$.

Denote $\hat{B} = [b_1, b_2, \dots, b_m]$. According to the definition of controllability indices, and by rearranging the order of columns in $\hat{B}$, without loss of generality we assume that $\hat{B}$ is in its original order such that the matrix

$$M = [b_1, \hat{E}b_1, \dots, \hat{E}^{\sigma_1^c - 1} b_1,\ b_2, \hat{E}b_2, \dots, \hat{E}^{\sigma_2^c - 1} b_2,\ \dots,\ b_m, \hat{E}b_m, \dots, \hat{E}^{\sigma_m^c - 1} b_m]$$

is nonsingular.

Denote $d_i = \sum_{j=1}^{i} \sigma_j^c$, $i = 1, 2, \dots, m$, and T_k is the d_ith row in M^{-1}, $k = 1, 2, \dots, m$. We construct

$$T = [T_1/T_1\hat{E}/\dots/T_1\hat{E}^{\sigma_1^c - 1}/T_2/T_2\hat{E}/\dots/T_2\hat{E}^{\sigma_2^c - 1}/\dots/T_m/T_m\hat{E}/\dots/T_m\hat{E}^{\sigma_m^c - 1}]. \tag{7-3.10}$$

By the constructing method, we know that T satisfies $|TM| = (-1)^d$, d is a positive scalar, showing the nonsingularity of T. For the nonsingular matrix T chosen in such a way, we define

$$\tilde{E} = T\hat{E}T^{-1} = (\tilde{E}_{ij}), \qquad \tilde{B} = T\hat{B}, \qquad \tilde{C} = CT^{-1}. \tag{7-3.11}$$

Then

$$\tilde{E}_{ii} = \begin{bmatrix} 0 & 1 & 0 & \dots & 0 \\ 0 & 0 & 1 & \dots & 0 \\ & \dots & & \dots & \\ 0 & 0 & 0 & \dots & 1 \\ * & * & * & \dots & * \end{bmatrix} \in \mathbb{R}^{\sigma_i^c \times \sigma_j^c}$$

$$\tilde{E}_{ij} = \begin{bmatrix} 0 & 0 & 0 & \dots & 0 \\ 0 & 0 & 0 & \dots & 0 \\ & \dots & & \dots & \\ 0 & 0 & 0 & \dots & 0 \\ * & * & * & \dots & * \end{bmatrix} \in \mathbb{R}^{\sigma_i^c \times \sigma_j^c}, \quad i \neq j \tag{7-3.12}$$

$$\tilde{B} = \begin{bmatrix} 0 & 0 & 0 & \dots & 0 \\ & \dots & & \dots & \\ 0 & 0 & 0 & \dots & 0 \\ 1 & * & * & \dots & * \\ 0 & 0 & 0 & \dots & 0 \\ & \dots & & \dots & \\ 0 & 1 & * & \dots & * \\ & \dots & & \dots & \\ 0 & 0 & 0 & \dots & 1 \end{bmatrix} \in \mathbb{R}^{n \times m}, \tag{7-3.13}$$

where "*" represents the possibly nonzero elements.

On the other hand, equation (7-3.7) shows

$$G_F(s) = C(\mu\hat{E} - I - \hat{B}\hat{K}_1 - K_2\mu)^{-1}\hat{B}F,$$

where $\mu = s+\alpha$, $K_1 = K_1 - K_2\alpha$.

Let $\tilde{E}_m \in \mathbb{R}^{m\times n}$ be the matrix formed by the d_i rows of $\tilde{E}$, $\tilde{B}_m$ be an mxm nonsingular matrix formed by the d_i rows of $\tilde{B}$,

$$\tilde{K}_1 = \hat{K}_1 T^{-1}, \qquad \tilde{K}_2 = K_2 T^{-1}$$

$$\tilde{S}(\mu) = \mathrm{diag}(\begin{bmatrix} \mu^{\sigma_i^c - 1} \\ \vdots \\ \mu \\ 1 \end{bmatrix}, \ i = 1, 2, \dots , m). \qquad (7\text{-}3.14)$$

It is easy to verify that

$$(\tilde{E}\mu - (I+\tilde{B}\tilde{K}_1+\tilde{B}\tilde{K}_2\mu))\tilde{S}(\mu) = \tilde{B}\tilde{B}_m^{-1}((\tilde{E}_m\mu - \tilde{B}_m\tilde{K}_1 - \tilde{B}_m\tilde{K}_2\mu)\tilde{S}(\mu) - I).$$

Thus we have

$$(\tilde{E}\mu - (I+\tilde{B}\tilde{K}_1+\tilde{B}\tilde{K}_2\mu))^{-1}\tilde{B} = \tilde{S}(\mu)((\tilde{E}_m - \tilde{B}_m\tilde{K}_2)\mu\tilde{S}(\mu) - \tilde{B}_m\tilde{K}_1\tilde{S}(\mu) - I)^{-1}\tilde{B}_m.$$

Rearranging the columns in $(\tilde{E}_m - \tilde{B}_m\tilde{K}_2)\mu\tilde{S}(\mu) - \tilde{B}_m\tilde{K}_1\tilde{S}(\mu) - I$ so as to be written as

$$(\tilde{E}_m - \tilde{B}_m\tilde{K}_2)\mu\tilde{S}(\mu) - \tilde{B}_m\tilde{K}_1\tilde{S}(\mu) - I = KS(\mu),$$

where

$$S(\mu) = \mathrm{diag}(\ [\mu^{\sigma_i^c}, \ \dots \ , \mu \ , \ 1]^{\tau}, \ i = 1, 2, \ \dots \ , m).$$

Paying attention to the form of $S(\mu)$ and the nonsingularity of $\tilde{B}_m$, from (7-3.14) we assure that K is determined uniquely by K_1, K_2, and, conversely, K_1, K_2 may be selected to satisfy (7-3.17) once K is fixed.

Thus

$$G_F(s) = \tilde{C}\tilde{S}(\mu)(KS(\mu))^{-1}\tilde{B}_m F.$$

Denoting $R(s) = \tilde{C}\tilde{S}(\mu)$, $N(s) = \bar{K}S(\mu)$, $\bar{K} = (\tilde{B}_m F)^{-1}K$, we have

$$G_F(s) = R(s)N^{-1}(s). \qquad (7\text{-}3.15)$$

Therefore, from this process we see that $R(s)$ is determined completely by the system itself and $N(s)$ may take any nonsingular matrix whose order of ith column is not greater than σ_i^c by suitable selection of K_1, K_2, F.

In summary we have proven the following theorem.

<u>Theorem 7-3.2.</u> Assume that system (7-3.1) is controllable, rank$B = m$. Then under P-D state feedback (7-3.2), the closed-loop transfer matrix $G_F(s)$ has the following

structure:

$$C_F(s) = R(s)N^{-1}(s) \tag{7-3.16}$$

where R(s) is determined by the system itself and N(s) may take any nonsingular polynomial matrix whose order of ith column is not greater than σ_i^c.

For multivariable system (7-3.1), it is called a pure prediction system (PPS) if there exists a matrix K such that

$$y(s) = Ku(s).$$

A direct result of Theorem 7-3.2 is the following theorem.

Theorem 7-3.3. Assume that system (7-3.1) is controllable, rankB = m. There exist feedback gain matrices K_1, K_2, and F such that the closed-loop system (7-3.3) is a PPS if and only if a matrix $K \in \mathbb{R}^{r\times m}$ exists satisfying $R(s) = \tilde{C}\tilde{S}(s) = KT(s)$, where T(s) is nonsingular.

Example 7-3.2. For system (7-3.9), $\alpha = 0$ satisfies $|\alpha E+A| \neq 0$, and $\sigma_1^c = 1$, $\sigma_2^c = 2$. Example 7-3.1 gives

$$\hat{E} = \begin{bmatrix} \frac{1}{2} & -\frac{1}{2} & 0 \\ 0 & 0 & 0 \\ \frac{1}{2} & \frac{1}{2} & 0 \end{bmatrix}, \qquad \hat{B} = \begin{bmatrix} 0 & -\frac{1}{2} \\ 0 & 1 \\ 1 & \frac{1}{2} \end{bmatrix} = [b_1, b_2]$$

$$M = [b_1, b_2, \hat{E}b_2] = \begin{bmatrix} 0 & -\frac{1}{2} & -\frac{3}{4} \\ 0 & 1 & 0 \\ 1 & \frac{1}{2} & \frac{1}{4} \end{bmatrix}$$

and

$$M^{-1} = \begin{bmatrix} \frac{1}{3} & -\frac{1}{3} & 1 \\ 0 & 1 & 0 \\ -\frac{4}{3} & -\frac{2}{3} & 0 \end{bmatrix}.$$

Thus we have $T_1 = [\frac{1}{3}, -\frac{1}{3}, 1]$, $T_2 = [-\frac{4}{3}, -\frac{2}{3}, 0]$, and

$$T = \begin{bmatrix} T_1 \\ T_2 \\ T_2\hat{E} \end{bmatrix} = \begin{bmatrix} \frac{1}{3} & -\frac{1}{3} & 1 \\ -\frac{4}{3} & -\frac{2}{3} & 0 \\ -\frac{2}{3} & \frac{2}{3} & 0 \end{bmatrix}, \qquad T^{-1} = \begin{bmatrix} 0 & -\frac{1}{2} & -\frac{1}{2} \\ 0 & -\frac{1}{2} & 1 \\ 1 & 0 & \frac{1}{2} \end{bmatrix}.$$

Direct computation shows that

$$\tilde{E} = T\hat{E}T^{-1} = \begin{bmatrix} 0 & -\frac{1}{2} & 0 \\ 0 & 0 & 1 \\ 0 & 0 & \frac{1}{2} \end{bmatrix}, \qquad \tilde{B} = \begin{bmatrix} 1 & 0 \\ 0 & 0 \\ 0 & 1 \end{bmatrix}$$

$$\tilde{C} = CT^{-1} = [1 \;\; 0 \;\; \tfrac{1}{2}].$$

Therefore, we have $R(s) = [\,1 \;\; \frac{1}{2}]$.

By Theorem 7-3.2, we know that for this system the general structure of its closed-loop system under feedback $u = K_1x + K_2\dot{x} + Fv$ is

$$G_F(s) = [\,1 \;\; \tfrac{1}{2}]N^{-1}(s),$$

where $N(s)$ is nonsingular and

$$N(s) = \begin{bmatrix} a_{11}+a_{12}s & a_{21}+a_{22}s+a_{23}s^2 \\ a_{31}+a_{32}s & a_{41}+a_{42}s+a_{43}s^2 \end{bmatrix}.$$

Choosing $a_{11} = a_{41} = 1$, and the others are zero, we simply have $G(s) = [\,1, \;\; \frac{1}{2}]$.

In viewing of the preceding results, we can now consider the model-matching problem.

The so-called model-matching problem may be stated in such a way: For a given transfer matrix $G_m(s)$ (the model), we find the feedback matrices K_1, K_2, F such that $G_F(s) = G_m(s)$, or the closed-loop system takes $G_m(s)$ as its transfer matrix.

Let $G_m(s)$ be a given model. We make decomposition on the matrix $G_m(s)$

$$G_m(s) = R_m(s)N_m^{-1}(s) \qquad (7\text{-}3.17)$$

where $R_m(s)$, $N_m(s)$ are $r\times m$ and $m\times m$ polynomial matrices, $N_m(s)$ is nonsingular, and $R_m(s)$ and $N_m(s)$ are right coprime, i.e.,

$$\operatorname{rank}\begin{bmatrix} R_m(s) \\ N_m(s) \end{bmatrix} = m, \qquad \forall\; s \in \mathbb{C},\; s \text{ finite}.$$

<u>Theorem 7-3.4</u>. Let system (7-3.1) be controllable, $\operatorname{rank}B = m$, $G_m(s) = R_m(s)N_m^{-1}(s)$ be an $r\times m$ transfer matrix. Then there exist feedback gain matrices K_1, K_2, F such that

$$G_F(s) = G_m(s)$$

if and only if there exists a nonsingular polynomial matrix $H(s) \in \mathbb{R}^{m\times m}$ such that

1. $R(s) = R_m(s)H(s)$.
2. The order of the ith column of $B_m(s)H(s)$ is not greater than σ_i^c, $i = 1, 2, \ldots, m$.

Proof. Sufficiency: For any $N_m(s)H(s)$ satisfying condition 2, there exists a constant matrix M such that $N_m(s)H(s) = MS(s)$. Let $\bar{K} = M$. Then K_1, K_2, F may be chosen from $\bar{K}$. And we have

$$G_F(s) = R(s)N^{-1}(s) = R_m(s)H(s)(N_m(s)H(s))^{-1} = G_m(s).$$

Necessity: Let K_1, K_2, F be gain matrices such that

$$G_F(s) = R(s)N^{-1}(s) = R_m(s)N_m^{-1}(s).$$

Then we have

$$R(s) = R_m(s)N_m^{-1}(s)N(s) = R_m(s)\mathrm{adj}(N_m(s))N(s)/|N_m(s)| \tag{7-3.18}$$

Furthermore, under the decomposition, $R_m(s)$ and $N_m(s)$ are right coprime. There must exist two matrices $D_1(s)$ and $D_2(s)$ such that

$$D_1(s)R_m(s) + D_2(s)N_m(s) = I_m.$$

Right multiplying both sides of this equation by $\mathrm{adj}(N_m(s))N(s)$ yields

$$D_1(s)R_m(s)\mathrm{adj}(N_m(s))N(s) + D_2(s)N_m(s)\mathrm{adj}(N_m(s))N(s) = \mathrm{adj}(N_m(s))N(s).$$

Noticing the fact $N_m(s)\mathrm{adj}(N_m(s)) = |N_m(s)|I$ and (7-3.18), we know

$$\mathrm{adj}(N_m(s))N(s) = H(s)|N_m(s)|,$$

where $H(s) = D_1(s)R(s) + D_2(s)N(s)$, indicating that

$$N_m^{-1}(s)N(s) = \mathrm{adj}(N_m(s))N(s)/|N_m(s)| = H(s)$$

is nonsingular. Hence

$$N(s) = N_m(s)H(s)$$

and

$$R(s) = R_m(s)N_m^{-1}(s)N(s) = R_m(s)H(s).$$

This is condition 1. Condition 2 is easy to obtain according to Theorem 7-3.2. Q.E.D.

7-4. Singular Input Function Observers

Consider the singular system (7-3.1):

$$\begin{aligned} E\dot{x}(t) &= Ax(t) + Bu(t) \\ y(t) &= Cx(t) \end{aligned} \tag{7-4.1}$$

where $x(t) \in \mathbb{R}^n$, $u(t) \in \mathbb{R}^m$, $y(t) \in \mathbb{R}^r$; rank$E < n$, and system (7-4.1) is regular.

Definition 7-4.1. Consider system (7-4.1) with initial condition $x(0)$; $y(t)$ is its measure output when $t \geqslant 0$. If there exist matrices E_c, A_c ($|sE_c - A_c| \neq 0$), B_c, C_c, and initial condition $x_c(0)$ such that

$$\begin{aligned} E_c\dot{x}_c(t) &= A_c x_c(t) + B_c y(t) \\ u(t) &= C_c x_c(t) \end{aligned} \tag{7-4.2}$$

where $x_c \in \mathbb{R}^{n_c}$, and equation (7-4.2) holds for any control input, we will call system (7-4.1) input function observable, and (7-4.2) is an input function observer (IFO).

The IFO, or inverse system, is a system to construct precisely (observe) input function $u(t)$ using the measurement of original system as its input.

The IFOs may be described by Figure 7-4.1.

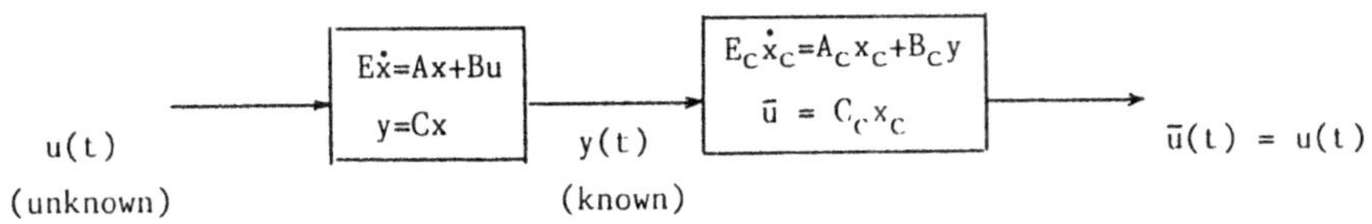

Figure 7-4.1. Input Function Observer

Let

$$G(s) = C(sE-A)^{-1}B \tag{7-4.3}$$

and

$$G_c(s) = C_c(sE_c-A_c)^{-1}B_c \tag{7-4.4}$$

be the transfer matrices of (7-4.1) and (7-4.2), respectively. Then we have the following theorem.

Theorem 7-4.1. System (7-4.1) is input function observable if and only if there exist constant matrices E_c, A_c, B_c, and C_c, such that

$$G_c(s)G(s) = I_m \ . \tag{7-4.5}$$

Proof. Necessity: Assume that system (7-4.1) is input function observable. There exist constant matrices E_c, A_c, B_c, C_c, and initial condition $x_c(0)$ such that (7-4.2) holds for any $u(t)$. Taking Laplace transformation on both sides of (7-4.2) we obtain

$$u(s) = C_c(sE_c-A_c)^{-1}(E_c x_c(0) + B_c y(s)). \tag{7-4.6}$$

Moreover, from (7-4.1) we have $y(s) = C(sE-A)^{-1}(Ex(0)+Bu(s))$. Substituting it into (7-4.6) yields

$$u(s) = C_c(sE_c-A_c)^{-1}[E_c x_c(0) + B_c C(sE-A)^{-1}(Ex(0)+Bu(s))].$$

Since this holds for any $u(s)$, it must be that

$$C_c(sE_c-A_c)^{-1}B_c C(sE-A)^{-1}B = I_m,$$

which is (7-4.5).

Sufficiency: Sufficiency is proven constructively. Let E_c, A_c, B_c, and C_c exist satisfying (7-4.5). We construct the system

$$\begin{aligned} E_c\dot{z}_1 &= A_c z_1 - B_c C z_2 + B_c y, \quad z_1(0) = 0 \\ E\dot{z}_2 &= A z_2, \quad z_2(0) = x(0) \\ \hat{u} &= C_c z_1 . \end{aligned}$$

By taking Laplace transformation on both sides of these equations we know that

$$\begin{aligned} \hat{u}(s) &= C_c z_1(s) = C_c(sE_c-A_c)^{-1}(B_c y(s)-B_c C z_2(s)) \\ &= C_c(sE_c-A_c)^{-1}[B_c C(sE-A)^{-1}(Ex(0)+Bu(s))- B_c C(sE-A)^{-1}Ex(0)] \\ &= G_c(s)G(s)u(s) = u(s). \end{aligned}$$

$\hat{u}$ and u are identical. Thus the system

$$\begin{bmatrix} E_c & 0 \\ 0 & E \end{bmatrix}\begin{bmatrix} \dot{z}_1 \\ \dot{z}_2 \end{bmatrix} = \begin{bmatrix} A_c & -B_c C \\ 0 & A \end{bmatrix}\begin{bmatrix} z_1 \\ z_2 \end{bmatrix} + \begin{bmatrix} B_c \\ 0 \end{bmatrix} y$$

$$z_1(0) = 0, \qquad z_2(0) = x(0) \tag{7-4.7}$$

$$u = [C_c \quad 0][z_1/z_2]$$

is an IFO for system (7-4.1). Hence, this system is input function observable. Q.E.D.

Obviously, the substate z_2 in (7-4.7) is unobservable and the transfer matrix of (7-4.7) is $G_c(s)$.

As shown in Chapter 2, any rational matrix has a singular system realization. Combining the two preceding theorem, we have the following result.

Theorem 7-4.2. System (7-4.1) is input function observable iff its transfer matrix G(s) is left invertible.

Further from Theorem 7-4.2, we have the following corollary.

Corollary 7-4.1. If system (7-4.1) is input function observable, $r \geqslant m$, or in other words, to determine inputs from measurements, there must not be fewer measures than inputs.

Example 7-4.1. Consider the following four-order system

$$\begin{bmatrix} 1 & 0 & 0 & 0 \\ 0 & 1 & 0 & 0 \\ 0 & 0 & 0 & 0 \\ 0 & 0 & 0 & 0 \end{bmatrix} \dot{x} = \begin{bmatrix} 2 & 0 & 0 & 0 \\ 0 & 2 & 0 & 0 \\ 0 & 0 & 1 & 0 \\ 0 & 0 & 0 & 1 \end{bmatrix} x + \begin{bmatrix} 1 & 0 \\ 0 & 1 \\ -1 & 0 \\ 0 & -1 \end{bmatrix} u$$

$$y = \begin{bmatrix} -6 & 3 & -2 & 1 \\ 0 & 0 & 0 & 0 \\ 2 & -1 & 1 & 0 \end{bmatrix} x \tag{7-4.8}$$

with transfer matrix

$$G(s) = C(sE-A)^{-1}B = \begin{bmatrix} \frac{-2(s+1)}{s-2} & \frac{s+1}{s-2} \\ 0 & 0 \\ \frac{s}{s-2} & \frac{-1}{s-2} \end{bmatrix},$$

which is left invertible. By Theorem 7-4.2, its IFO exists.

It has been shown in Example 7-1.2 that

$$G_c(s) = \begin{bmatrix} \frac{1}{s+1} & \frac{1}{s+2} & 1 \\ \frac{s}{s+1} & \frac{s+1}{s+2} & 2 \end{bmatrix}$$

is a left inverse of G(s). $G_c(s)G(s) = I_m$; m = 2. Choosing

$$E_c = \begin{bmatrix} 1 & 0 & 0 & 0 \\ 0 & 1 & 0 & 0 \\ 0 & 0 & 0 & 0 \\ 0 & 0 & 0 & 0 \end{bmatrix}, \quad A_c = \begin{bmatrix} -1 & 0 & 0 & 0 \\ 0 & -2 & 0 & 0 \\ 0 & 0 & 1 & 0 \\ 0 & 0 & 0 & 1 \end{bmatrix}, \quad B_c = \begin{bmatrix} 1 & 0 & 0 \\ 0 & 1 & 0 \\ 0 & 0 & 1 \\ 1 & 1 & 2 \end{bmatrix},$$

$$C_c = \begin{bmatrix} 1 & 0 & -1 & 0 \\ -1 & -1 & 0 & -1 \end{bmatrix}$$

it will be $G_c(s) = C_c(sE_c-A_c)^{-1}B_c$. Thus, from Theorem 7-4.1 we may construct the following IFO for system (7-4.8):

$$\begin{bmatrix} 1&0&0&0&&&& \\ 0&1&0&0&&0&& \\ 0&0&0&0&&&& \\ 0&0&0&0&&&& \\ &&&&1&0&0&0 \\ &0&&&0&1&0&0 \\ &&&&0&0&0&0 \\ &&&&0&0&0&0 \end{bmatrix}\begin{bmatrix}\dot z_1\\ \dot z_2\end{bmatrix} = \begin{bmatrix} -1&0&0&0&-6&3&-2&1 \\ 0&-2&0&0&0&0&0&0 \\ 0&0&1&0&2&-1&1&0 \\ 0&0&0&1&-2&1&0&1 \\ &&&&2&0&0&0 \\ &&0&&0&2&0&0 \\ &&&&0&0&1&0 \\ &&&&0&0&0&1 \end{bmatrix}\begin{bmatrix}z_1\\ z_2\end{bmatrix} + \begin{bmatrix} 0&0&0 \\ 0&1&0 \\ 0&0&1 \\ 1&1&2 \\ && \\ &0& \\ && \\ && \end{bmatrix} y$$

$$z_1(0) = 0, \quad z_2(0) = x(0) \tag{7-4.9}$$

$$u = \begin{bmatrix} 1&1&-1&0&0&0&0&0 \\ -1&-1&0&-1&0&0&0&0 \end{bmatrix}\begin{bmatrix}z_1\\ z_2\end{bmatrix}$$

which is of order 8.

IFOs obtained in such a way are generally of high order. But it has been shown (El-Tohami et al., 1985) that a process may be imposed on it to obtain IFOs of lower order.

Furthermore, from

$$\text{rank}(s\begin{bmatrix}E&0\\0&0\end{bmatrix}-\begin{bmatrix}A&B\\C&0\end{bmatrix}) = \text{rank}\begin{bmatrix}sE-A&-B\\-C&0\end{bmatrix}\begin{bmatrix}I&-(sE-A)^{-1}B\\0&-I\end{bmatrix}$$

$$= n + \text{rank}C(sE-A)^{-1}B$$

and Theorem 7-4.2, we have the following corollary.

Corollary 7-4.2. System (7-4.1) is input function observable if and only if

$$\text{rank}(s\begin{bmatrix}E&0\\0&0\end{bmatrix}+\begin{bmatrix}A&B\\C&0\end{bmatrix}) = n + m$$

is false for no more than n+m s.

This condition may be easily verified by a slight modification of the "shuffle" algorithm in Section 1-2.

7-5. Normal Input Function Observers

For the difficulty in the realization of singular systems, as pointed out before, singular IFOs are difficult to realize physically. Now we will consider the existence and design methods for normal IFOs, which are easier to realize than singular ones. The normal IFO is an IFO in the form of

$$\begin{aligned} \dot{x}_c &= A_c x_c + B_c y \\ u &= F_c x_c + Fy. \end{aligned} \tag{7-5.1}$$

Generally, a singular system does not necessarily have normal IFOs. However, there exist IFOs for some singular systems. It has been proven that only singular systems may have normal IFOs. There definitely exist no normal IFOs for normal systems.

<u>Theorem 7-5.1</u>. If there exists a normal IFO (7-5.1) for system (7-4.1), E must be singular.

<u>Proof</u>. Let (7-5.1) be a normal IFO for system (7-4.1). According to Theorem 7-4.1, we have

$$[F + F_c(sI-A_c)^{-1}B_c][C(sE-A)^{-1}B] = I_m.$$

If E is nonsingular, without loss of generality, we assume $E = I_n$, and the preceding equation becomes

$$[F + F_c(sI-A_c)^{-1}B_c]C(sI-A)^{-1}B = I_m.$$

which holds for any s. But this is obviously false if we set $s \to \infty$ in the preceding equation. This conflict shows E is singular. Q.E.D.

Thus only singular systems have the possibility to possess a normal IFO. Since the transfer matrix is determined by the controllable and observable subsystem, a singular system may have a normal IFO only if its controllable and observable subsystem is singular.

For any regular ssytem (7-4.1), there exist two nonsingular matrices Q and P such that it is r.s.e. to EF1:

$$\begin{aligned} \dot{x}_1 &= A_1 x_1 + B_1 u \\ N\dot{x}_2 &= x_2 + B_2 u \\ y &= C_1 x_1 + C_2 x_2 \end{aligned} \tag{7-5.2}$$

where $P^{-1}x = [x_1/x_2]$, $x_1 \in \mathbb{R}^{n_1}$, $x_2 \in \mathbb{R}^{n_2}$, $n_1 + n_2 = n$, and

$$QEP = \mathrm{diag}(I_{n_1}, N), \qquad QAP = \mathrm{diag}(A_1, I_{n_2}),$$

$$QB = [B_1/B_2], \qquad CP = [C_1 \;\; C_2],$$

$N \in \mathbb{R}^{n_2 x n_2}$ is nilpotent with a nilpotent index h.

Subject to EF1 (7-5.2), the transfer matrix G(s) is

$$G(s) = C(sE-A)^{-1}B = C_1(sI-A_1)^{-1}B_1 + C_2(sN-I)^{-1}B_2. \tag{7-5.3}$$

Let us take observability decomposition on (N,C_2) to obtain

$$TNT^{-1} = \begin{bmatrix} N_{11} & 0 \\ N_{21} & N_{22} \end{bmatrix}, \qquad C_2T^{-1} = [C_{21} \;\; 0]$$

where T is nonsingular and (N_{11},C_{21}) is observable. By denoting

$$Tx_2 = \begin{bmatrix} x_{21} \\ x_{22} \end{bmatrix}, \qquad x_{2i} \in \mathbb{R}^{n_{2i}}, \; i = 1, 2,$$

$$TB_2 = \begin{bmatrix} B_{21} \\ B_{22} \end{bmatrix}, \qquad B_{2i} \in \mathbb{R}^{n_{2i} x m}, \quad i = 1, 2,$$

$n_{21} + n_{22} = n_2$, system (7-5.2) (thus system (7-4.1)) is r.s.e. to

$$\begin{aligned} \dot{x}_1 &= A_1x_1 + B_1u \\ N_{11}\dot{x}_{21} &= x_{21} + B_{21}u \\ N_{21}\dot{x}_{21} + N_{22}\dot{x}_{22} &= x_{22} + B_{22}u \\ y &= C_1x_1 + C_{21}x_{21}. \end{aligned} \tag{7-5.4}$$

Theorem 7-5.2. System (7-4.1) has a normal IFO if and only if there exists an $M \in \mathbb{R}^{m x n_{21}}$ such that

1. $MB_{21} = I_m$, 2. $MN_{11} = 0$.

Proof. Necessity: Let (7-5.1) be a normal IFO. Then by Theorem 7-4.1 it will be

$$[F + F_c(sI-A_c)^{-1}B_c]C(sE-A)^{-1}B = I_m.$$

Expanding it we obtain

$$\begin{aligned} I_m &= [F + F_c(sI-A_c)^{-1}B_c][C_1(sI-A_1)^{-1}B_1 + C_{21}(sN_{11}-I)^{-1}B_{21}] \\ &= [F + F_c(sI-A_c)^{-1}B_c]C_1(sI-A_1)^{-1}B_1 \end{aligned}$$

$$- FC_{21} \sum_{i=0}^{h-1} s^i N_{11}^i B_{21} - F_c(sI-A_c)^{-1} B_c C_{21} \sum_{i=0}^{h-1} s^i N_{11}^i B_{21}. \tag{7-5.5}$$

Since

$$s^i(sI-A_c)^{-1} = s^{i-1}I + s^{s-2}A_c + \dots + A_c^i(sI-A_c)^{-1}, \tag{7-5.6}$$

equation (7-5.5) becomes

$$I_m = [F+F_c(sI-A_c)^{-1}B_c]C_1(sI-A_1)^{-1}B_1 - FC_{21} \sum_{i=0}^{h-1} s^i N_{11}^i B_{21} - F_c(sI-A_c)^{-1}B_cC_{21}B$$

$$- \sum_{i=1}^{h-1} F_c[s^{i-1}I+s^{i-2}A_c+ \dots + A_c^i(sI-A\ )^{-1}B_cC_{21}N_{11}^iB_{21}$$

$$- F_c(sI-A_c)^{-1}B_cC_{21}B_{21}.$$

Equaling the coefficients of corresponding terms at both sides of this equation yields

$$- FC_{21}B_{21} - \sum_{i=1}^{h-1} F_cA_c^{i-1}B_cC_{21}N_{11}^iB_{21} = I_m. \tag{7-5.7}$$

On the other hand, by combining (7-4.1) and (7-5.1), we know that

$$\begin{aligned} u(s) &= F_cx_c(s) + Fy(s) \\ &= F_c(sI-A_c)^{-1}x_c(0) + [F+F_c(sI-A_c)^{-1}B_c]C(sE-A)^{-1}Ex(0) \\ &\quad + [F+F_c(sI-A_c)^{-1}B_c]C(sE-A)^{-1}Bu(s) \end{aligned}$$

holds for any $x(0)$, $x_c(0)$, and $u(s)$. Particularly, setting $u(s) = 0$ results in

$$F_c(sI-A_c)^{-1}x_c(0) + [F+F_c(sI-A_c)^{-1}B_c]C(sE-A)^{-1}Ex(0) = 0.$$

Keeping in mind that the preceding equation holds for any $x(0)$, by using (7-5.6) to expand this equation, and equaling the constant terms on both sides of it, we have

$$FC_2N + F_cB_cC_2N^2 + \dots + F_cA_c^{h-2}B_cC_2N^h = 0,$$

or equivalently,

$$FC_{21}N_{11} + F_cB_cC_{21}N_{11}^2 + \dots + F_cA_c^{h-2}B_cC_{21}N_{11}^h = 0. \tag{7-5.8}$$

Let

$$M = -(FC_{21} + F_cB_cC_{21}N_{11} + \dots + F_cA_c^{h-2}B_cC_{21}N_{11}^{h-1}).$$

Equations (7-5.7) and (7-5.8) are conditions 1 and 2.

Sufficiency: Assume that matrix M satisfies conditions 1 and 2. Left multiplying both sides of the second equation in (7-5.4) with M we obtain

$$u = -Mx_{21}. \tag{7-5.9}$$

On the other hand, system (7-5.4) implies that the substate

$$\tilde{x} = \begin{bmatrix} x_1 \\ x_{21} \end{bmatrix}$$

satisfies the dynamic equation:

$$\begin{bmatrix} I & 0 \\ 0 & N_{11} \\ C_1 & C_{21} \end{bmatrix} \dot{\tilde{x}} = \begin{bmatrix} A_1 & -B_1M \\ 0 & I-B_{21}M \\ 0 & 0 \end{bmatrix} \tilde{x} + \begin{bmatrix} 0 \\ 0 \\ I \end{bmatrix} \dot{y}. \tag{7-5.10}$$

By decomposition, (N_{11}, C_{21}) is observable. Therefore the matrices

$$\begin{bmatrix} N_{11} \\ C_{21} \end{bmatrix} \quad \text{and} \quad \begin{bmatrix} I & 0 \\ 0 & N_{11} \\ C_1 & C_{21} \end{bmatrix}$$

are of full column ranks. Thus, a matrix H may be chosen so that

$$H \begin{bmatrix} I & 0 \\ 0 & N_{11} \\ C_1 & C_{21} \end{bmatrix} = I.$$

Left multiplying both sides of (7-4.10) by H we have

$$\dot{\tilde{x}} = A_c\tilde{x} + \bar{B}_c\dot{y} \tag{7-5.11}$$

where

$$A_c = H \begin{bmatrix} A_1 & -B_1M \\ 0 & I-B_{21}M \\ 0 & 0 \end{bmatrix}, \qquad \bar{B}_c = H \begin{bmatrix} 0 \\ 0 \\ I \end{bmatrix}.$$

Denoting $x_c = \tilde{x} - \bar{B}_c y$, (7-5.11) and (7-5.9) respectively become

$$\begin{aligned} \dot{x}_c &= A_c x_c + B_c y, \qquad x_c(0) = \tilde{x}(0) - \bar{B}_c y(0) \\ u &= F_c x_c + Fy \end{aligned} \tag{7-5.12}$$

where $B_c = A_c\bar{B}_c$, $F_c = -M[0 \quad I]$, $F = -M[0 \quad I]\bar{B}_c$. Thus (7-5.12) is a normal IFO for system (7-4.1). Q.E.D.

Some more convenient conditions may be deduced from the conditions in this theorem.

Let T_1 and T_2 be any two nonsingular matrices satisfying

$$T_1N_{11}T_2 = \text{diag}(I_{\bar{q}}, 0), \qquad \bar{q} = \text{rank}N_{11}$$

and denote $T_1B_{21} = [\breve{B}_1/\tilde{B}_2]$. Theorem 7-5.2 implies the following.

Corollary 7-5.1. System (7-5.1) has a normal IFO in the form of (7-5.1) if and only if $\text{rank}\tilde{B}_2 = m$.

Or, in other words, the dimension of the observable subsystem is no less than the number of inputs.

Corollary 7-5.1 further shows

Corollary 7-5.2. Dual normalizable system (7-4.1) has a normal IFO (7-5.1) if and only if rank[E B] = rankE + m, which is convenient for testification.

The design of normal IFOs may follow the sufficient proof in Theorem 7-5.2.

Example 7-5.1. Example 7-4.1 has shown that system (7-4.8) has a singular IFO (7-4.9).

For system (7-4.8), $m = r = 2$, $n = 4$, rankE = 2, and rank[E/C] = 4. Thus it is dual normalizable, and rank[E B] = 4 = rankE + m. Hence, by Corollary 7-5.2, this system has a normal IFO in the form of (7-5.1).

In fact, if we let $x = [x_1/x_2]$, $x_1, x_2 \in \mathbb{R}^2$, system (7-4.8) may be written as

$$\begin{aligned} \dot{x}_1 &= \begin{bmatrix} 2 & 0 \\ 0 & 2 \end{bmatrix} x_1 + \begin{bmatrix} 1 & 0 \\ 0 & 1 \end{bmatrix} u \\ 0 &= x_2 - u \\ y &= \begin{bmatrix} -6 & 3 \\ 0 & 0 \\ 2 & -1 \end{bmatrix} x_1 + \begin{bmatrix} -2 & 1 \\ 0 & 0 \\ 1 & 0 \end{bmatrix} x_2 . \end{aligned} \tag{7-5.13}$$

Clearly,

$$u = x_2 . \tag{7-5.14}$$

Moreover, system (7-4.8) shows that x satisfies

$$\begin{bmatrix} 1 & 0 & 0 & 0 \\ 0 & 1 & 0 & 0 \\ 0 & 0 & 0 & 0 \\ 0 & 0 & 0 & 0 \\ -6 & 2 & -2 & 1 \\ 0 & 0 & 0 & 0 \\ 2 & -1 & 1 & 0 \end{bmatrix} \dot{x} = \begin{bmatrix} 2 & 0 & 1 & 0 \\ 0 & 2 & 0 & 1 \\ 0 & 0 & 1 & 0 \\ 0 & 0 & 0 & 1 \\ & & & \\ & 0 & & \\ & & & \end{bmatrix} x + \begin{bmatrix} & & \\ & 0 & \\ & & \\ & & \\ 1 & 0 & 0 \\ 0 & 1 & 0 \\ 0 & 0 & 1 \end{bmatrix} \dot{y} . \tag{7-5.15}$$

Left multiplying both sides of (7-5.15) by

$$H = \begin{bmatrix} 1 & 0 & 0 & 0 & 0 & 0 & 0 \\ 0 & 1 & 0 & 0 & 0 & 0 & 0 \\ 0 & 0 & 0 & 0 & 0 & 0 & 1 \\ 0 & 0 & 0 & 0 & 1 & 0 & 2 \end{bmatrix}$$

and rewriting, we obtain

$$\dot{x}_c = \begin{bmatrix} 2 & 0 & 1 & 0 \\ 0 & 2 & 0 & 1 \\ 0 & 0 & 0 & 0 \\ 0 & 0 & 0 & 0 \end{bmatrix} x_c + \begin{bmatrix} 0 & 0 & 1 \\ 1 & 0 & 2 \\ 0 & 0 & 0 \\ 0 & 0 & 0 \end{bmatrix} y$$

$$x_c(0) = x(0) - \begin{bmatrix} 0 & 0 & 0 \\ 0 & 0 & 0 \\ 0 & 0 & 1 \\ 1 & 0 & 2 \end{bmatrix} y(0) \qquad (7\text{-}5.16)$$

$$u = \begin{bmatrix} 0 & 0 & 1 & 0 \\ 0 & 0 & 0 & 1 \end{bmatrix} x_c + \begin{bmatrix} 0 & 0 & 1 \\ 1 & 0 & 2 \end{bmatrix} y$$

where

$$x_c = x - \begin{bmatrix} 0 & 0 & 0 \\ 0 & 0 & 0 \\ 0 & 0 & 1 \\ 1 & 0 & 2 \end{bmatrix} y,$$

which is apparently a normal IFO for system (7-4.8).

Furthermore, (7-5.13) and (7-5.14) show that if we use the matrix $\bar{\Pi}$

$$\bar{\Pi} = \begin{bmatrix} 0 & 0 & 1 \\ 1 & 0 & 2 \end{bmatrix}$$

to left multiply both sides of the third equation in (7-5.13) it will be

$$u = -\begin{bmatrix} 2 & -1 \\ -2 & 1 \end{bmatrix} x_c + \begin{bmatrix} 0 & 0 & 1 \\ 1 & 0 & 2 \end{bmatrix} y \qquad (7\text{-}5.17)$$

where $x_c = x_1$. The substitution of it into the first equation in (7-5.13) yields

$$\dot{x}_c = \begin{bmatrix} 0 & 1 \\ 2 & 1 \end{bmatrix} x_c + \begin{bmatrix} 0 & 0 & 1 \\ 1 & 0 & 2 \end{bmatrix} y$$
$$x_c(0) = x_1(0) \qquad (7\text{-}5.18)$$

which, together with (7-5.17), forms another normal IFO for system (7-4.8) of order 2, which is lower than the preceding one.

7-6. Decoupling control

In a multivariable control system, every scalar output is generally affected by several control inputs, which results in a complicated input-output relationship in control system. Decoupling control is a kind of control strategy that can make the closed-loop system have a simple input-output relationship. One output may be completely determined by only one control input. Such systems are called decoupled. Obviously, as long as the decoupling is achieved, the multivariable system is split into several independent single input, single output systems. The advantage of decoupling control lies in its simplicity and reliability, especially when the control is partially executed by a human factor.

7-6.1. Dynamic decoupling control

Definition 7-6.1. A singular system is called (dynamically) decoupled if its transfer matrix is diagonally nonsingular.

Consider the multivariable singular system

$$\begin{aligned} E\dot{x} &= Ax + Bu \\ y &= Cx \end{aligned} \tag{7-6.1}$$

where $x \in \mathbb{R}^n$, $u \in \mathbb{R}^m$, $y \in \mathbb{R}^r$ are its state, control input, and measure output, respectively; E, $A \in \mathbb{R}^{n\times n}$, $B \in \mathbb{R}^{n\times m}$, $C \in \mathbb{R}^{r\times n}$ are constant matrices; $\operatorname{rank}E < n$. System (7-6.1) is regular.

Let the feedback control be the P-D state feedback

$$u = K_1x + K_2\dot{x} + Fv \tag{7-6.2}$$

where K_1, $K_2 \in \mathbb{R}^{m\times n}$, $F \in \mathbb{R}^{m\times m}$ are constant matrices.

System (7-6.1) and control (7-6.2) form the closed-loop system:

$$\begin{aligned} (E-BK_2)\dot{x} &= (A+BK_1)x + BFv \\ y &= Cx \end{aligned} \tag{7-6.3}$$

with the transfer matrix

$$G_F(s) = C[s(E-BK_2) - (A+BK_1)]^{-1}BF. \tag{7-6.4}$$

Since $G_F(s)$ is an $r\times m$ matrix, to guarantee that $G_F(s)$ is diagonal we stipulate $r = m \leqslant n$ and F is nonsingular.

In this section, system (7-6.1) is assumed both controllable and observable.

By definition, the decoupling problem is to find feedback matrices K_1, K_2, F such that $G_F(s)$ is diagonally nonsingular, i.e., $G_F(s) = \operatorname{diag}(g_1(s), g_2(s), \ldots, g_m(s))$, $g_i(s)$ is nonzero polynomial, $i = 1, 2, \ldots, m$.

We have seen in Theorem 7-3.2 that if system (7-6.1) is controllable and $\operatorname{rank}B = m$, the closed-loop transfer matrix under P-D state feedback (7-6.2) has the form:

$$G_F(s) = R(s)N^{-1}(s) \tag{7-6.5}$$

where $R(s)$ is independent of the selection of K_1, K_2, and F. $N(s)$ may take any nonsingular polynomial matrix whose order of ith column is not greater than σ_i^c. Let $N(s)$

be denoted by $N_0(s)$ when $K_1 = K_2 = 0$ and $F = I$. Then

$$G(s) = C(sE-A)^{-1}B = R(s)N_0^{-1}(s). \tag{7-6.6}$$

Theorem 7-6.1. For system (7-6.1), there exists feedback control (7-6.2) such that the closed-loop system (7-6.3) is decoupled iff its transfer matrix $G(s)$ is nonsingular.

Proof. Necessity: Let K_1, K_2, F be matrices such that (7-6.3) is decoupled. Then $G_F(s)$ is diagonally nonsingular, implying the nonsingularity of $G(s)$ by (7-6.6).

Sufficiency: If $G(s)$ is nonsingular, from (7-6.6) we know $R(s)$ is nonsingular. Note that the order of ith column of $R(s)$ is not greater than σ_i^c. There exist K_1, K_2, F such that $N(s) = R(s)$, and $G_F(s) = I$ is diagonally nonsingular. Q.E.D.

Thus a necessary condition for decoupling is that rankB = rankC = m = r.

If system (7-6.1) may be decoupled via feedback control (7-6.2), the preceding theorem shows that the transfer matrix $G(s) = C(sE-A)^{-1}B$ is nonsingular. Note that for any α satisfying $|\alpha E-A| \neq 0$, and $|G(\alpha)| = |C(\alpha E-A)^{-1}B| \neq 0$,

$$G_F(s) = C(s(E-BK_2) - (A+BK_1))^{-1}B = C(sE-A)^{-1}B[I - (K_1+sK_2)(sE-A)^{-1}B]^{-1}F \tag{7-6.7}$$

and

$$C(sE-A)^{-1}B = C(\alpha E-A)^{-1}B + \alpha C(\alpha E-A)^{-1}(sE-A)^{-1}B - sC(\alpha E-A)^{-1}(sE-A)^{-1}B.$$

By setting $K_1 = -\alpha FC(\alpha E-A)^{-1}$, $K_2 = FC(\alpha E-A)^{-1}$, $F = [C(\alpha E-A)^{-1}B]^{-1}$, we have $G_F(s) = I$, implying the closed-loop system is a PPS.

By noticing m = r, from Corollary 7-4.2 and Theorem 7-6.1, we have the following.

Corollary 7-6.1. There exists state feedback control (7-6.2) such that the closed-loop system (7-6.3) is decoupled if and only if the following matrix pencil

$$\left(\begin{bmatrix} E & 0 \\ 0 & 0 \end{bmatrix}, \begin{bmatrix} A & B \\ C & 0 \end{bmatrix}\right)$$

is regular, which is easy to testify via the "shuffle" algorithm.

Particularly, when $E = I_n$, Theorem 7-6.1 is applicable to the normal system

$$\dot{x} = Ax + Bu$$
$$y = Cx. \tag{7-6.9}$$

This means that if the state feedback $u = Kx + v$ is extended to the general form (7-6.2), the decoupling problem may have a wider feasibility so as to find a feedback control such that the closed-loop system is a PPS.

7-6.2. Static decoupling

Now we will consider the static decoupling problem.

Definition 7-6.2. System (7-6.1) is called static decoupling if it is stable and

$$\lim_{s\to 0} G(s) = \lim_{s\to 0} C(sE-A)^{-1}B \tag{7-6.10}$$

is a diagonally nonsingular matrix; $\lim_{s\to 0} G(s)$ is assumed to exist finitely.

Under the control (7-6.2), the static decoupling problem is to choose gain matrices K_1, K_2, F such that

$$\lim_{s\to 0} G_F(s) = \lim_{s\to 0} C(s(E-BK_2) - (A+BK_1))^{-1}BF = \mathrm{diag}(g_1, g_2, \ldots, g_m)$$

where $g_i \neq 0$ is scalar, $i = 1, 2, \ldots, m$.

The static decoupling property states an asymptotical decoupling behavior, without mention the decoupling property for any finite time period.

Theorem 7-6.2. For system (7-6.1), there exists a feedback control (7-6.2) such that the closed-loop system (7-6.3) is static decoupling if and only if it is stabilizable and

$$\mathrm{rank}\begin{bmatrix} A & B \\ C & 0 \end{bmatrix} = n + m. \tag{7-6.11}$$

Proof. Assume that there exist feedback matrices K_1, K_2, and F such that (7-6.3) is static decoupling. Therefore, $\sigma(E-BK_2, A-BK_1) \subset \mathbb{C}^-$. The closed-loop system is stable, and system (7-6.1) is stabilizable. $A-BK_1$ is nonsingular. In this case,

$$\lim_{s\to 0} G_F(s) = \lim_{s\to 0} C[s(E-BK_2) - (A+BK_1)]^{-1}BF = -C(A+BK_1)^{-1}BF$$

is nonsingular, i.e.,

$$\mathrm{rank}\ C(A+BK_1)^{-1}B = m. \tag{7-6.12}$$

On the other hand, for such a matrix K_1,

$$\mathrm{rank}\begin{bmatrix} A & B \\ C & 0 \end{bmatrix} = \mathrm{rank}\begin{bmatrix} A+BK_1 & B \\ C & 0 \end{bmatrix} = \mathrm{rank}\begin{bmatrix} A+BK_1 & 0 \\ C & -C(A+BK_1)^{-1}B \end{bmatrix}. \tag{7-6.13}$$

Combining it with (7-6.12) we conclude that (7-6.11) holds.

Conversely, if (7-6.11) is true and system (7-6.1) is stabilizable, we may choose a matrix K_1 such that

$$\sigma(E, A+BK_1) \subset \mathbb{C}^- \tag{7-6.14}$$

and thus both $A+BK_1$ and $C(A+BK_1)^{-1}B$ are nonsingular according to (7-6.13). Setting

$$K_2 = 0, \qquad F = -[C(A+BK_1)^{-1}B]^{-1},$$

we have

$$\lim_{s\to 0} G_F(s) = \lim_{s\to 0} C[s(E-BK_2) - (A+BK_1)]^{-1}BF = I_m.$$

therefore, for such matrices K_1, K_2, and F the closed-loop system (7-6.3) is static decoupling. Q.E.D.

Remark 1. A necessary condition for static decoupling is rankB = rankC = m.

Remark 2. K_1 and K_2 are used only to stabilize the closed-loop system. F is chosen to assure the static decoupling.

Remark 3. As seen in the proof process, there exists a feedback (7-6.2) such that closed-loop system (7-6.3) is static decoupling if and only if a P-state feedback

$$u = K_1x + Fv \tag{7-6.14}$$

may be chosen such that the closed-loop system

$$\begin{aligned} E\dot{x} &= (A+BK_1)x + BFv \\ y &= Cx \end{aligned} \tag{7-6.15}$$

is static decoupling.

Thus, for static decoupling problem we prefer using the P-state feedback (7-6.14) for its simplicity.

Example 7-6.1. Consider the singular system

$$\begin{aligned} \begin{bmatrix} 1 & 0 & 0 \\ 0 & 1 & 0 \\ 0 & 0 & 0 \end{bmatrix} \dot{x} &= \begin{bmatrix} 0 & 1 & 0 \\ 0 & 0 & 0 \\ 0 & 1 & 1 \end{bmatrix} x + \begin{bmatrix} 1 & 0 \\ 0 & 1 \\ 0 & 0 \end{bmatrix} u \\ y &= \begin{bmatrix} 1 & 0 & 0 \\ 0 & 0 & 1 \end{bmatrix} x, \end{aligned} \tag{7-6.16}$$

which is stabilizable. The feedback gain matrix

$$K_1 = \begin{bmatrix} 0 & 0 & 0 \\ -1 & -2 & 0 \end{bmatrix}$$

satisfies $\sigma(E, A+BK_1) = \{-1, -1\} \subset \mathbb{C}^-$.

In this system n = 3, m = r = 2, and

$$\text{rank}\begin{bmatrix} A & B \\ C & 0 \end{bmatrix} = 5 = n + m.$$

Thus, by Theorem 7-6.2, there exists a (7-6.14) such that the closed-loop system (corresponding to (7-6.15)) is static decoupling.

In fact, if we select

$$F = [C(A+BK_1)^{-1}B]^{-1}$$

$$= \begin{bmatrix} 0 & -1 \\ -1 & 2 \end{bmatrix}$$

and define

$$u = \begin{bmatrix} 0 & 0 & 0 \\ -1 & -2 & 0 \end{bmatrix} x + \begin{bmatrix} 0 & -1 \\ -1 & 2 \end{bmatrix} v$$

the closed-loop system is

$$\begin{bmatrix} 1 & 0 & 0 \\ 0 & 1 & 0 \\ 0 & 0 & 0 \end{bmatrix} \dot{x} = \begin{bmatrix} 0 & 1 & 0 \\ -1 & -2 & 0 \\ 0 & 1 & 1 \end{bmatrix} x + \begin{bmatrix} 0 & -1 \\ -1 & 2 \\ 0 & 0 \end{bmatrix} v$$

$$y = \begin{bmatrix} 1 & 0 & 0 \\ 0 & 0 & 1 \end{bmatrix} x,$$

which is static decoupling. This may be verified by selecting v to be constant input, in which case

$$\lim_{t \to \infty} y(t) = v.$$

Concerning the relationship between static decoupling and dynamic decoupling, the following has been proven (Dai, 1988f):

<u>Theorem 7-6.3</u>. System (7-6.1) may be decoupled via P-D state feedback (7-6.2) if it may be static decoupled via state feedback (7-6.2).

Thus, when (7-6.11) holds, there are two design ways for our selection: dynamic decoupling or static decoupling. The former decouples the system instantly without guaranteeing the stability for the closed-loop system, while the latter, assuring the closed-loop stability, cannot achieve the instant decoupling property.

It is worth pointing out that the inverse of Theorem 7-6.4 is false.

<u>Example 7-6.5</u>. In the system

$$\begin{bmatrix} 1 & 0 & 0 \\ 0 & 1 & 0 \\ 0 & 1 & 0 \end{bmatrix} \dot{x} = \begin{bmatrix} 0 & 1 & 0 \\ 0 & 0 & 0 \\ 0 & 0 & 1 \end{bmatrix} x + \begin{bmatrix} 1 & 0 \\ 0 & 1 \\ 0 & 0 \end{bmatrix} u$$

$$y = \begin{bmatrix} 1 & 0 & 0 \\ 0 & 0 & 1 \end{bmatrix} x \qquad (7\text{-}6.17)$$

$m = r = 2$, $n = 3$. Since

$$\text{rank}\begin{bmatrix} A & B \\ C & 0 \end{bmatrix} = 4 < 5 = n + m,$$

this system cannot be static decoupled via (7-6.2).

However, choosing $\alpha = 1$ we have

$$\operatorname{rank} C(\alpha E+A)^{-1}B = \operatorname{rank}\begin{bmatrix} 1 & -1 \\ 0 & -1 \end{bmatrix} = 2 = m.$$

Thus the transfer matrix of (7-6.17) is nonsingular and by Theorem 7-6.1 the system may be (dynamically) decoupled via feedback (7-6.2).

7-7. Notes and References

Closed-loop structures under P-D state feedback control are based on Wolovich (1974) and Dai (1986b). In this chapter, we theoretically discussed the IFO problem for singular systems based on Dai (1988c). This problem was investigated earlier by El-Tohami et al. (1984, 1985) using a matrix pseudo-inverse method. Decoupling problems for singular systems are first explored by Christodoulou (1986), and further studied by Zhou et al. (1987), Mertzios and Christodoulou (1986), and Dai (1988l,h). In Dai (1988h) a strong decoupling problem is investigated. The strong decoupling problem emphasizes the closed-loop not only (dynamically or static) decoupled but also ensures no infinite pole, or no impulse terms exist in closed-loop state response. The primary purpose of this chapter is to show the difference and similarity between singular and normal systems.

CHAPTER 8

INTRODUCTION TO DISCRETE-TIME SINGULAR SYSTEMS

Discrete-time singular systems, or sampled digital singular systems, are one kind of system in which the variables take their value at instantaneous time points. Discrete-time systems defer themselves from continuous-time ones in that their signals are in the form of sampled data.

With the development of the digital computer, the discrete-time system theory plays an important role in control theory. In real systems, the discrete-time system often appears when it is the result of sampling the continuous-time system; or when only discrete data are available for use; or when computers are involved in the control loop. When computers are used in a control system, they process the sampled data to determine the control law for the next step, then via analog digital converter, realizing the control via the excutive components. Discrete-time systems exist tremendously in social systems, time series analysis, etc. The Leontief dynamic economic system model in the introduction of Chapter 1 is a discrete-time singular system.

The general discrete-time singular system may be described by the following model:

$$\begin{aligned} &f_k(x(k+1),x(k),\ldots,x(0),u(k),u(k-1),\ldots,u(0)) = 0 \\ &y(k) = g_k(x(k),x(k-1),\ldots,x(0),u(k),u(k-1),\ldots,u(0)) \end{aligned} \qquad (8\text{-}0.1)$$

where the state $x(k) = x(t_k)$ (or $x(Tk)$, where T is the sampling period), $x(k) \in \mathbb{R}^n$, the control input $u(k) \in \mathbb{R}^m$, the measure output $y(k) \in \mathbb{R}^r$, all at the kth sampling time point. f_k and g_k are vector functions of appropriate dimensions in $x(k+1)$, $x(k)$, $\ldots$, $x(0)$, $u(k)$, $u(k-1)$, $\ldots$, $u(0)$.

A special case of (8-0.1) is

$$\begin{aligned} Ex(k+1) &= Ax(k) + Bu(k) \\ y(k) &= Cx(k), \qquad\qquad k = 0, 1, 2, \ldots \end{aligned}$$

where $E, A \in \mathbb{R}^{n\times n}$, $B \in \mathbb{R}^{n\times m}$, $C \in \mathbb{R}^{r\times n}$ are constant matrices. This is the discrete-time system we are going to study in this chapter.

Although much difference lies between discrete-time and continuous-time systems, discrete-time systems are in most aspects akin to continuous-time systems. Thus, to avoid tedious repetition, results for discrete-time systems are only listed without any proof unless necessary.

8-1. Finite Discrete-Time Series

Consider the finite time series described by

$$\begin{aligned} Ex(k+1) &= Ax(k) + Bu(k) \\ y(k) &= Cx(k), \qquad k = 0, 1, \dots, L \end{aligned} \tag{8-1.1}$$

where $x(k) \in \mathbb{R}^n$, $u(k) \in \mathbb{R}^m$, $y(k) \in \mathbb{R}^r$, are its state, control input, and measure output, respectively; E, $A \in \mathbb{R}^{nxn}$, $B \in \mathbb{R}^{nxm}$, $C \in \mathbb{R}^{rxn}$ are constant matrices; $L \geqslant n$ is the fixed terminal point; and rankE < n. We view (8-1.1) as a system of finite time series.

For the input series u(0), u(1), ... , u(L), the state x(0), x(1), ... , x(L) of system (8-1.1) satisfies and is determined by the following equation

$$\begin{bmatrix} -A & E & & & \\ & -A & E & & \\ & & \ddots & \ddots & \\ & & & & E \\ & & & -A & E \end{bmatrix} \begin{bmatrix} x(0) \\ x(1) \\ \vdots \\ x(L) \end{bmatrix} = \begin{bmatrix} Bu(0) \\ Bu(1) \\ \vdots \\ Bu(L) \end{bmatrix} \tag{8-1.2}$$

Note that the coefficient in (8-1.2) is an nLxn(L+1) matrix. Thus it has n independent solutions if they exist. If there exists a condition defined by initial and terminal conditions such that different solutions determined by this condition are different at least at one(initial or terminal) point, the condition will be called a complete condition.

Luenberger (1977) has pointed that under the regularity assumption for (8-1.1), i.e., $|zE-A| \not\equiv 0$, a complete condition may be selected from x(0) and x(L) such that any state x(k), $0 \leqslant k \leqslant L$, is uniquely determined by this condition and inputs u(k), k = 0, 1, ... , L.

For normal systems, the complete condition is the initial state condition.

To guarantee the uniqueness of the state solution, system (8-1.1) is always assumed regular.

Subject to the regularity, there exist two nonsingular matrices Q and P such that

$$QEP = \text{diag}(I_{n_1}, N), \qquad QAP = \text{diag}(A_1, I_{n_2}) \tag{8-1.3}$$

where $A_1 \in \mathbb{R}^{n_1 x n_1}$, $N \in \mathbb{R}^{n_2 x n_2}$, N is nilpotent with nilpotent index h, and $n_1 + n_2 = n$. Let

$$x(k) = P\begin{bmatrix} x_1(k) \\ x_2(k) \end{bmatrix} \begin{matrix} \} n_1 \\ \} n_2 \end{matrix}, \qquad QB = \begin{bmatrix} B_1 \\ B_2 \end{bmatrix} \begin{matrix} \} n_1 \\ \} n_2 \end{matrix}, \qquad CP = [C_1 \quad C_2].$$

System (8-1.1) is r.s.e. to

$$x_1(k+1) = A_1x_1(k) + B_1u(k) \tag{8-1.4a}$$

$$Nx_2(k+1) = x_2(k) + B_2u(k) \tag{8-1.4b}$$

$$y(k) = C_1x_1(k) + C_2x_2(k) \tag{8-1.4c}$$

$$k = 0,\ 1,\ \dots\ L.$$

In these equations, (8-1.4a) is a forward recurrent equation whose state is determined uniquely by initial state $x_1(0)$ and $u(k)$, $k = 0, 1, \dots, L$,

$$x_1(k) = A_1^k x_1(0) + \sum_{i=0}^{k-1} A_1^{k-i-1}B_1u(i). \tag{8-1.5}$$

While (8-1.4b) is a backward recurrence whose state is uniquely determined by terminal state $x_2(L)$ and $u(k)$, $k = 0, 1, \dots, L$ according to

$$x_2(k) = N^{L-k}x_2(L) - \sum_{i=0}^{L-k-1} N^iB_2u(k+i). \tag{8-1.6}$$

Equations (8-1.5) and (8-1.6) show that the initial substate $x_1(0)$ and the terminal substate $x_2(L)$ form a complete condition, under which the state and output are

$$x(k) = P\begin{bmatrix} I \\ 0 \end{bmatrix}(A_1^k x_1(0) + \sum_{i=0}^{k-1} A_1^{k-i-1}B_1u(i) + P\begin{bmatrix} 0 \\ I \end{bmatrix}(N^{L-k}x_2(L) - \sum_{i=0}^{L-k-1} N^iB_2u(k+i))$$

$$y(k) = Cx(k)$$

$$k = 0,\ 1,\ \dots,\ L.$$

This is the general state represention. Therefore, the state at any time point k for a finite time series (8-1.1) is generally related with not only the initial state and former inputs, as in the normal system case, but also terminal state and future inputs up to time point L.

For simplicity, we always assume that system (8-1.1) is in its r.s.e. form (8-1.4).

Example 8-1.1. Consider the finite time series

$$\begin{bmatrix} 1 & 0 & 0 & 0 \\ 0 & 1 & 0 & 0 \\ 0 & 0 & 0 & 1 \\ 0 & 0 & 0 & 0 \end{bmatrix} x(k+1) = \begin{bmatrix} 1 & 1 & 0 & 0 \\ 0 & 1 & 0 & 0 \\ 0 & 0 & 1 & 0 \\ 0 & 0 & 0 & 1 \end{bmatrix} x(k) + \begin{bmatrix} 0 \\ 1 \\ 1 \\ -1 \end{bmatrix} u(k)$$

$$y(k) = [0\ \ 1\ \ 1\ \ 0]x(k) \tag{8-1.7}$$

$$k = 0,\ 1,\ \dots,\ L.$$

By writing $x(k) = [x_1(k)/x_2(k)]$, $x_1(k)$, $x_2(k) \in \mathbb{R}^2$, system (8-1.7) may be written as

$$x_1(k+1) = \begin{bmatrix} 1 & 1 \\ 0 & 1 \end{bmatrix} x_1(k) + \begin{bmatrix} 0 \\ 1 \end{bmatrix} u(k) \tag{8-1.8a}$$

$$\begin{bmatrix} 0 & 1 \\ 0 & 0 \end{bmatrix} x_2(k+1) = x_2(k) + \begin{bmatrix} 1 \\ -1 \end{bmatrix} u(k)$$

$$y(k) = [0 \quad 1]x_1(k) + [1 \quad 0]x_2(k) \qquad (8\text{-}1.8b)$$

$$k = 0, 1, \dots, L.$$

From (8-1.5)-(8-1.6), its solution is

$$x_1(k) = A_1^k x_1(0) + \sum_{i=0}^{k-1} A_1^{k-i-1} B_1 u(i)$$

$$= \begin{bmatrix} 1 & k \\ 0 & 1 \end{bmatrix} x_1(0) + \sum_{i=0}^{k-1} \begin{bmatrix} k-i-1 \\ 1 \end{bmatrix} u(i), \qquad 0 \leqslant k \leqslant L;$$

$$x_2(k) = N^{L-k} x_2(L) - \sum_{i=0}^{L-k-1} N^i B_2 u(k+i)$$

$$= \begin{cases} \begin{bmatrix} 0 & 1 \\ 0 & 0 \end{bmatrix} x_2(L) - \begin{bmatrix} 1 \\ -1 \end{bmatrix} u(k) & \text{when } k = L-1 \\ \begin{bmatrix} 1 \\ 0 \end{bmatrix} u(k+1) - \begin{bmatrix} 1 \\ -1 \end{bmatrix} u(k) & \text{when } 0 \leqslant k \leqslant L-2 \end{cases}$$

Clearly, $x_2(k)$ is independent of the terminal state when $k \leqslant L-2$.

As seen in (8-1.5)-(8-1.6), the substate $x_1(k)$ is determined completely by initial condition $x_1(0)$ and the former inputs, $u(i)$, $i = 0, 1, \dots, k-1$. This relationship between state $x(k)$ and $u(k)$ is called causality.

For example, when $N = 0$ in (8-1.6), $x_2(k) = -B_2u(k)$. This is a causal relationship.

Generally, for system (8-1.1) we define the following.

Definition 8-1.1. The finite time series (8-1.1) is called causal if its state $x(k)$ $(0 \leqslant k \leqslant L)$ at any time point k is determined completely by initial condition $x(0)$ and former (the kth step is included) inputs $u(0)$, $u(1)$, ...,,$u(k)$; otherwise, it is termed noncausal.

Causality is an important relationship. Thus the time series possessing such relationships is an important class of systems with the advantage of easy realizability. Discrete-time normal systems have this property.

Theorem 8-1.1. Time series (8-1.1) is causal iff $\deg(|zE-A|) = \operatorname{rank}E$, or $N = 0$, the system has no infinite poles.

Proof. The result is easy to obtain through the definition and state represention (8-1.6). Q.E.D.

Next we will introduce the concepts of controllability and observability for finite time series.

8-1.1. Controllability

Definition 8-1.2. Finite time series (8-1.1) is called controllable if for any complete condition $[x_1(0)/x_2(L)]$ and $w \in \mathbb{R}^n$ there exists a time point k_1, $0 \leqslant k_1 \leqslant L$ and control inputs $u(0), u(1), \ldots, u(L)$ such that $x(k_1) = w$.

Thus, the controllability is pointwise. Subject to the controllability assumption, state $x(k)$ may pervade the whole space vector.

Theorem 8-1.2. Time series (8-1.1) is controllable iff

$$\mathrm{rank}[B_1, A_1B_1, \ldots, A_1^{n_1-1}B_1] = n_1$$

$$\mathrm{rank}[B_2, NB_2, \ldots, N^{n_2-1}B_2] = n_2$$

where the items are determined in (8-1.4).

Proof. Necessity: Let $x_1(0) = 0$, $x_2(L) = 0$. Under the controllability assumption, for any $w \in \mathbb{R}^n$, there exists a time point k_1 and $u(i)$, $i = 0, 1, \ldots, L$, such that $x(k_1) = w$. On the other hand, from the state represention, we have

$$w = x(k_1) = [x_1(k_1)/x_2(k_1)]$$

$$= \left[\begin{array}{cccc|cccc} A_1^{k_1-1}B_1, & \ldots, & A_1B_1, & B_1 & 0, & 0, & \ldots, & 0 \\ 0, & \ldots, & 0, & 0 & B_2, & NB_2, & \ldots, & N^{L-k_1-1}B_2 \end{array}\right]\begin{bmatrix} u(0) \\ \vdots \\ u(L) \end{bmatrix}. \tag{8-1.9}$$

From (8-1.9) and the arbitrariness of w we know that the conclusion holds.

Sufficiency: For any complete condition $[x_1(0)/x_2(L)] \in \mathbb{R}^n$, from (8-1.5)-(8-1.6), the state has the form

$$x(k) = \left[\begin{array}{ccc|ccc} A_1^{k-1}B_1, \ldots, & A_1B_1, & B_1, & 0, & & 0 \\ & 0 & & B_2, & NB_2, \ldots, & N^{L-k-1}B_2 \end{array}\right]\begin{bmatrix} u(0) \\ \vdots \\ u(L) \end{bmatrix} + \begin{bmatrix} A_1^k x_1(0) \\ N^{L-k}x_2(L) \end{bmatrix}. \tag{8-1.10}$$

Under the sufficient assumption, the matrix

$$M = \mathrm{diag}([A_1^{n_1-1}B_1, \ldots, A_1B_1, B_1], [B_2, NB_2, \ldots, N^{L-n_1-1}B_2])$$

is of full row rank ($L \geqslant n$). Therefore, for any $w \in \mathbb{R}^n$ we choose $k_1 = n_1$ and

$$\begin{bmatrix} u(0) \\ \vdots \\ u(L-1) \end{bmatrix} = M^{\tau}(MM^{\tau})^{-1}\left(w - \begin{bmatrix} A_1^{n_1}x_1(0) \\ \\ N^{L-n_1}x_2(L) \end{bmatrix}\right).$$

The inputs determined here will satisfy $x(n_1) = w$. Q.E.D.

Therefore, system (8-1.1) is controllable if and only if its forward and backward recurrenses are controllable. Since $x_1(k)$ is governed by $u(0)$, $u(1)$, ..., $u(k-1)$, and $x_2(k)$ by $u(k)$, $u(k+1)$, ..., $u(L-1)$, we may choose control inputs respectively for the control purpose of $x_1(k)$ and $x_2(k)$.

According to this theorem, time series (8-1.7) is controllable.

8-1.2. R-controllability

For any fixed terminal condition $x_2(L) \in \mathbb{R}^{n_2}$, we use $R(x_2(L))$ to denote the reachable (state) set of (8-1.1) starting from any initial condition, which is simply called the initial reachable set:

$$R(x_2(L)) = \{ w \mid w \in \mathbb{R}^n, \text{ there exist } x_1(0),\ 0 \leqslant k_1 \leqslant L, \text{ and } u(0), \\ u(1), \ldots, u(L), \text{ such that } x(k_1) = w \}.$$

It is clear that the initial reachable set $R(x_2(L))$ is dependent of $x_2(L)$. For different $x_2(L)$, $R(x_2(L))$ may be different.

<u>Example 8-1.2</u>. Since system (8-1.7) is controllable, for any terminal condition $x_2(L)$, $R(x_2(L)) = \mathbb{R}^n$.

<u>Example 8-1.3</u>. Consdier the finite tiem series of order 3:

$$\begin{bmatrix} 1 & 0 & 0 \\ 0 & 0 & 1 \\ 0 & 0 & 0 \end{bmatrix}\begin{bmatrix} x_1(k+1) \\ x_2(k+1) \end{bmatrix} = \begin{bmatrix} 2 & 0 & 0 \\ 0 & 1 & 0 \\ 0 & 0 & 1 \end{bmatrix}\begin{bmatrix} x_1(k) \\ x_2(k) \end{bmatrix} + \begin{bmatrix} 1 \\ 0 \\ 0 \end{bmatrix}u(k)$$

$$y(k) = [1 \quad 0 \quad 0][x_1(k)/x_2(k)] \tag{8-1.11}$$

where $x_1(k) \in \mathbb{R}$, $x_2(k) \in \mathbb{R}^2$. For any complete condition $[x_1(0)/x_2(L)]$, its state is

$$x_1(k) = 2^k x_1(0) + \sum_{i=0}^{k-1} 2^{k-i-1}u(i)$$

$$x_2(k) = \begin{cases} \begin{bmatrix} 0 & 1 \\ 0 & 0 \end{bmatrix}x_2(L) & \text{when } k = L-1 \\ 0 & \text{when } 0 \leqslant k < L-1, \end{cases}$$

showing that for any terminal condition $x_2(L)$, the initial reachable set is

$$R(x_2(L)) = \mathbb{R} \oplus ((\begin{bmatrix} 0 & 1 \\ 0 & 0 \end{bmatrix} x_2(L)) \cup x_2(L))$$

which is obviously dependent of $x_2(L)$.

Definition 8-1.3. Finite time series (8-1.1) is called controllable in initial reachable set (R-controllable) if for any fixed terminal condition $x_2(L)$ the state starting from any initial condition may be controlled by inputs to reach any state in $R(x_2(L))$ at a certain time point.

R-controllability guarantees the controllability of any state in the initial reachable set.

Theorem 8-1.3. Time series (8-1.1) is R-controllable iff

$$\text{rank}[B_1, A_1B_1, \ldots, A_1^{n_1-1}B_1] = n_1,$$

i.e., subsystem (8-1.4a) is controllable.

It is easy to testify that both time series (8-1.7) and (8-1.11) are R-controllable.

8-1.3. Y-controllability

We have seen that under the feedback control

$$u(k) = Kx(k) + v(k), \qquad k = 0, 1, \ldots, L \tag{8-1.12}$$

where $K \in \mathbb{R}^{m\times n}$ is a constant matrix, $v(k)$ is the new control input, when applied to (8-1.1), the closed-loop system is

$$Ex(k+1) = (A+BK)x(k) + Bv(k) \tag{8-1.13}$$

Definition 8-1.4. Time series (8-1.1) is called causality controllable (Y-controllable) if there exists a feedback (8-1.12) such that its closed-loop system (8-1.13) is causal. Here Y comes from the first letter of the Chinese word for "Causality."

In most real systems, noncausality is often unexpected, otherwise, it may cause many problems in controlling, identification, and estimation. The Y-controllability assures the ability to control the causality via state feedback control (8-1.12).

Apparently, from Theorems 8-1.1 and 3-2.1 we have the following theorem.

Theorem 8-1.4. Time series (8-1.1) is Y-controllable if and only if there exists a matrix $K \in \mathbb{R}^{m\times n}$ so that $\deg(|zE - (A+BK)|) = \text{rank}E$, or equivalently,

$$\text{rank}\begin{bmatrix} E & 0 & 0 \\ A & E & B \end{bmatrix} = \text{rank}E.$$

8-1.4. Observability, R-observability and Y-observability

What is characterized by the concepts of controllability, R-controllability, and Y-controllability is the controllability of inputs on the state. Their dual concepts are the three observabilities that describe the ability to reconstruct the state from measurements $y(k)$, $k = 0, 1, \ldots, L$.

As shown before, state $x(k)$ at any time point k, $0 \leqslant k \leqslant L$, is completely determined by the complete condition $[x_1(0)/x_2(L)] \in \mathbb{R}^n$ and inputs $u(k)$, $k = 0, 1, \ldots, L$. Since $u(k)$ is a known vector, the observabilities for (8-1.1) actually are the ability of reconstructing the complete condition $[x_1(0)/x_2(L)] \in \mathbb{R}^n$ from measurement $y(k)$.

Definition 8-1.5.

1. Time series (8-1.1) is called observable if its state $x(k)$ at any time point k is uniquely determined by $\{u(i), y(i), i = 0, 1, \ldots, L\}$.

2. It is called R-observable if it is observable in the initial reachable set $R(x_2(L))$ for any fixed terminal condition $x_2(L) \in \mathbb{R}^{n_2}$.

3. It is called causal observable (Y-observable) if its state $x(k)$ at any time point k is uniquely determined by initial condition $x_1(0)$ and the former (k is included) inputs $u(i)$, together with former measurements $y(i)$, $i = 0, 1, \ldots, k$.

Theorem 8-1.5. Let the items be as determined in (8-1.4).

1. Time series (8-1.1) is observable iff

$$\mathrm{rank}[C_1/C_1A_1/\ldots/C_1A_1^{n_1-1}] = n_1 \quad \text{and} \quad \mathrm{rank}[C_2/C_2N/\ldots/C_2N^{n_2-1}] = n_2.$$

2. Time series (8-1.1) is R-observable iff

$$\mathrm{rank}[C_1/C_1A_1/\ldots/C_1A_1^{n_1-1}] = n_1.$$

3. Time series (8-1.1) is Y-observable iff

$$\mathrm{rank}\begin{bmatrix} E & A \\ 0 & E \\ 0 & C \end{bmatrix} = n + \mathrm{rank}E.$$

Example 8-1.4. Consider the time series (8-1.8) in which

$$A_1 = \begin{bmatrix} 1 & 1 \\ 0 & 1 \end{bmatrix}, \quad N = \begin{bmatrix} 0 & 1 \\ 0 & 0 \end{bmatrix}, \quad C_1 = [0 \quad 1], \quad C_2 = [1 \quad 0]$$

and

$$\mathrm{rank}\begin{bmatrix} C_1 \\ C_1A_1 \end{bmatrix} = \mathrm{rank}\begin{bmatrix} 0 & 1 \\ 0 & 1 \end{bmatrix} = 1 < 2 = n_1,$$

$$\text{rank}\begin{bmatrix} E & A \\ 0 & E \\ 0 & C \end{bmatrix} = 7 = n + \text{rank}E.$$

Thus, from Theorem 8-1.5 it is neither observable nor R-observable, but it is Y-observable.

Example 8-1.5. In the time series (8-1.11),

$$A_1 = 2, \quad N = \begin{bmatrix} 0 & 1 \\ 0 & 0 \end{bmatrix}, \quad C_1 = 1, \quad C_2 = 0,$$

and

$$\text{rank}\begin{bmatrix} C_1 \\ C_1A_1 \end{bmatrix} = \text{rank}\begin{bmatrix} 1 \\ 2 \end{bmatrix} = 1, \qquad \text{rank}\begin{bmatrix} C_2 \\ C_2N \end{bmatrix} = 0 < 2$$

$$\text{rank}\begin{bmatrix} E & A \\ 0 & E \\ 0 & C \end{bmatrix} = 4 < 5 = n + \text{rank}E.$$

Thus, it is neither observable nor Y-observable, but it is R-observable.

8-2. Solution, Controllability and Observability in Discrete-Time Singular Systems

Now we turn to the discrete-time singular system

$$\begin{aligned} Ex(k+1) &= Ax(k) + Bu(k) \\ y(k) &= Cx(k), \end{aligned} \qquad k = 0, 1, 2, \ldots \tag{8-2.1}$$

which will be discussed in this chapter. In (8-2.1), the state $x(k) \in \mathbb{R}^n$, the control input $u(k) \in \mathbb{R}^m$ and the measure output $y(k) \in \mathbb{R}^r$; $E, A \in \mathbb{R}^{nxn}$, $B \in \mathbb{R}^{nxm}$, $C \in \mathbb{R}^{rxn}$ are constant matrices. It is assumed that system (8-2.1) is regular and rank$E < n$.

The difference between singular system (8-2.1) and time series (8-1.1) lies in that (8-1.1) is a finite time series, thus has both initial and terminal conditions, and (8-2.1) is an infinite time series, having only the initial condition.

First of all, for singular system (8-2.1), there exist two nonsingular matrices Q and P such that (8-2.1) is r.s.e. to its EF1:

$$\begin{aligned} x_1(k+1) &= A_1x_1(k) + B_1u(k) \\ y_1(k) &= C_1x_1(k) \end{aligned} \tag{8-2.2a}$$

$$Nx_2(k+1) = x_2(k) + B_2u(k)$$
$$y_2(k) = C_2x_2(k) \tag{8-2.2b}$$

$$y(k) = y_1(k) + y_2(k) \tag{8-2.2c}$$
$$k = 0,\ 1,\ 2,\ \ldots$$

in which the coordinate transformation is

$$x(k) = P\begin{bmatrix} x_1(k) \\ x_2(k) \end{bmatrix}, \qquad x_1(k) \in \mathbb{R}^{n_1}, \quad x_2(k) \in \mathbb{R}^{n_2}$$

$n_1 + n_2 = n$, and

$$QEP = \mathrm{diag}(I_{n_1}, N), \qquad QAP = \mathrm{diag}(A_1, I_{n_2})$$
$$QB = [B_1/B_2], \qquad CP = [C_1 \quad C_2] \tag{8-2.3}$$

where $A_1 \in \mathbb{R}^{n_1 x n_1}$, $N \in \mathbb{R}^{n_2 x n_2}$ is nilpotent with nilpotent index h.

In EF1, subsystems (8-2.2a) and (8-2.2b) are called the forward and backward subsystems, respectively. In (8-2.2), the measurements y_1, y_2 are only in form. They are not available.

Forward subsystem (8-2.2a) is a discrete-time normal system whose state is

$$x_1(k) = A_1^k x_1(0) + \sum_{i=0}^{k-1} A_1^{k-i-1} B_1 u(i) \tag{8-2.4}$$

in which there exists the causal relationship between state and inputs.

The backward subsystem (8-2.2b) is a backward recurrence of state. By repeatedly left multiplying both sides of it with $N^0 = I$, N^1, N^2, ... , N^{h-1}, we have

$$x_2(k) = Nx_2(k+1) - B_2u(k)$$
$$Nx_2(k) = N^2x_2(k+2) - NB_2u(k+1)$$
$$N^2x_2(k+2) = N^3x_2(k+3) - N^2B_2u(k+2)$$
$$\ldots \qquad \ldots$$
$$N^{h-1}x_2(k+h-1) = N^hx_2(k+h) - N^{h-1}B_2u(k+h-1).$$

The addition of these equations, and noticing $N^h = 0$, yields the substate

$$x_2(k) = -\sum_{i=0}^{h-1} N^i B_2 u(k+i) \tag{8-2.5}$$

from which we see that to determine the substate $x_2(k)$ future inputs $u(i)$, $k \leqslant i \leqslant k+$

h-1, are needed.

Comparing (8-1.6) with (8-2.5) we see that they are identical whenever $k \leqslant L-h$. The only difference happens at the time point $L-h < k \leqslant L$. Therefore, if we view (8-2.1) as a limit case of (8-1.1), equation (8-2.5) is a limit case of (8-1.6).

Combining (8-2.3)-(8-2.5), we obtain the general state for system (8-2.1);

$$
\begin{aligned}
x(k) &= P[I/0]x_1(k) + P[0/I]x_2(k) \\
&= P\begin{bmatrix} I \\ 0 \end{bmatrix}(A_1^k x_1(0) + \sum_{i=0}^{k-1} A_1^{k-i-1}B_1u(i)) - P\begin{bmatrix} 0 \\ I \end{bmatrix}\sum_{i=0}^{h-1} N^i B_2u(k+i) \\
&= P\begin{bmatrix} I \\ 0 \end{bmatrix}(A_1^k\begin{bmatrix} I \\ 0 \end{bmatrix}P^{-1}x(0) + \sum_{i=0}^{k-1} A_1^{k-i-1}B_1u(i)) - P\begin{bmatrix} 0 \\ I \end{bmatrix}\sum_{i=0}^{h-1} N^i B_2u(k+i) \\
y(k) &= Cx(k),
\end{aligned}
\tag{8-2.6}
$$

which is the represention of state x(k) and measure output at any time point k.

<u>Example 8-2.1</u>. Consider the following discrete-time singular system

$$
\begin{bmatrix} 1 & 0 & 0 & 0 \\ 0 & 1 & 0 & 0 \\ 0 & 0 & 0 & 1 \\ 0 & 0 & 0 & 0 \end{bmatrix} x(k+1) = \begin{bmatrix} 1 & 1 & 0 & 0 \\ 0 & 1 & 0 & 0 \\ 0 & 0 & 1 & 0 \\ 0 & 0 & 0 & 1 \end{bmatrix} x(k) + \begin{bmatrix} 0 \\ 1 \\ 1 \\ -1 \end{bmatrix} u(k)
$$

$$
y(k) = [0 \quad 1 \quad 1 \quad 0]x(k) \qquad k = 0, 1, 2, \ldots \tag{8-2.7}
$$

This system has the same form as (8-1.7) with the difference that it is an infinite process and (8-1.7) is finite.

From the state represention in Example 8-1.1, we obtain its state

$$x(k) = [x_1(k)/x_2(k)]$$

$$x_1(k) = \begin{bmatrix} 1 & k \\ 0 & 1 \end{bmatrix} x_1(0) + \sum_{i=0}^{k-1} \begin{bmatrix} k-i-1 \\ 1 \end{bmatrix} u(i)$$

$$x_2(k) = \begin{bmatrix} u(k+1)-u(k) \\ u(k) \end{bmatrix},$$

in which the first variable of $x_2(k)$ is the jump amplitude, while the second one is the input.

While it is wellknown that causality holds in discrete-time normal systems, the discrete-time singular system is not always causal. To determine its state, future inputs are often required. The fact of noncausality indeed exists in real systems. For eaxmple, the Leontief dynamic model in economic system is described by

$$x(k) = Ax(k) + B(x(k+1) - x(k)) + d(k)$$

where d(k) is input, including items such as consuming. The purpose of production is consuming. But a time delay exists between these two aspects. Therefore, a future consuming (input) goal is often used to guide the present production (state). Non-causal systems exist when the variables evolve in space rather than in time. The non-causality, on one hand, also characterizes the discrete-time singular system.

Example 8-2.2. In this example we will study the causality in a singular system of the form (8-2.1).

I. Causality between state and inputs.

Without loss of generality, system (8-2.1) is assumed to be in its EF1 (8-2.2). Obviously, causality exists in system (8-2.2) iff the backward subsystem (8-2.2b) is causal.

From (8-2.5), it is clear that causality exists between state and inputs iff

$$NB_2 = 0. \qquad (8\text{-}2.8)$$

II. Causality between measure output and inputs.

Since the input-output relationship is determined only by the controllable and observable subsystem, it doesn't lose any generality if we assume that (N,B_2,C_2) is controllable and observable. Since causality exists between y(k) and u(k) iff such a relationship exists between $y_2(k)$ and u(k), and

$$y_2(k) = C_2x_2(k) = -\sum_{i=0}^{h-1} C_2N^iB_2u(k+i), \qquad (8\text{-}2.9)$$

the causal relationship exists between $y_2(k)$ and inputs (thus between y(k) and inputs) if and only if

$$C_2N^iB_2 = 0, \qquad i = 1, 2, \dots , h-1,$$

i.e.,

$$[C_2/C_2N/\dots/C_2N^{h-1}]N[B_2, NB_2, \dots , N^{h-1}B_2] = 0. \qquad (8\text{-}2.10)$$

The controllability and observability of (N,B_2,C_2) indicates that N = 0.

Another difference between discrete-time singular systems and normal ones is that singular systems don't always have solutions for any initial condition. This point may be easily seen in (8-2.6), in which when k = 0 it will be

$$x(0) = P\begin{bmatrix} I \\ 0 \end{bmatrix}[I \quad 0]P^{-1}x(0) - P\begin{bmatrix} 0 \\ I \end{bmatrix}\sum_{i=0}^{h-1} N^iB_2u(i)$$

or, in another form

$$[0 \quad I]P^{-1}x(0) = \sum_{i=0}^{h-1} N^i B_2 u(i), \tag{8-2.11}$$

which is the so-called admissible initial condition satisfied by the initial state $x(0)$.

Recalling (8-2.5), we see that the admissibility of the initial state is brought out by the backward subsystem. Since the singular system can describe a kind of large-scale system composed of interconnected subsystems, its substate $x_2(k)$ is a kind of pseudo-state in the sense that it only reflects the interconnection between subsystems. Naturally, its initial state $x_2(0)$ loses its meaning as a normal initial state. It is the interconnection between subsystems at the initial time.

For example, assume $N = 0$. Then (8-2.5) becomes $x_2(k) = -B_2u(k)$, which is only a linear combination of input $u(k)$.

We use I_o to denote the set of admissible initial state:

$$I_o = \{ x(0) \in \mathbb{R}^n \mid [0 \quad I]P^{-1}x(0) = \sum_{i=0}^{h-1} N^i B_2 u(i) \}. \tag{8-2.12}$$

When $u(k) = 0$, the free response of (8-2.1) is

$$\begin{aligned} x_1(k) &= A_1^k x_1(0) \\ x_2(k) &= 0. \end{aligned} \tag{8-2.13}$$

$$k = 0, 1, 2, \ldots$$

The substate $x_2(k)$ of the backward subsystem is identically zero.

Equations (8-2.6) and (8-2.11) not only show the difference between discrete-time singular system and a normal one, but also reveal the difference between discrete-time singular systems and continuous-time ones.

As seen in Chapter 1, the continuous-time system

$$\begin{aligned} E\dot{x}(t) &= Ax(t) + Bu(t) \\ y(t) &= Cx(t) \end{aligned} \tag{8-2.14}$$

has the state response:

$$x(t) = P\begin{bmatrix} I \\ 0 \end{bmatrix}[e^{A_1 t}x_1(0) + \int_0^t e^{A_1(t-\tau)}B_1u(\tau)d\tau]$$

$$- P\begin{bmatrix} 0 \\ I \end{bmatrix}\sum_{i=0}^{h-1} \delta^{(i)}(t)N^{i+1}x_2(0) - P\begin{bmatrix} 0 \\ I \end{bmatrix}\sum_{i=0}^{h-1} N^i B_2 u^{(i)}(t). \tag{8-2.15}$$

Therefore, the comparison of (8-2.6) with (8-2.15) reveals that while no impulse terms are involved in the state of discrete-time systems, noncausality happens in such systems; and, although causality exists in continuous-time systems, impulse terms and derivatives of input are contained in the state response. Furthermore, the term $u(k+i)$ in the state of discrete-time system is just in correspondence with the term $u^{(i)}(t)$ in continuous-time case.

As for system (8-2.1), we give the following definitions.

Definition 8-2.1. Consider ssytem (8-2.1).

1. It is called controllable (R-controllable and Y-controllable) if for any sufficiently large $L \geqslant n$ the finite time series (8-1.1) is controllable (R-controllable and Y-controllable).

2. It is called observable (R-observable and Y-observable) if for any sufficiently large $L \geqslant n$ the finite time series (8-1.1) is observable (R-observable and Y-observable).

Clearly, these concepts are natural generalizations of those for normal systems. They are determined by system (8-2.1) and are independent of L. Duality exists among these concepts.

A direct result of Theorems 8-1.2 - 8-1.5 and Definition 8-2.1 is the following theorem.

Theorem 8-2.1.

1. Discrete-time system (8-2.1) is controllable (R-controllable, Y-controllable) if and only if continuous-time system (8-2.14) is controllable (R-controllable, impulse controllable).

2. Discrete-time system (8-2.1) is observable (R-observable, Y-observable) if and only if continuous-time system (8-2.14) is observable (R-observable, impulse observable).

By this theorem, it is easy to testify the controllabilities or observabilities for system (8-2.1) by using the criteria for system (8-2.14).

For system (8-2.1), these concepts have the same physical meaning as for time series (8-1.1). For example, if system (8-2.1) is controllable, for any initial condition $x_1(0)$ (or admissible initial condition $x(0)$) and $w \in \mathbb{R}^n$, there exists a time point k_1 and input series $u(0)$, $u(1)$, ... , $u(k_1+h-1)$ such that $x(k_1) = w$. Thus, the controllability guarantees the ability to control state by inputs. The other concepts also have their corresponding physical senses.

Example 8-2.4. The same coefficient continuous-time system for system (8-2.7) is

$$\begin{bmatrix} 1 & 0 & 0 & 0 \\ 0 & 1 & 0 & 0 \\ 0 & 0 & 0 & 1 \\ 0 & 0 & 0 & 0 \end{bmatrix} \dot{x}(t) = \begin{bmatrix} 1 & 1 & 0 & 0 \\ 0 & 1 & 0 & 0 \\ 0 & 0 & 1 & 0 \\ 0 & 0 & 0 & 1 \end{bmatrix} x(t) + \begin{bmatrix} 0 \\ 1 \\ 1 \\ -1 \end{bmatrix} u(t)$$

$$y(t) = [0 \quad 1 \quad 1 \quad 0]x(t)$$

By the criteria in Chapter 2 it may be testified that it is controllable, R-controllable, impulse controllable, and impulse observable. Thus, system (8-2.7) is controllable, R-controllable, Y-controllable and Y-observable.

8-3. State Feedback and Pole Placement in Discrete-Time Singular Systems

Consider system (8-2.1):

$$\begin{aligned} Ex(k+1) &= Ax(k) + Bu(k) \\ y(k) &= Cx(k). \end{aligned} \tag{8-3.1}$$

Definition 8-3.1. Discrete-time system (8-3.1) is termed stable if its free response satisfies

$$\|x(k)\| \leqslant \alpha\beta^k \|x(0)\|, \qquad k = 0, 1, \dots, \quad \alpha > 0,\ 0 < \beta < 1,$$

for any admissible initial condition $x(0)$ and $k \geqslant 0$.

According to the definition, if system (8-3.1) is stable, its free response satisfies $\lim_{k\to\infty} x(k) = 0$. This is an asymptotical stability in Lyapunov's sense.

For regular system (8-3.1), its free response is (8-2.13). It is stable iff its forward subsystem is stable.

Theorem 8-3.1. Discrete-time system (8-3.1) is stable iff its finite pole set $\sigma(E,A)$ is within the unit circle on the complex plane, which is denoted by U^+.

Example 8-3.1. In system (8-2.7), $\sigma(E,A) = \{1, 1\}$. Thus it is unstable.

Example 8-3.2. Consider the system

$$\begin{aligned} \begin{bmatrix} 1 & 0 & -1 \\ 0 & 1 & 0 \\ 0 & 0 & 0 \end{bmatrix} x(k+1) &= \begin{bmatrix} -\frac{1}{2} & 0 & 2 \\ -1 & \frac{1}{3} & -1 \\ 0 & 0 & 1 \end{bmatrix} x(k) + \begin{bmatrix} 1 \\ 0 \\ -1 \end{bmatrix} u(k) \\ y(k) &= [0 \quad 1 \quad 0] x(k) \end{aligned} \tag{8-3.2}$$

in which

$$E = \begin{bmatrix} 1 & 0 & -1 \\ 0 & 1 & 0 \\ 0 & 0 & 0 \end{bmatrix}, \qquad A = \begin{bmatrix} -\frac{1}{2} & 0 & 2 \\ -1 & \frac{1}{3} & -1 \\ 0 & 0 & 1 \end{bmatrix}.$$

Since the pole set is $\sigma(E,A) = \{-\frac{1}{2}, \frac{1}{3}\} \subset U^+$. Thus this system is stable and its free response satisfies

$$\lim_{k\to\infty} x(k) = 0, \quad \forall\ x(0).$$

Although stability is required by real systems, this property sometimes doesn't hold. Now we consider stabilizing a system by feedback control.

Let the feedback be

$$u(k) = Kx(k) + v(k) \tag{8-3.3}$$

where $K \in \mathbb{R}^{m\times n}$, $v(k)$ is the new input. State feedback (8-3.3) and system (8-3.1) form the closed-loop system

$$\begin{aligned} Ex(k+1) &= (A+BK)x(k) + Bv(k) \\ y(k) &= Cx(k). \end{aligned} \tag{8-3.4}$$

To guarantee the existence and uniqueness of solution for any control, closed-loop ssytem (8-3.4) is always assumed regular.

<u>Definition 8-3.2</u>. Discrete-time system (8-3.1) is called stabilizable if there exists a feedback gain $K \in \mathbb{R}^{m\times n}$ such that closed-loop system (8-3.4) is stable; it is called detectable if its dual system

$$\begin{aligned} E^{\tau}x(k+1) &= A^{\tau}x(k) + C^{\tau}u(k) \\ y(k) &= B^{\tau}x(k) \end{aligned}$$

is stabilizable.

Concerning the stabilizability and detectability we have the following theorem.

<u>Theorem 8-3.2</u>. System (8-3.1) is stabilizable (detectable) iff

$$\mathrm{rank}[zE-A,\ \ B] = n, \quad \forall\ z \in \mathbb{C}, \quad |z| \geqslant 1,\ z \text{ finite} \tag{8-3.5}$$
$$(\mathrm{rank}[zE-A/C] = n, \quad \forall\ z \in \mathbb{C}, \quad |z| \geqslant 1,\ z \text{ finite})$$

<u>Proof</u>. We choose to prove the criterion for stabilizability. The result for detectability is ready by duality.

Necessity: Assume that system (8-3.1) is stabilizable. Then a amtrix $K \in \mathbb{R}^{m\times n}$ exists so that $\sigma(E, A+BK) \subset U^{+}$. Therefore,

$$\mathrm{rank}[zE-(A+BK)] = n, \quad \forall\ z \in \mathbb{C}, \quad |z| \geqslant 1,\ z \text{ finite}.$$

On the other hand, we know

$$\mathrm{rank}[zE-(A+BK)] = \mathrm{rank}[zE-A,\ \ B]\begin{bmatrix} I \\ -K \end{bmatrix}.$$

The combination of these two equations results in $\mathrm{rank}[zE-A,\ B] = n$, $\forall\ z \in \mathbb{C}$, $|z| \geqslant 1$, z finite.

Sufficiency: For any regular ssytem (8-3.1), there exist two nonsingular matrices Q and P such that

$$QEP = \mathrm{diag}(I_{n_1}, N), \quad QAP = \mathrm{diag}(A_1, I_{n_2}), \quad QB = [B_1/B_2]$$

where $n_1 + n_2 = n$, $N \in \mathbb{R}^{n_2 \times n_2}$ is nilpotent. In this case,

$$\mathrm{rank}[zE-A,\ B] = \mathrm{rank}[zQEP-QAP,\ QB]$$

$$= \mathrm{rank}\begin{bmatrix} zI-A_1 & 0 & B_1 \\ 0 & zN-I & B_2 \end{bmatrix} = n_2 + \mathrm{rank}[zI-A_1,\ B_1], \quad \forall\ z \in \mathbb{C},\ |z| \geqslant 1, \ z \text{ finite}.$$

Subject to assumption (8-3.5), we have

$$\mathrm{rank}[zI-A_1,\ B_1] = n_1, \quad \forall\ z \in \mathbb{C}, \quad |z| \geqslant 1,\ z \text{ finite}.$$

Hence, a matrix K_1 may be selected so that $\sigma(A_1+B_1K_1) \subset U^+$. Let the feedback gain matrix be $K = [K_1,\ 0]P^{-1}$. It is easy to verify that $\sigma(E,A+BK) = \sigma(A_1+B_1K_1)$ is within the unit circle on the complex plane. Thus system (8-3.1) is stabilizable. Q.E.D.

The proof process shows that system (8-3.1) is stabilizable (detectable) iff its forward subsystem is stabilizable (detectable). Or equivalently, the uncontrollable poles of the forward subsystem is within the unit circle on the complex plane.

Example 8-3.3. System (8-2.7) has been shown to be unstable. But, since

$$\mathrm{rank}[zE-A,\ B] = 4, \quad \forall\ z \in \mathbb{C},\ z \text{ finite},$$

it is stabilizable. In fact, if we choose $K = [-0.25 \quad -2 \quad 0 \quad 0]$, it will be $\sigma(E,A+BK) = \{0.5, -0.5\}$, which is within the unit circle on the complex plane.

Relationships among these concepts are

controllability → R-controllability → stabilizability
controllability ⇄ Y-controllability
controllability ⇄ R-controllability + rank[E B] = n

observability → R-observability → detectability
observability → Y-observability
observability ⇄ R-observability + rank[E/C] = n

Concerning the pole placement problem, we have the following theorem.

Theorem 8-3.3. Consider system (8-3.1).

1. For any feedback control (8-3.3), closed-loop system (8-3.4) has at most rankE finite poles; and there exists a (8-3.3) such that (8-3.4) has the most rankE finite poles (or no infinite poles) iff this system is Y-controllable.

2. Feedback control (8-3.3) may be selected so that its closed-loop system is stable and has rankE stable finite poles (or no infinite poles) iff the system is stabilizable and Y-controllable.

Therefore, the Y-controllability guarantees the ability to drive infinite poles to finite positions, which corresponds to the controllability on causality.

By this theorem, for any feedback gain matrix K closed-loop system (8-3.4) has at most rankE finite poles, and furthermore, it has been proven that such rankE poles may be arbitrarily assigned if the system is both R-controllable and Y-controllable.

Furthermore, there exists feedback control (8-3.3) such that the closed-loop system is both stable and causal if and only if system (8-3.1) is stabilizable and Y-controllable.

Example 8-3.4. Direct testification shows that system (8-2.7) is stabilizable and Y-controllable. Let the feedback gain matrix be

$$K = [-1, \ -\frac{41}{12}, \ -1, \ -\frac{7}{3}].$$

Then $\sigma(E,A+BK) = \{0.5, -0.5, -\frac{1}{3}\}$. The closed-loop system has no infinite poles. Thus, under the feedback control $u(k) = Kx(k) + v(k)$, closed-loop ssytem (8-2.7) is not only stable but also causal.

By duality, it is easy to know from Theorem 8-3.3 that the following holds.

Theorem 8-3.4. There exists a matrix $G \in \mathbb{R}^{n\times r}$ so that $\sigma(E,A+GC) \subset U^+$ and $\deg(|zE-(A+GC)|) = \text{rank}E$ iff system (8-3.1) is detectable and Y-observable.

This result will be needed later in the design of state observers and dynamic compensators.

The next result shows that by feedback control not only can drive infinite poles to finite positions, but also can assign finite poles to infinite positions.

Let

$$\begin{aligned} x_1(k+1) &= A_1x_1(k) + B_1u(k) \\ Nx_2(k+1) &= x_2(k) + B_2u(k) \\ y(k) &= C_1x_1(k) + C_2x_2(k) \end{aligned} \tag{8-3.6}$$

be the EF1 for system (8-3.1). Here Q and P are nonsingular, $x(k) = P[x_1(k)/x_2(k)]$, $x_1(k) \in \mathbb{R}^{n_1}$, $x_2(k) \in \mathbb{R}^{n_2}$, $n_1 + n_2 = n$, and

$$QEP = \text{diag}(I_{n_1}, N), \qquad QAP = \text{diag}(A_1, I_{n_2}),$$

$$QB = [B_1/B_2], \qquad CP = [C_1, C_2].$$

Theorem 8-3.5.(Dai, 1988a). There exists a feedback gain matrix $K \in \mathbb{R}^{m\times n}$ such that closed-loop system (8-3.4) has no finite poles, i.e.,

$$|zE - (A+BK)| = \alpha \qquad z \in \mathbb{C} \tag{8-3.7}$$

$\alpha \neq 0$ is a constant, if and only if system (8-3.1) is R-controllable and either $B_2 \neq 0$ or $n_1 = 0$.

Now we will examine the physical sense of this theorem.

Let the matrix K satisfy (8-3.7). Then the closed-loop system (8-3.4) has only infinite poles. Thus two nonsingular matrices $\tilde{Q}$ and $\tilde{P}$ may be chosen such that (8-3.4) is r.s.e. to

$$\tilde{N}\tilde{x}(k+1) = \tilde{x}(k) + \tilde{B}v(k) \tag{8-3.8}$$

where $x(k) = \tilde{P}\tilde{x}(k)$, and $\tilde{Q}E\tilde{P} = \tilde{N}$, $\tilde{Q}(A+BK)\tilde{P} = I_n$, $\tilde{Q}B = \tilde{B}$, $\tilde{N}$ is nilpotent and its nilpotent index is denoted by $\tilde{h}$. According to (8-2.5), the state response of (8-3.8) is

$$x(k) = \tilde{P}\tilde{x}(k) = -\sum_{i=0}^{h-1} \tilde{P}\tilde{N}^i\tilde{B}v(k+i)$$

showing that x(k) is independent of former inputs; x(k) indicates the control function of present and future inputs v(k), v(k+1), ... , $v(k+\tilde{h}-1)$, revealing some properties of future inputs. Especially, when {v(k)} is not available in advance, its properties may be conjectured from that of output {y(k)}. This is one special feature of discrete-time singular systems.

The free response of (8-3.8) is $\tilde{x}(k) \equiv 0$, $k \geqslant 0$.

The following two results are useful.

Corollary 8-3.1. If system (8-3.1) is controllable and $n_2 > 0$, there always exists a matrix that satisfies (8-3.7).

Corollary 8-3.2. If system (8-3.1) is observable and $n_2 > 0$ (rankE < n), there always exists a matrix G such that $\deg(|zE-(A-GC)|) = 0$.

Example 8-3.5. Consider system (8-2.7):

$$\begin{bmatrix} 1 & 0 & 0 & 0 \\ 0 & 1 & 0 & 0 \\ 0 & 0 & 0 & 1 \\ 0 & 0 & 0 & 0 \end{bmatrix} x(k+1) = \begin{bmatrix} 1 & 1 & 0 & 0 \\ 0 & 1 & 0 & 0 \\ 0 & 0 & 1 & 0 \\ 0 & 0 & 0 & 1 \end{bmatrix} x(k) + \begin{bmatrix} 0 \\ 1 \\ 1 \\ -1 \end{bmatrix} u(k). \tag{8-3.9}$$

$$y(k) = [0 \quad 1 \quad 1 \quad 0]x(k)$$

It is easy to testify that the system is controllable and $n_2 = 2 > 0$.

For any matrix $K = [k_1, k_2, k_3, k_4]$, we have

$$|zE-(A+BK)| = -k_3z^3 + (1+3k_3-k_4)z^2 - (2+3k_3-2k_4-k_2)z + 1+k_3-k_1-k_2-k_4.$$

Thus, if we choose $k_1 = -1$, $k_2 = 0$, $k_3 = 0$, $k_4 = 1$, it will be $|zE-(A+BK)| = 1$.

8-4. State Observation for Discrete-Time Singular Systems

Consider the singular system

$$\begin{aligned} Ex(k+1) &= Ax(k) + Bu(k) \\ y(k) &= Cx(k) \end{aligned} \tag{8-4.1}$$

where $x(k) \in \mathbb{R}^n$, $u(k) \in \mathbb{R}^m$, $y(k) \in \mathbb{R}^r$, $E, A \in \mathbb{R}^{n\times n}$, $B \in \mathbb{R}^{n\times m}$, $C \in \mathbb{R}^{r\times n}$ are constant matrices, rank$E < n$ is assumed.

The state observation problem for system (8-4.1) is to construct a dynamic system of the form

$$\begin{aligned} E_cx_c(k+1) &= A_cx_c(k) + B_cu(k) + Gy(k) \\ w(k) &= F_cx_c(k) + Fu(k) + Hy(k) \end{aligned} \tag{8-4.2}$$

where $x_c(k) \in \mathbb{R}^{n_c}$, $w(k) \in \mathbb{R}^n$, $E_c, A_c \in \mathbb{R}^{n_c\times n_c}$, B_c, C, F_c, F, H are constant matrices, and $|zE_c-A_c| \not\equiv 0$, such that the asymptotical state reconstruction is achieved:

$$\lim_{k\to\infty} (w(k) - x(k)) = 0, \quad \forall\ x(0), x_c(0).$$

In general, the state observers for discrete-time singular systems may be constructed in ways analogous to those of the continuous-time case.

For system (8-4.1), we consider the following dynamic system

$$E\hat{x}(k+1) = A\hat{x}(k) + Bu(k) + G(y(k)-C\hat{x}(k)) \tag{8-4.3}$$

In (8-4.3), the first two terms on the right-hand side model the dynamic equation of system (8-4.1), while the last term is employed to amend the mismatch in the modeling process. If $\hat{x}(0) = x(0)$, the states of (8-4.3) and (8-4.1) are identical, i.e., $\hat{x}(k) = x(k)$, $k \geqslant 0$.

Let $e(k) = x(k) - \hat{x}(k)$ be the estimation error between $x(k)$ and its estimation $\hat{x}(k)$. Then $e(k)$ satisfies the dynamics

$$Ee(k+1) = (A-GC)e(k), \quad e(0) = x(0) - \hat{x}(0) \tag{8-4.4}$$

If system (8-4.1) is detectable, a matrix G may be selected so that $\sigma(E,A-GC) \subset$

U^+. Thus system (8-4.4) is stable, $\lim_{k\to\infty} e(k) = 0$, $\forall\ e(0)$, implying that $\hat{x}(k)$ asymptotically reconstructs the state $x(k)$.

In summary we have proven the following.

Theorem 8-4.1. If system (8-4.1) is detectable, there must exist a matrix $G \in \mathbb{R}^{n\times r}$ such that (8-4.3) is an observer for system (8-4.1).

Furthermore, if rank$E < n$ and system (8-4.1) is observable, from Corollary 8-3.2 there exists a matrix $G \in \mathbb{R}^{n\times r}$ satisfying $|zE - (A-GC)| = $ constant. In this case two nonsingular matrices Q_1 and P_1 may be chosen so that (8-4.4) is r.s.e. to

$$\tilde{N}\tilde{e}(k+1) = \tilde{e}(k), \qquad k = 0,\ 1,\ 2,\ \ldots$$

where $Q_1EP_1 = \tilde{N}$, $Q_1(A-GC)P_1 = I_n$, $e(k) = P_1\tilde{e}(k)$ and $\tilde{N}$ is nilpotent. Therefore,

$$e(k) = P_1\tilde{e}(k) = 0, \qquad k = 0,\ 1,\ 2,\ \ldots$$

or in other words, $x(k) = \hat{x}(k)$, $k = 0, 1, \ldots$, $x(k)$ and $\hat{x}(k)$ are identical. This property is termed exact reconstruction. Hence we have proven the following.

Theorem 8-4.2. If system (8-4.1) is observable, rank$E < n$, there exists a matrix G such that the state of system (8-4.1) is reconstructed exactly by an observer (8-4.3).

What is emphasized in this theorem is that without any knowledge of initial conditions, the state of system (8-4.1) may be reconstructed exactly at the very beginning by an observer of the form (8-4.3). This reflects a special feature of singular systems. Normal systems don't have such a property.

Example 8-4.1. Consider the system

$$\begin{bmatrix} 1 & 0 & 0 & 0 \\ 0 & 1 & 0 & 0 \\ 0 & 0 & 0 & 0 \\ 0 & 0 & 1 & 0 \end{bmatrix} x(k+1) = \begin{bmatrix} 1 & 0 & 0 & 0 \\ 1 & 1 & 0 & 0 \\ 0 & 0 & 1 & 0 \\ 0 & 0 & 0 & 1 \end{bmatrix} x(k) + \begin{bmatrix} 0 \\ 1 \\ 0 \\ 2 \end{bmatrix} u(k)$$

$$y(k) = [0 \quad 1 \quad 1 \ -1]x(k) \tag{8-4.5}$$

for which the matrix $G = [1 \quad 0 \quad 0 \ -1]^\tau$ satisfies $|zE - (A-GC)| = 1 = $ constant. By Theorem 8-4.2, the observer

$$\begin{bmatrix} 1 & 0 & 0 & 0 \\ 0 & 1 & 0 & 0 \\ 0 & 0 & 0 & 0 \\ 0 & 0 & 1 & 0 \end{bmatrix} \hat{x}(k+1) = \begin{bmatrix} 1 & -1 & -1 & 1 \\ 1 & 1 & 0 & 0 \\ 0 & 0 & 1 & 0 \\ 0 & 1 & 1 & 0 \end{bmatrix} \hat{x}(k) + \begin{bmatrix} 0 \\ 1 \\ 0 \\ 2 \end{bmatrix} u(k) + \begin{bmatrix} 1 \\ 0 \\ 0 \\ -1 \end{bmatrix} y(k)$$

reconstructs exactly the state of system (8-4.5) at every step.

Obviously, if (8-2.2) is the EF1 for system (8-4.1), Theorem 8-4.2 holds, provided system (8-4.1) is R-observable, rank$E < n$ and $C_2 \neq 0$.

As seen in Section 8-2, a realization of discrete-time system needs future inputs. Thus the singular observer, though it exists theoretically, is difficult, or even impossible, to physically realize, especially the state reconstructor. We use the term state reconstructor to refer an observer that reconstructs exactly the state for a singular system. Thus, we now consider the design of normal observers.

The normal observer for system (8-4.1) is a kind of observer (8-4.2) when $E_c = I$. Analogous to that in the continuous-time case, we may prove the following two theorems (Dai, 1988a).

Theorem 8-4.3. If system (8-4.1) is detectable, dual normalizable, rank$C = r$, it has a reduced-order normal observer of order n-r of the form:

$$x_c(k+1) = A_c x_c(k) + B_c u(k) + Gy(k)$$
$$w(k) = F_c x_c(k) + Fy(k)$$

where $x_c(k) \in \mathbb{R}^{n-r}$, and $\lim_{k\to\infty}(w(k)-x(k)) = 0$, $\forall\ x(0),\ x_c(0)$.

Such observers are characterized by their measure outputs $w(k)$ which don't include input signal $u(k)$ visibly. As pointed out in Section 4-3, if (A_c, F_c) is observable, such observers are of the lowest order of n-r.

Theorem 8-4.4. Assume that system (8-4.1) is detectable and Y-observable. Then it has a state observer of order not greater than rankE of the following form:

$$x_c(k+1) = A_c x_c(k) + B_c u(k) + Gy(k)$$
$$w(k) = F_c x_c(k) + Fy(k) + Hu(k)$$

in which $x_c(k) \in \mathbb{R}^d$, $d \leqslant$ rankE, $w(k) \in \mathbb{R}^n$, such that

$$\lim_{k\to\infty} (w(k) - x(k)) = 0, \quad \forall\ x(0),\ x_c(0).$$

As mentioned earlier, the causal relationship generally doesn't exist between state and input in discrete-time singular systems. Any state $x(k)$ at time k cannot be determined by initial condition and inputs $u(0)$, $u(1)$, ... , $u(k)$, as in the case of normal systems. However, an interesting phenomenon is shown in Theorems 8-4.3 and 8-4.4. Under some conditions, the state $x(k)$ of discrete-time singular system (8-4.1) may be asymptotically estimated by former outputs $y(0)$, $y(1)$,..., $y(k)$, together with former inputs $u(0)$, $u(1)$, ..., $u(k)$. And the estimation error may be made arbitrarily small when time step k is sufficiently large.

These two kinds of observers may be designed in the same way as in the continuous-time case.

Example 8-4.2. Now we will consider the design of normal observers for system (8-4.5).

In this system, rankC = 1 and the nonsingular matrices

$$Q = \begin{bmatrix} 0 & 0 & 1 & 0 \\ 1 & 0 & 0 & 0 \\ 0 & 1 & 0 & 0 \\ 0 & 0 & 0 & 1 \end{bmatrix}, \qquad P = \begin{bmatrix} 0 & 1 & 0 & 0 \\ 0 & 0 & 1 & 0 \\ 0 & 0 & 0 & 1 \\ -1 & 0 & 1 & 1 \end{bmatrix}$$

satisfy

$$QEP = \mathrm{diag}(0,\ I_3), \qquad CP = [1 \mid 0 \quad 0 \quad 0]$$

$$QAP = \left[\begin{array}{c|ccc} 0 & 0 & 0 & 1 \\ \hline 0 & 1 & 0 & 0 \\ 0 & 1 & 1 & 0 \\ -1 & 0 & 1 & 1 \end{array}\right], \qquad QB = \left[\begin{array}{c} 0 \\ \hline 0 \\ 1 \\ 2 \end{array}\right].$$

By denoting

$$P^{-1}x(k) = \begin{bmatrix} x_1(k) \\ x_2(k) \end{bmatrix}, \qquad x_1(k) \in \mathbb{R}, \quad x_2(k) \in \mathbb{R}^3,$$

system (8-4.5) is r.s.e. to

$$x_1(k) = y(k)$$

$$x_2(k+1) = \begin{bmatrix} 1 & 0 & 0 \\ 1 & 1 & 0 \\ 0 & 1 & 1 \end{bmatrix} x_2(k) + \begin{bmatrix} 0 \\ 1 \\ 2 \end{bmatrix} u(k) + \begin{bmatrix} 0 \\ 0 \\ -1 \end{bmatrix} y(k) \tag{8-4.6}$$

$$\tilde{y} \triangleq [0 \quad 0 \quad 1] x_2(k) = 0.$$

Let

$$G_2 = [\tfrac{1}{6},\ 1,\ \tfrac{11}{6}]^\tau.$$

Then G_2 satisfies

$$\sigma\left(\begin{bmatrix} 1 & 0 & 0 \\ 1 & 1 & 0 \\ 0 & 1 & 1 \end{bmatrix} - \begin{bmatrix} 1/6 \\ 1 \\ 11/6 \end{bmatrix} [0 \quad 0 \quad 1] \right) = \{0,\ \tfrac{1}{2},\ \tfrac{2}{3}\},$$

which is within the unit circle on the complex plane. hence we may construct the observer for substate $x_2(k)$:

$$\hat{x}_2(k+1) = \left(\begin{bmatrix} 1 & 0 & 0 \\ 1 & 1 & 0 \\ 0 & 1 & 1 \end{bmatrix} - G_2[0 \quad 0 \quad 1] \right) \hat{x}_2(k) + \begin{bmatrix} 0 \\ 1 \\ 2 \end{bmatrix} u(k) + \begin{bmatrix} 0 \\ 0 \\ -1 \end{bmatrix} y(k) + G_2 \tilde{y} \tag{8-4.7}$$

such that $\lim_{k \to \infty} (x_2(k) - \hat{x}_2(k)) = 0$, $\forall\ x(0),\ \hat{x}_2(0)$. Combining it with (8-4.6) we thus obtain the normal observer for system (8-4.5) of the form:

$$x_c(k+1) = \begin{bmatrix} 1 & 0 & -\frac{1}{6} \\ 1 & 1 & -1 \\ 0 & 1 & -\frac{5}{6} \end{bmatrix} x_c(k) + \begin{bmatrix} 0 \\ 1 \\ 2 \end{bmatrix} u(k) + \begin{bmatrix} 0 \\ 0 \\ -1 \end{bmatrix} y(k)$$

$$w(k) = \begin{bmatrix} 1 & 0 & 0 \\ 0 & 1 & 0 \\ 0 & 0 & 1 \\ 0 & 1 & 1 \end{bmatrix} x_c(k) + \begin{bmatrix} 0 \\ 0 \\ 0 \\ -1 \end{bmatrix} y(k) \tag{8-4.8}$$

where $x_c(k) = \hat{x}_2(k)$, such that $\lim_{k\to\infty}(w(k)-x(k)) = 0$, $\forall\, x(0), x_c(0)$. This is a reduced-order observer of order three in whose measure output the input u(k) is not visibly involved.

Furthermore, we can prove that the state at any time point k may be exactly reconstructed from former measurements y(0), y(1), ... , y(k), together with former inputs u(0), u(1),..., u(k). This is the minimal-time state reconstruction problem discussed later.

Assume that system (8-4.1) is observable. Then a matrix $G_1 \in \mathbb{R}^{n\times r}$ may be chosen so that

$$\deg(|zE - (A-G_1C)|) = \text{rank}E \tag{8-4.9}$$

from which we know that there exist nonsingular matrices Q_1 and P_1 such that

$$\begin{aligned} Q_1EP_1 &= \text{diag}(I_q, 0), & Q_1(A-G_1C)P_1 &= \text{diag}(\hat{A}_1, I), \quad q = \text{rank}E \\ Q_1B &= [\hat{B}_1/\hat{B}_2], & CP_1 &= [\hat{C}_1, \hat{C}_2] \end{aligned} \tag{8-4.10}$$

Let ν_o be the observability index of $(\hat{A}_1,\hat{C}_1)$, which is independent of the selection of Q_1, P_1, G_1, provided (8-4.9)-(8-4.10) hold (Theorem 3-4.1).

<u>Theorem 8-4.5.</u> If system (8-4.1) is observable, it has a normal observer of the form:

$$\begin{aligned} x_c(k+1) &= A_cx_c(k) + B_cu(k) + Gy(k) \\ w(k) &= F_cx_c(k) + Fy(k) + Hu(k) \end{aligned} \tag{8-4.11}$$

where $x_c(k) \in \mathbb{R}^{n_c}$, A_c, B_c, G, F_c, F, and H are constant matrices, $n_c \leqslant$ rankE, such that the state x(k) is reconstructed exactly in ν_o steps.

<u>Proof.</u> As mentioned earlier, for any $G \in \mathbb{R}^{n\times r}$ satisfying $\deg(|zE-(A-GC)|) = \text{rank}E$, the system

$$E\hat{x}(k+1) = (A-GC)\hat{x}(k) + Bu(k) + Gy(k) \tag{8-4.12}$$

is causal. To construct the minimum-time state observer so that the state x(k) is reconstructed out in finite steps, we use $e(k) = x(k) - \hat{x}(k)$ to denote the construction error between x(k) and $\hat{x}(k)$. Then e(k) has the dynamics

$$Ee(k+1) = (A-GC)e(k) \tag{8-4.13}$$

Let G_1 satisfy (8-4.9). For any matrix $G \in \mathbb{R}^{n\times r}$, if we decompose

$$e(k) = P_1\begin{bmatrix} e_1(k) \\ e_2(k) \end{bmatrix}, \qquad Q_1(G-G_1) = \begin{bmatrix} \hat{G}_{11} \\ \hat{G}_{12} \end{bmatrix}$$

the error dynamics (8-4.13) is r.s.e. to

$$\begin{aligned} e_1(k+1) &= (\hat{A}_1-\hat{G}_{11}\hat{C}_1)e_1(k) - \hat{G}_{11}\hat{C}_2e_2(k) \\ 0 &= -\hat{G}_{12}\hat{C}_1e_1(k) + (I-\hat{G}_{12}\hat{C}_2)e_2(k). \end{aligned} \tag{8-4.14}$$

Since G satisfies $\deg(|zE-(A-GC)|) = \mathrm{rank}E$, $(I-\hat{G}_{12}\hat{C}_2)^{-1}$ exists. Hence (8-4.14) becomes

$$\begin{aligned} e_1(k+1) &= (\hat{A}_1-\hat{G}_{11}\hat{C}_1-\hat{G}_{11}\hat{C}_2(I-\hat{G}_{12}\hat{C}_2)^{-1}\hat{G}_{12}\hat{C}_1)e_1(k) \\ e_2(k) &= (I-\hat{G}_{12}\hat{C}_2)^{-1}\hat{G}_{12}\hat{C}_1e_1(k). \end{aligned} \tag{8-4.15}$$

Therefore, the problem of a number k_o existing such that $e(k) = 0$, $k \geqslant k_o$, is changed into finding a matrix $G \in \mathbb{R}^{n\times r}$ satisfying $\deg(|zE-(A-GC)|) = \mathrm{rank}E$ such that

$$e_1(k) = 0, \quad e_2(k) = 0, \qquad k \geqslant k_o.$$

And the minimum-time state reconstruction problem is consequently changed into setting system (8-4.15) such that its state reaches and keeps at zero in minimum steps.

Without loss of generality, we assume $\mathrm{rank}\hat{C}_1 = r$. According to the results in Section 7-3 and by duality, there exists a nonsingular matrix T such that

$$\tilde{A}_1 = T\hat{A}_1T^{-1} = \left[\begin{array}{cccc|cccc|c|cccc}
0 & \cdots & & * & & & & * & & & & & * \\
1 & 0 & & * & & & & * & & & & & * \\
 & 1 & \ddots & \vdots & & 0 & & \vdots & & & \cdots & & \vdots \\
 & & 1 & * & & & & * & & & & & * \\
\hline
 & & & * & 0 & & & * & & & & & * \\
 & & & * & 1 & 0 & & * & & & & & * \\
 & 0 & & \vdots & & 1 & \ddots & \vdots & & & \cdots & & \vdots \\
 & & & * & & & 1 & * & & & & & * \\
\hline
 & & & & & & & & \ddots & & & & \\
\hline
 & & & * & & & & * & & 0 & & & * \\
 & & & * & & & & * & & 1 & 0 & & * \\
 & 0 & & \vdots & & 0 & & \vdots & & & 1 & \ddots & \vdots \\
 & & & * & & & & * & & & & 1 & *
\end{array}\right]$$

$$\underbrace{\qquad\qquad}_{d_1}\uparrow_{\sigma_1^o} \qquad \underbrace{\qquad\qquad}_{d_2}\uparrow_{\sigma_2^o} \qquad \underbrace{\qquad\qquad}_{d_r}\uparrow_{\sigma_r^o}$$

$$\tilde{C}_1 = \hat{C}_1 T^{-1} = \left[\begin{array}{cccc|ccc|c|ccc} 0 & 0 & & 1 & 0 & & 0 & & 0 & & 0 \\ 0 & 0 & \cdots & * & 0 & \cdots & 1 & \cdots & 0 & \cdots & 0 \\ 0 & 0 & & * & 0 & & * & & 0 & & 0 \\ & \cdots & \cdots & & & & & & \cdots & \cdots & \\ 0 & 0 & \cdots & * & 0 & \cdots & * & \cdots & 0 & \cdots & 1 \end{array}\right]$$

$$\underbrace{\qquad\qquad}_{d_1}\ \uparrow\sigma_1^{\circ}\quad \underbrace{\qquad\quad}_{d_2}\ \uparrow\sigma_2^{\circ}\qquad\qquad \underbrace{\qquad\quad}_{d_r}\ \uparrow\sigma_r^{\circ}$$

where "$*$" represents the possibly nonzero elements, $q = \text{rank}E = \sum_{i=1}^{r} d_i$, $\nu_o = \max\{d_i\}$.

Let $\bar{A}_o$ be the qxr matrix formed by the σ_i°th columns of $\tilde{A}_1$ and $\bar{C}_o$ be the rxr nonsingular matrix formed by the σ_i°th columns of $\tilde{C}_1$. If we set

$$G_o = \bar{A}_o\bar{C}_o^{-1}, \quad \hat{G}_{12} = 0, \quad \hat{G}_{11} = TG_o,$$

direct calculation shows

$$\tilde{A}_1 - G_o\tilde{C}_1 = \left[\begin{array}{ccccc|ccccc|c|ccccc} 0 & & & & 0 & & & & & & & & & & & \\ 1 & 0 & & & 0 & & & & & & & & & & & \\ & 1 & \ddots & & \vdots & & & & & & & & & 0 & & \\ & & \ddots & 0 & \vdots & & & & & & & & & & & \\ & & & 1 & 0 & & & & & & & & & & & \\ \hline & & & & & 0 & & & & 0 & & & & & & \\ & & & & & 1 & 0 & & & 0 & & & & & & \\ & & & & & & 1 & \ddots & & \vdots & & & & & & \\ & & & & & & & \ddots & 0 & \vdots & & & & & & \\ & & & & & & & & 1 & 0 & & & & & & \\ \hline & & & & & & & & & & \ddots & & & & & \\ \hline & & & & & & & & & & & 0 & & & & 0 \\ & & 0 & & & & & & & & & 1 & 0 & & & 0 \\ & & & & & & & & & & & & 1 & \ddots & & \vdots \\ & & & & & & & & & & & & & \ddots & 0 & \vdots \\ & & & & & & & & & & & & & & 1 & 0 \end{array}\right]$$

Thus, $\tilde{A}_1 - G_o\tilde{C}_1$ is nilpotent, $(\tilde{A}_1 - G_o\tilde{C}_1)^{\nu_o} = 0$. For such a G_o we have

$$(\hat{A}_1 - \hat{G}_{11}\hat{C}_1 - \hat{G}_{11}\hat{C}_2(I - \hat{G}_{12}\hat{C}_2)^{-1}\hat{G}_{12}\hat{C}_1)^{\nu_o} = (\hat{A}_1 - \hat{G}_{11}\hat{C}_1)^{\nu_o} = T(\tilde{A}_1 - G_o\tilde{C}_1)^{\nu_o}T^{-1} = 0.$$

Therefore, $k_o = \nu_o$, and

$$G = G_1 + Q_1^{-1}[\hat{G}_{11}/\hat{G}_{12}] = G_1 + Q_1^{-1}[TG_o/0].$$

Thus observer (8-4.12) reconstructs exactly the state of system (8-4.1) in ν_o steps.

Moreover, by construction the matrix G satisfies $\deg(|zE-(A-GC)|) = \text{rank}E$. Hence, there exist nonsingular matrices Q_2 and P_2 such that

$$Q_2EP_2 = \text{diag}(I_q, 0), \qquad Q_2(A-GC)P_2 = \text{diag}(A_o, I).$$

As a result, under the coordinate transformation

$$\hat{x}(k) = P_2\begin{bmatrix}\hat{x}_1(k)\\ \hat{x}_2(k)\end{bmatrix}, \qquad Q_2B = \begin{bmatrix}\bar{B}_1\\ \bar{B}_2\end{bmatrix}, \qquad Q_2G = \begin{bmatrix}\bar{G}_1\\ \bar{G}_2\end{bmatrix}, \qquad CP_2 = [\bar{C}_1, \ \bar{C}_2]$$

observer (8-4.12) is r.s.e. to

$$\hat{x}_1(k+1) = A_o\hat{x}_1(k) + \bar{B}_1u(k) + \bar{G}_1y(k)$$
$$0 = \hat{x}_2(k) + \bar{B}_2u(k) + \bar{G}_2y(k),$$

which may be written in another form:

$$x_c(k+1) = A_ox_c(k) + \bar{B}_1u(k) + \bar{G}_1y(k)$$
$$w(k) = P_2\begin{bmatrix}I\\ 0\end{bmatrix}x_c(k) + P_2\begin{bmatrix}0\\ -B_2\end{bmatrix}u(k) + P_2\begin{bmatrix}0\\ -G_2\end{bmatrix}y(k),$$

where $x_c(k) = \hat{x}_1(k)$, such that

$$w(k) - x(k) = 0, \qquad k \geqslant k_o, \qquad \forall \ x(0), x_c(0).$$

Therefore, this observer is what is required. This establishes the theorem. Q.E.D.

Clearly, this observer is a minimum-time state reconstruction observer under a design method such as Theorem 7-4.5.

Example 8-4.3. For the singular system

$$\begin{bmatrix}1&0&0&0\\0&1&0&0\\0&0&0&0\\0&0&1&0\end{bmatrix}x(k+1) = \begin{bmatrix}1&0&0&0\\1&1&0&0\\0&0&1&0\\0&0&0&1\end{bmatrix}x(k) + \begin{bmatrix}0\\1\\0\\2\end{bmatrix}u(k)$$
$$y(k) = [0 \quad 1 \quad 1 \ -1]x(k) \tag{8-4.16}$$

under the coordinate transformation

$$Q = \begin{bmatrix}1&0&0&0\\0&1&0&0\\0&0&-1&1\\0&0&1&0\end{bmatrix}, \qquad P = \begin{bmatrix}1&0&0&0\\0&1&0&0\\0&0&1&0\\0&1&1&1\end{bmatrix}$$

and

$$P^{-1}x(k) = \begin{bmatrix}x_1(k)\\ x_2(k)\end{bmatrix}$$

it is r.s.e. to

$$x_1(k+1) = \begin{bmatrix}1&0&0\\1&1&0\\0&1&0\end{bmatrix}x_1(k) + \begin{bmatrix}0\\1\\2\end{bmatrix}u(k) + \begin{bmatrix}0\\0\\-1\end{bmatrix}y(k) \tag{8-4.17a}$$

$$x_2(k) = -y(k)$$
$$0 = [0 \quad 0 \quad 1]x_1(k). \tag{8-4.17b}$$

The substate $x_2(k)$ is given by $-y(k)$. Now we consider constructing the observer for substate $x_1(k)$.

Let $G = [1 \quad 3 \quad 2]^{\tau}$. Then

$$\left(\begin{bmatrix} 1 & 0 & 0 \\ 1 & 1 & 0 \\ 0 & 1 & 0 \end{bmatrix} - G[0 \quad 0 \quad 1]\right)^3 = \begin{bmatrix} 1 & 0 & -1 \\ 1 & 1 & -3 \\ 0 & 1 & -2 \end{bmatrix}^3 = 0.$$

Therefore, we may construct the following state observer for substate $x_1(k)$:

$$x_c(k+1) = \begin{bmatrix} 1 & 0 & -1 \\ 1 & 1 & -3 \\ 0 & 1 & -2 \end{bmatrix} x_c(k) + \begin{bmatrix} 0 \\ 1 \\ 2 \end{bmatrix} u(k) + \begin{bmatrix} 0 \\ 0 \\ -1 \end{bmatrix} y(k)$$

such that

$$x_1(k) - x_c(k) = 0, \quad k \geqslant 3, \quad \forall\ x(0),\ x_c(0).$$

Combining it with (8-4.17), and $x(k) = P[x_1(k)/x_2(k)]$, it is obvious that the system

$$x_c(k+1) = \begin{bmatrix} 1 & 0 & -1 \\ 1 & 1 & -3 \\ 0 & 1 & -2 \end{bmatrix} x_c(k) + \begin{bmatrix} 0 \\ 1 \\ 2 \end{bmatrix} u(k) + \begin{bmatrix} 0 \\ 0 \\ -1 \end{bmatrix} y(k)$$
$$w(k) = \begin{bmatrix} 1 & 0 & 0 \\ 0 & 1 & 0 \\ 0 & 0 & 1 \\ 0 & 1 & 1 \end{bmatrix} x_c(k) + \begin{bmatrix} 0 \\ 0 \\ 0 \\ -1 \end{bmatrix} y(k) \tag{8-4.18}$$

is a normal state observer for system (8-4.16). And the state $x(k)$ may be reconstructed exactly by (8-4.18) in three steps.

Moreover, if system (8-4.1) is observable, there exists a matrix $G \in \mathbb{R}^{n \times r}$ so that $|E+GC| \neq 0$. Let $\hat{A} = (E+GC)^{-1}A$ and μ_0 be the observability index of $(\hat{A}, C)$, which has been proven to be a fixed value independent of the selection of $G \in \mathbb{R}^{n \times r}$ satisfying $|E+GC| \neq 0$.

Then we can construct state observers for system (8-4.1) of another form.

<u>Theorem 8-4.6</u>. If system (8-4.1) is observable, it has a normal state observer

$$x_c(k+1) = A_c x_c(k) + B_c u(k) + G_c y(k)$$
$$w(k) = F_c x_c(k) + Fy(k) \tag{8-4.19}$$

such that its state $x(k)$ may be reconstructed exactly in μ_0 steps.

Such observers are characterized by their output in which the inputs are not used directly, which are different from observers of the form (8-4.11).

8-5. Dynamic Compensation for Discrete-Time Singular Systems

For discrete-time singular system (8-4.1):

$$\begin{aligned} Ex(k+1) &= Ax(k) + Bu(k) \\ y(k) &= Cx(k) \end{aligned} \qquad (8\text{-}5.1)$$

its state compensation problem os to construct the dynamic compensators of the form

$$\begin{aligned} E_c x_c(k+1) &= A_c x_c(k) + B_c y(k) \\ u(k) &= F_c x_c(k) + Fy(k) \end{aligned} \qquad (8\text{-}5.2)$$

where $x_c(k) \in \mathbb{R}^{n_c}$, A_c, B_c, E_c, F_c, and F are constant matrices and system (8-5.2) is regular, such that the closed-loop system

$$\begin{bmatrix} E & 0 \\ 0 & E_c \end{bmatrix}\begin{bmatrix} x(k+1) \\ x_c(k+1) \end{bmatrix} = \begin{bmatrix} A+BFC & BF_c \\ B_cC & A_c \end{bmatrix}\begin{bmatrix} x(k) \\ x_c(k) \end{bmatrix} \qquad (8\text{-}5.3)$$

$$y(k) = [C \quad 0][x(k)/x_c(k)]$$

is stable, i.e.,

$$\sigma\left(\begin{bmatrix} E & 0 \\ 0 & E_c \end{bmatrix}, \begin{bmatrix} A+BFC & BF_c \\ B_cC & A_c \end{bmatrix}\right) \subset U^+.$$

Thus closed-loop system (8-5.3) is asymptotically stable. Systems (8-5.2) and (8-5.3) are guaranteed to be regular.

If E_c is nonsingular, we will term (8-5.2) a normal (dynamic) compensator; otherwise, it is a singular compensator.

For the discrete-time singular system (8-5.1), its compensators may be designed in a similar way to that for the continuous-time case, in a form of observer-based controllers.

Another problem studied in this section is the deadbeat control problem: Find control inputs u(0), u(1), ... such that the state x(k) of system (8-5.1) at any instant k is transferred to zero in finite steps. The deadbeat control problem, in a sense, is the dual problem of the exact state reconstruction problem in the last section.

8-5.1. Singular compensators

Following are two results on singular compensators for system (8-5.1).

<u>Theorem 8-5.1</u>. System (8-5.1) has a singular compensator of the form of (8-5.2) if

and only if it is stabilizable and detectable.

Theorem 8-5.2. Assume that system (8-5.1) is controllable and observable, rank$E < n$. Then it has a deadbeat controller in the form of (8-5.2) such that its state from arbitrary initial condition is driven to and kept at zero from the initial instant.

Proof. Subject to the controllability and observability assumptions for system (8-5.1), rank$E < n$, from Corollary 8-3.1 we know that there exist two matrices $K \in \mathbb{R}^{m\times n}$, $G \in \mathbb{R}^{n\times r}$ so that

$$|zE - (A+BK)| = \text{constant} \neq 0$$

$$|zE - (A-GC)| = \text{constant} \neq 0.$$

For system (8-5.1) we now construct the control

$$u(k) = Kx_c(k), \tag{8-5.4}$$

where $x_c(k)$ is the output of its singular observer of the following form:

$$Ex_c(k+1) = (A-GC)x_c(k) + Bu(k) + Gy(k). \tag{8-5.5}$$

In this case, the closed-loop system formed by (8-5.4) and (8-5.5) is

$$\begin{bmatrix} E & 0 \\ 0 & E \end{bmatrix}\begin{bmatrix} x(k+1) \\ x_c(k+1) \end{bmatrix} = \begin{bmatrix} A & BK \\ GC & A-GC+BK \end{bmatrix}\begin{bmatrix} x(k) \\ x_c(k) \end{bmatrix}, \tag{8-5.6}$$

which has the characteristic polynomial

$$\left|\begin{bmatrix} E & 0 \\ 0 & E \end{bmatrix} - \begin{bmatrix} A & BK \\ GC & A-GC+BK \end{bmatrix}\right| = |zE-(A+BK)||zE-(A-GC)| = \text{constant} \neq 0.$$

Combining it with the discussion in Section 8-3 we know that the closed-loop system (8-5.6) has the state response

$$x(k) = 0, \quad x_c(k) = 0, \quad k = 0, 1, 2, \ldots \tag{8-5.7}$$

Thus the compensator

$$Ex_c(k+1) = (A-GC+BK)x_c(k) + Gy(k)$$
$$u(k) = Kx_c(k)$$

determined by (8-5.4) and (8-5.5) is a deadbeat controller for system (8-5.1). Q.E.D.

It is wellknown that for a normal system the deadbeat controller drives its state from an arbitrary initial condition to zero in $k > 0$ steps, or at least one step. However, this theorem shows that, different from the normal case, the deadbeat controller, which is in the form of a singular compensator, can transfer the state of singular system (8-5.1) from arbitrary initial condition to zero in zero steps, or from the initial instant.

Example 8-5.1. Consider the singular system

$$\begin{bmatrix} 1 & 0 & 0 \\ 1 & 1 & 0 \\ 0 & 0 & 0 \end{bmatrix} x(k+1) = \begin{bmatrix} 2 & 1 & 0 \\ 0 & 3 & 0 \\ 0 & 0 & 1 \end{bmatrix} x(k) + \begin{bmatrix} 0 \\ 1 \\ -1 \end{bmatrix} u(k)$$

$$y(k) = [1 \ \ -1 \ \ 2]x(k) \qquad (8\text{-}5.8)$$

which is both controllable and observable.

It is easy to testify that the two matrices

$$K = [1 \ \ 0 \ \ 1], \qquad G = [1 \ \ 2 \ \ 0.5]^{\tau}$$

satisfy

$$|zE-(A+BK)| = 1, \quad |zE-(A-GC)| = -3.$$

From Theorem 8-5.1 we can construct the following deadbeat controller

$$\begin{bmatrix} 1 & 0 & 0 \\ 1 & 1 & 0 \\ 0 & 0 & 1 \end{bmatrix} x_c(k+1) = \begin{bmatrix} 1 & 2 & -2 \\ -1 & 5 & -3 \\ -1.5 & 0.5 & -1 \end{bmatrix} x_c(k) + \begin{bmatrix} 1 \\ 2 \\ 0.5 \end{bmatrix} y(k)$$

$$u(k) = [1 \ \ 0 \ \ 1]x_c(k)$$

such that the state of system (8-5.8) is driven to and kept at zero from the initial instant.

8-5.2. Normal compensators

Now we will consider the normal compensators for system (8-5.1):

$$\begin{aligned} x_c(k+1) &= A_c x_c(k) + B_c y(k) \\ u(k) &= F_c x_c(k) + Fy(k) \end{aligned} \qquad (8\text{-}5.9)$$

where $x_c(k) \in \mathbb{R}^{n_c}$, A_c, B_c, F_c, and F are constant matrices of appropriate dimensions.

<u>Theorem 8-5.3</u>. System (8-5.1) has a normal compensator in the form of (8-5.9) if and only if it is stabilizable and detectable.

Thus, the existence conditions for normal and singular systems are the same. But, for the noncausality in singular systems, the causal normal compensators have an advantage for practical system design. Therefore, normal compensators are often adopted for control system design.

It has been proven that although, under certain conditions, system (8-5.1) may have a deadbeat controller in the form of (8-5.9), the state of the closed-loop system could not be identically zero. Generally, we can design a normal compensator (8-5.9) to transfer the state of system (8-5.1) to and keep it at zero in finite steps.

To consider this problem, we note that the compensators are often designed based on state feedback realized via a state observer. Thus, let us consider the observer-based normal compensator:

$$\begin{aligned} x_c(k+1) &= A_c x_c(k) + B_c u(k) + Gy(k) \\ w(k) &= F_c x_c(k) + Fy(k) \\ u(k) &= Kw(k) \end{aligned} \tag{8-5.10}$$

in which $w(k)$ is the output of observer such that $\lim_{k\to\infty}(w(k)-x(k)) = 0$, $\forall\, x(0)$, $x_c(0)$.

Let system (8-5.1) be controllable and (A_c, F_c) be observable. Then, from Theorem 4-3.4 we know that the necessary and sufficient conditions for $w(k)$ to be an observer for state $x(k)$ are

1. $B_c = MB$.
2. $MA - A_c ME = GC$.
3. $I = F_c ME + FC$.
4. $\sigma(A_c) \subset U^+$.

M is a constant matrix.

Denoting $e(k) = x_c(k) - MEx(k)$, from (8-5.1) and (8-5.10) we have

$$\begin{bmatrix} E & 0 \\ 0 & I \end{bmatrix}\begin{bmatrix} x(k+1) \\ e(k+1) \end{bmatrix} = \begin{bmatrix} A+BK & BKF_c \\ 0 & A_c \end{bmatrix}\begin{bmatrix} x(k) \\ e(k) \end{bmatrix}, \tag{8-5.11}$$

i.e.,

$$Ex(k+1) = (A+BK)x(k) + BKF_c e(k) \tag{8-5.12a}$$

$$e(k+1) = A_c e(k), \qquad e(0) = x_c(0) - MEx(0). \tag{8-5.12b}$$

Let K be chosen such that

$$|zE - (A+BK)| = \text{constant} \neq 0. \tag{8-5.13}$$

Then there exist nonsingular matrices Q_2 and P_2 such that (8-5.12a) is r.s.e. to

$$\tilde{N}\tilde{x}(k+1) = \tilde{x}(k) + \tilde{B}e(k) \tag{8-5.14}$$

where

$$\tilde{N} = Q_2 E P_2, \quad Q_2(A+BK)P_2 = I_n, \quad Q_2 BKF_c = \tilde{B}, \quad x(k) = \tilde{P}\tilde{x}(k).$$

Let $\tilde{h}$ be the nilpotent index of $\tilde{N}$. From (8-5.13) and (8-5.14) we know that $x(k)$ and $e(k)$ are given by

$$x(k) = -P_2 \sum_{i=0}^{\tilde{h}-1} \tilde{N}^i \tilde{B} e(k+i)$$

$$e(k) = A_c^i e(0).$$

Thus, to transfer the state $x(k)$ to zero in minimum time is to find the matrix A_c

so that e(k) reaches and keeps at zero in minimum time, or to determine matrix A_c so that the ν satisfying $A_c^\nu = 0$ is minimum. As a result, the deadbeat controller problem via compensator (8-5.10) is equivalent to finding matrices K and A_c subject to w(k) being an observer for x(k).

Particularly, if system (8-5.1) is observable, the observability index μ_o of $((E+G_1C)^{-1}A,C)$ is a fixed value independent of the selection of matrix G_1 provided $G_1 \in \mathbb{R}^{n\times r}$ satisfies $|E+G_1C| \neq 0$. Thus we can prove the following.

Theorem 8-5.4. Let system (8-5.1) be controllable and observable, rank$E < n$. Then it has a normal compensator in the form of (8-5.10) such that the state of system (8-5.1) is driven to and kept at zero from an arbitrary initial condition in μ_o steps.

Proof. Under the assumption of observability, a matrix $K \in \mathbb{R}^{m\times n}$ may be chosen so that

$$|zE - (A+BK)| = \text{constant} \neq 0.$$

Hence, as long as the feedback control $u(k) = Kx(k)$ is applied to (8-5.1), the state of the closed-loop system is identically zero. Since x(k) is inaccessible, we next realize it via state observer.

Note the assumption that system (8-5.1) is observable. There exists a matrix G_1 satisfying $|E+G_1C| \neq 0$. Via a similar procedure as that used in the last section, a matrix G_o may be chosen so that

$$((E+G_1C)^{-1}A - G_oC)^{\mu_o} = 0.$$

For such G_1, G_o, we choose $G_2 = (E+G_1C)G_o$ and construct the state observer

$$(E+G_2C)\hat{x}(k+1) = (A-G_2C)\hat{x}(k) + Bu(k) + G_2y(k+1) + G_1y(k). \tag{8-5.15}$$

Direct computation shows that

$$x(k) = \hat{x}(k), \quad k \geqslant \mu_o.$$

By defining $x_c(k) = \hat{x}(k) - (E+G_1C)^{-1}G_2y(k)$, system (8-5.15) may be written as

$$\begin{aligned} x_c(k+1) &= A_cx_c(k) + B_cu(k) + G_cy(k) \\ w(k) &= \hat{x}(k) = F_cx_c(k) + Fy(k) \end{aligned} \tag{8-5.16}$$

where

$$A_c = (E+G_1C)^{-1}(A-G_2C),$$

$$B_c = (E+G_1C)^{-1}B,$$

$$G_c = (E+G_1C)^{-1}G_2 + A_c(E+G_1C)^{-1}G_1$$

$$F = (E+G_1C)^{-1}G_1, \qquad F_c = I.$$

According to the previous discussion, we know that the compensator

$$x_c(k+1) = (A_c+B_cK)x_c(k) + (G_c+B_cKF)y(k)$$

$$u(k) = Kx_c(k) + KFy(k)$$

formed by feedback $u(k) = Kw(k)$ and observer (8-5.16) is a deadbeat controller for system (8-5.1) such that its state from arbitrary initial condition is driven to and kept at zero in μ_o steps. Q.E.D.

Example 8-5.2. As shown in Example 8-5.1, system (8-5.8) is controllable and observable, $\text{rank}E = 2 < n = 3$. The matrix $K = [1\ 0\ 1]$ satisfies $|zE-(A+BK)| = 1$ and the matrix $G_1 = [0\ \ 0\ \ 1]^\tau$ satisfies $|E+G_1C| \neq 0$. Direct computation shows

$$(E+G_1C)^{-1}A = \begin{bmatrix} 2 & 1 & 0 \\ -2 & 2 & 0 \\ -2 & 0.5 & 0.5 \end{bmatrix}$$

and the observability index of $((E+G_1C)^{-1}A,C)$ is 3. Let $G_o = [-4\ \ 16\ \ 12.25]^\tau$. It will be

$$((E+G_1C)^{-1}A - G_oC)^3 = 0.$$

For these matrices chosen as earlier, we have

$$G_2 = (E+G_1C)G_o = [-4\ \ 12\ \ 4.5]$$

$$A_c = (E+G_1C)^{-1}(A-G_2C) = (E+G_1C)^{-1}A - G_oC$$

$$= \begin{bmatrix} 6 & -3 & 8 \\ -18 & 18 & -32 \\ 14.25 & -12.75 & -24 \end{bmatrix}$$

$$B_c = (E+G_1C)^{-1}B = [0\ \ 1\ \ 0]^\tau$$

$$G_c = (E+G_1C)^{-1}G_2 + A_c(E+G_1C)^{-1}G_1 = [0\ \ 0\ \ 0.25]^\tau$$

$$F = (E+G_1C)^{-1}G_1 = [0\ \ 0\ \ 0.5]^\tau$$

$$A_c+B_cK = \begin{bmatrix} 6 & -3 & 8 \\ -17 & 18 & -31 \\ 14.25 & -12.75 & -24 \end{bmatrix}$$

$$G_c+B_cKF = [0\ \ 0.5\ \ 0.25]^\tau$$

$$KF = 0.5.$$

By Theorem 8-5.4, the compensator

$$x_c(k+1) = \begin{bmatrix} 6 & -3 & 8 \\ -17 & 18 & -31 \\ 14.25 & -12.75 & -24 \end{bmatrix} x_c(k) + \begin{bmatrix} 0 \\ 0.5 \\ 0.25 \end{bmatrix} y(k)$$

$$u(k) = [1\ \ 0\ \ 1]x_c(k) + 0.5y(k)$$

is a deabeat controller for singular system (8-5.8) such that

$$x(k) = 0, \quad x_c(k) = 0, \quad \forall\ x(0),\ x_c(0), \quad k \geqslant 3.$$

8-6. Notes and References

Discrete-time singular systems were first studied by Luenberger (1977). Sections 8-1 and 8-2 are based on Dai (1988d). Sections 8-4 and 8-5 are based on Dai (1988a) and Dai and Wang (1987a), respectively, and other papers such as Bender (1987), Campbell and Rodriguez (1985), El-Tohami et al. (1987), Lewis (1983a, 1984, 1985a), and Luenberger (1977, 1979, 1987).

The relationship between continuous-time and discrete-time singular systems such as discretization of continuous-time systems and control of discrete data on continuous-time systems deserves further work.

CHAPTER 9

OPTIMAL CONTROL

9-1. Introduction

So far, in the control system design we placed our design on one principle: to assign the closed-loop poles to arbitrarily prescribed positions. On one hand, such a design principle is difficult to handle since evaluting the performance of a system from pole distribution is not an easy task; on the other hand, in the pole assignment the feedback control is not unique. Thus, the freedom should be used to improve the closed-loop properties, such as fast response, minimum-energy consumption. These aims require choosing a best control law from all feasible strategies. This is the optimal control problem, loosely speaking.

It is worth pointing out that deviations always exist in a real system model, thus cause the optimal control to lose its expected property. In some cases, a nonoptimal control law may have better practical effect than a theoretically optimal one due to these reasons. However, the optimal method provides a more effective approach in control system design. In practice, this approach often proves reliable and useful.

For simplicity, we study only the simple linear singular system:

$$E\dot{x}(t) = Ax(t) + Bu(t) \tag{9-1.1}$$

where $x(t) \in \mathbb{R}^n$, $u(t) \in \mathbb{R}^m$ are its state and control input, respectively; E, $A \in \mathbb{R}^{n\times n}$, and $B \in \mathbb{R}^{n\times m}$ are constant matrices. It is assumed that system (9-1.1) is regular and $\operatorname{rank}E < n$.

In real systems, the control input $u(t)$ often subjects to a constraint π. We shall use U to denote the collection set of $u(t)$ subject to constraint π, which is called the constraint set, or simply, the constraint. Any $u(t) \in U$ is called admissible control. For singular systems, the admissible control should be first the piecewise sufficiently continuously differentiable function since derivatives of control input are included in the state of singular systems. The constraint π often is a vector function of state, control, and time t, or $\pi(x(t), u(t), t)$. We also suppose that a target $\mathbf{S}$ is given, which is a constraint on terminal state. For system (9-1.1), let J be a functional

$$J(x(t), u(t), t) \tag{9-1.2}$$

of state, control, and time, which is called a cost functional. The optimal control problem for system (9-1.1) with respect to the target **S**, constraint U, initial time t = 0, and initial state x(0) is to determine the admissible control $u^*(t) \in U$ such that the terminal state reaches **S** and the cost functional J is minimized (or maximized):

$$\min_{u \in U} J(x(t), u(t), t) = J(x^*(t), u^*(t), t) \qquad (9\text{-}1.3)$$

$$\left(\text{or} \ \max_{u \in U} J(x(t), u(t), t) = J(x^*(t), u^*(t), t)\ \right).$$

The cost J is chosen with respect to the problem in we are interested. Different Js represent different meaning of our optimal problem. Usually, it is the time-optimal control, the cheap control(minimum energy consumption), etc.

For normal systems, the optimal solution for general problems may be obtained from the wellknown maximum principle. However, there doesn't exist such a principle for general singular systems. Much effort has been made toward solving this problem (Lovass-Nagy et al., 1986; Lewis, 1985b; Bender and Laub, 1987b; Dai, 1988g). At present, the problems concerning optimal control are treated by changing them into optimal problems for normal systems, then solving via normal system theory.

9-2. Optimal Regulation with Quadratic Cost Functional

Consider the singular system

$$E\dot{x}(t) = Ax(t) + Bu(t) \qquad (9\text{-}2.1)$$

with the quadratic cost functional

$$J = \int_0^\infty [x^\tau(t)Fx(t) + u^\tau(t)Gu(t)]dt \qquad (9\text{-}2.2)$$

where F and G are symmetric positive definite. This is a problem with an infinite-interval cost functional. It is supposed that the admissible control is the piecewise sufficiently continuously differentiable function. For its simple form for analysis, this cost is often adopted in practice. Since imposing control means consuming of energy, the optimal control with quadratic cost functional is also called cheap control.

Cobb (1983a) has proven the following theorem.

Theorem 9-2.1. Consider (9-2.1) with quadratic cost functional (9-2.2). Then there exists an optimal control $u^*(t)$ such that cost functional (9-2.2) is minimized if and only if system (9-2.1) is stabilizable and impulse controllable.

We now consider solving the optimal control under the assumption of existence and uniqueness of the solution.

Assume that system (9-2.1) is stabilizable and detectable. Then there exists a matrix $K \in \mathbb{R}^{m\times n}$ satisfying

$$\deg(|sE - (A+BK)|) = \operatorname{rank}E. \tag{9-2.3}$$

Let the preliminary feedback control be

$$u = Kx + v, \tag{9-2.4}$$

where $v \in \mathbb{R}^m$ is the new input. When applied to system (9-2.1), the closed-loop system is

$$E\dot{x} = (A+BK)x + Bv. . \tag{9-2.5}$$

According to the selection of matrix K, equation (9-2.3) holds. Thus, there exist nonsingular matrices Q_1 and P_1 such that

$$Q_1EP_1 = \operatorname{diag}(I_q, 0), \quad Q_1(A+BK)P_1 = \operatorname{diag}(A_1, I), \quad q = \operatorname{rank}E.$$

By denoting

$$P_1^{-1}x = \begin{bmatrix} x_1 \\ x_2 \end{bmatrix}, \quad x_1 \in \mathbb{R}^q, \quad x_2 \in \mathbb{R}^{n-q}, \tag{9-2.6}$$

system (9-2.5) is r.s.e. to

$$\begin{aligned} \dot{x}_1 &= A_1x_1 + B_1v \\ 0 &= x_2 + B_2v. \end{aligned} \tag{9-2.7}$$

From (9-2.7), $x_2 = -B_2v$. Combining it with (9-2.6) we have

$$\begin{bmatrix} x \\ u \end{bmatrix} = \begin{bmatrix} I & 0 \\ K & I \end{bmatrix}\begin{bmatrix} x \\ v \end{bmatrix} = \begin{bmatrix} P_1 & 0 \\ KP_1 & I \end{bmatrix}\begin{bmatrix} x_1 \\ x_2 \\ v \end{bmatrix} = \begin{bmatrix} P_1 & 0 \\ KP_1 & I \end{bmatrix}\begin{bmatrix} I & 0 \\ 0 & -B_2 \\ 0 & I \end{bmatrix}\begin{bmatrix} x_1 \\ v \end{bmatrix}.$$

Thus cost functional (9-2.2) becomes

$$J = \int_0^\infty \begin{bmatrix} x \\ u \end{bmatrix}^\tau \begin{bmatrix} F & 0 \\ 0 & G \end{bmatrix}\begin{bmatrix} x \\ u \end{bmatrix} dt = \int_0^\infty \begin{bmatrix} x_1 \\ v \end{bmatrix}^\tau \begin{bmatrix} \bar{F} & H \\ H^\tau & \bar{G} \end{bmatrix}\begin{bmatrix} x_1 \\ v \end{bmatrix} dt \tag{9-2.8}$$

where

$$\begin{bmatrix} \bar{F} & H \\ H & \bar{G} \end{bmatrix} = \begin{bmatrix} I & 0 \\ 0 & -B_2 \\ 0 & I \end{bmatrix}^\tau \begin{bmatrix} P_1 & 0 \\ KP_1 & I \end{bmatrix}^\tau \begin{bmatrix} F & 0 \\ 0 & G \end{bmatrix}\begin{bmatrix} P_1 & 0 \\ KP_1 & I \end{bmatrix}\begin{bmatrix} I & 0 \\ 0 & -B_2 \\ 0 & I \end{bmatrix}. \tag{9-2.9}$$

Note that the matrix diag(F, G) is symmetric positive definite and matrix

$$\begin{bmatrix} P_1 & 0 \\ KP_1 & I \end{bmatrix}\begin{bmatrix} I & 0 \\ 0 & -B_2 \\ 0 & I \end{bmatrix}$$

is of full column rank. We know that matrix (9-2.9) is symmetric positive definite, and so are matrices $\bar{F}$ and $\bar{G}$. Therefore, cost functional (9-2.9) may be further written as

$$J = \int_0^\infty (x_1^\tau \tilde{F} x_1 + w^\tau \bar{G} w)dt \tag{9-2.10}$$

where

$$\tilde{F} = \bar{F} - H\bar{G}^{-1}H^\tau, \qquad w = v + \bar{G}^{-1}H^\tau x_1. \tag{9-2.11}$$

Paying attention to the relationship:

$$\begin{bmatrix} \tilde{F} & 0 \\ 0 & \bar{G} \end{bmatrix} = \begin{bmatrix} I & -\bar{F}^{-1}H \\ 0 & I \end{bmatrix}^\tau \begin{bmatrix} \bar{F} & H \\ H^\tau & \bar{G} \end{bmatrix}\begin{bmatrix} I & -\bar{F}^{-1}H \\ 0 & I \end{bmatrix}, \tag{9-2.12}$$

and nonsingularity of the matrix

$$\begin{bmatrix} I & -\bar{F}^{-1}H \\ 0 & I \end{bmatrix},$$

we know that matrix $\tilde{F}$ is symmetric positive definite.

Substituting $v = w - \bar{G}^{-1}H^\tau x_1$ into (9-2.7) we obtain the following system

$$\dot{x}_1 = \bar{A}_1 x_1 + B_1 w \tag{9-2.13}$$

where

$$\bar{A}_1 = A_1 - B_1\bar{G}^{-1}H^\tau. \tag{9-2.14}$$

Thus, the optimal control problem for singular system (9-2.1) is changed into the problem of finding the new control w for normal system (9-2.13) such that cost functional (9-2.10) is minimized. To solve this problem using the result in normal system theory, we need to prove the following lemma.

Lemma 9-2.3. $(\bar{A}_1, B_1)$ is stabilizable.

Proof. This is a direct result of the assumption of stabilizability for system (9-2.1), also noticing equation (9-2.14) and the stabilizability of (A_1, B_1). Q.E.D.

By applying the quadratic regulation problem for normal systems in Appendix C, the optimal control for system (9-2.13) minimizing the cost functional (9-2.10) is given by

$$w^* = -\bar{G}^{-1}B_1^\tau M x_1^* \tag{9-2.15}$$

where x_1^* is the optimal trajectory, the symmetric positive definite matrix M is the unique solution of the following Riccati equation

$$M\bar{A}_1 + \bar{A}_1^{\tau}M - MB_1\bar{G}^{-1}B_1^{\tau}M + \tilde{F} = 0, \tag{9-2.16}$$

and the minimum value of J is

$$\min J = x_1^{\tau}(0)Mx_1(0).$$

Therefore, by summing up this discussion, we obtain the optimal control $u^*(t)$ for singular system (9-2.1):

$$u^*(t) = Kx^*(t) + v^*(t) = Kx^*(t) + w^*(t) - \bar{G}^{-1}H^{\tau}x_1^*(t)$$

$$= (K - \bar{G}^{-1}(H^{\tau}+B_1^{\tau}M)[I \quad 0]P_1^{-1})x^*(t) \tag{9-2.17}$$

and the optimal trajectory $x^*(t)$:

$$E\dot{x}^*(t) = [A+B(K-\bar{G}^{-1}(H^{\tau}+B_1^{\tau}M)[I \quad 0]P_1^{-1})]x^*(t). \tag{9-2.18}$$

Moreover, direct computation shows that system (9-2.18) is stable and no impulse terms exist in optimal trajectory $x^*(t)$, the minimum value of J is

$$\min J = x^{\tau}(0)(P_1^{\tau})^{-1}\begin{bmatrix} I \\ 0 \end{bmatrix}M[I \quad 0]P_1^{-1}x(0). \tag{9-2.19}$$

Hence, in summary of this discussion, if the solution of optimal control problem for singular system (9-2.1) with quadratic cost functional (9-2.2) exists and is unique, the problem may be solved according to the following procedure.

1. Verify the stabilizability and inpulse controllability for system (9-2.1).

2. Choose matrix K so that $\deg(|sE-(A+BK)|) = \text{rank}E$.

3. Calculate the matrices A_1, B_1, B_2, Q_1, and P_1 according to (9-2.7).

4. Determine the matrices $\bar{F}$, H, $\bar{G}$, $\tilde{F}$, and $\bar{A}_1$ according to (9-2.9), (9-2.11), and (9-2.14).

5. Find the unique solution of Riccati equation (9-2.16).

6. Finally obtain the optimal control $u^*(t)$ and trajectory $x^*(t)$ using (9-2.17) and (9-2.18).

In the optimal regulation problem with quadratic cost functional for normal systems, to obtain the solution we need to solve a Riccati equation of order n, the same as the order of the system. However, as seen from (9-2.16), a Riccati equation of order $\text{rank}E < n$ is needed for this problem. This will be convenient when rankE is much smaller than n.

Optimal control (9-2.17) is in the form of state feedback.

Example 9-2.1. Consider the singular system

$$\begin{bmatrix} 1 & 0 & 0 \\ 0 & 0 & 1 \\ 0 & 0 & 0 \end{bmatrix} \dot{x} = \quad x + \begin{bmatrix} 1 \\ 0 \\ 1 \end{bmatrix} u \tag{9-2.20}$$

with cost functional

$$J = \int_0^{\infty} (x^{\tau}x + 2u^{\tau}u)dt. \tag{9-2.21}$$

This system is controllable, thus it is stabilizable amd impulse controllable. The procedure is applicable in this case. In this system, rankE = 2, n = 3, $F = I_3$, and G = 2. The matrix K = [0 1 0] satisfies deg(|sE-(A+BK)|) = rankE. Direct verification shows that under the nonsingular matrices

$$Q_1 = \begin{bmatrix} 1 & 0 & -1 \\ 0 & 1 & -1 \\ 0 & 0 & 1 \end{bmatrix}, \qquad P_1 = \begin{bmatrix} 1 & 0 & 0 \\ 0 & -1 & 1 \\ 0 & 1 & 0 \end{bmatrix}$$

we have

$$Q_1EP_1 = \text{diag}(I_2,\ 0), \qquad Q_1(A+BK)P_1 = \text{diag}(A_1,\ 1),$$

$$A_1 = \begin{bmatrix} 1 & -1 \\ 0 & -1 \end{bmatrix}, \qquad B_1 = \begin{bmatrix} 0 \\ -1 \end{bmatrix}, \qquad B_2 = 1.$$

For this system, equations (9-2.9), (9-2.11), and (9-2.14) give

$$\bar{F} = \text{diag}(1,\ 4), \quad \bar{G} = 1, \quad \tilde{F} = \text{diag}(1,\ 3)$$

$$\bar{A}_1 = \begin{bmatrix} 1 & -1 \\ 0 & 0 \end{bmatrix}, \qquad H = \begin{bmatrix} 0 \\ 1 \end{bmatrix}.$$

To solve Riccati equation (9-2.16), which here is

$$M\begin{bmatrix} 1 & -1 \\ 0 & 0 \end{bmatrix} + \begin{bmatrix} 1 & 0 \\ -1 & 0 \end{bmatrix}M + \begin{bmatrix} 1 & 0 \\ 0 & 3 \end{bmatrix} - M\begin{bmatrix} 0 & 0 \\ 0 & 1 \end{bmatrix}M = 0, \tag{9-2.22}$$

we denote

$$M = \begin{bmatrix} m_1 & m_2 \\ m_2 & m_3 \end{bmatrix}.$$

Performing the indicated matrix multiplication, from (9-2.22) we obtain the three equations:

$$2m_1 - m_2^2 + 1 = 0$$

$$m_2 - m_1 - m_2m_3 = 0$$

$$-2m_2 - m_3^2 + 3 = 0,$$

which has the unique solution

$$m_1 = 8+6\sqrt{2}, \qquad m_2 = -3-2\sqrt{2}, \qquad m_3 = 1+2\sqrt{2}.$$

Thus

$$M = \begin{bmatrix} 8+6\sqrt{2} & -3-2\sqrt{2} \\ -3-2\sqrt{2} & 1+2\sqrt{2} \end{bmatrix}.$$

According to (9-2.17) and (9-2.18), the optimal control and trajectory for this problem, respectively, is

$$u^*(t) = [K - \bar{G}^{-1}(H^\tau + B_1^\tau M)[I \;\; 0]P_1^{-1}]x^*(t) = [-3-2\sqrt{2},\ 1,\ 2\sqrt{2}]x^*(t)$$

and

$$\begin{bmatrix} 1 & 0 & 0 \\ 0 & 0 & 1 \\ 0 & 0 & 0 \end{bmatrix} \dot{x}^*(t) = \begin{bmatrix} -2-2\sqrt{2} & 1 & 2\sqrt{2} \\ 0 & 1 & 0 \\ -3-2\sqrt{2} & 1 & 1+2\sqrt{2} \end{bmatrix} x^*(t),$$

which has finite pole set $\{-\sqrt{2}, -\sqrt{2}\}$. Hence the closed-loop system is stable and no impulse terms exist in its optimal trajectory.

9-3. Time-Optimal Control

In the time-optimal problem, or fast control problem, the cost functional is taken as (the initial time is $t = 0$)

$$J = t, \tag{9-3.1}$$

which represents the time period taken to drive the state from one position to another.

Consider system (9-2.1):

$$E\dot{x} = Ax + Bu \tag{9-3.2}$$

with two fixed states: the initial state $x(0)$ and the terminal state $x(t_1)$. Its time-optimal control is to find the control $u^*(t)$ such that the state from $x(0)$ is driven to $x(t_1)$ in a minimum period, i.e.,

$$t_1 = \min J.$$

As shown in the concept of controllability in Chapter 2, if the control is unstrained the state of controllable system (9-3.2) may be driven from an arbitrary initial state to any terminal state in an arbitrarily short period. In this case our problem will lose interest. Therefore, the time-optimal problem is often studied with a constrained control. Let the constraints be bounded control:

$$|u_i| \leqslant a_i, \quad u = [u_1, u_2, \ldots, u_m], \ i = 1, 2, \ldots, m \tag{9-3.3}$$

and $u(t)$ is piecewise sufficiently continuously differentiable.

Furthermore, for the sake of simplicity, the terminal state is assumed to be zero, i.e., $x(t_1) = 0$.

For any regular system (9-3.2), there exist two nonsingular matrices Q and P such that it is r.s.e. to

$$\begin{aligned} \dot{x}_1 &= A_1x_1 + B_1u \\ N\dot{x}_2 &= x_2 + B_2u \end{aligned} \qquad (9\text{-}3.4)$$

where $A_1 \in \mathbb{R}^{n_1 \times n_1}$, $N \in \mathbb{R}^{n_2 \times n_2}$, $n_1 + n_2 = n$, N is nilpotent.

Now we will examine the time-optimal problem of two extreme cases.

9-3.1. The slow subsystem

Consider the slow subsystem

$$\dot{x}_1 = A_1x_1 + B_1u, \qquad x_1(0) = w_1. \qquad (9\text{-}3.5)$$

It is a normal system. From linear system theory we know that the optimal control under constraints (9-3.3) that drives any initial state of system (9-3.5) to zero in the minimum time is the switching control, which takes the extreme values.

9-3.2. The fast subsystem

For the fast subsystem

$$N\dot{x}_2 = x_2 + B_2u, \qquad x_2(0) = w_2, \qquad (9\text{-}3.6)$$

when $t > 0$, its state solution is

$$x_2(t) = -\sum_{i=0}^{h-1} N^iB_2u^{(i)}(t), \qquad t > 0, \qquad (9\text{-}3.7)$$

where h is the nilpotent index of N. Therefore if we choose $u(t) \equiv 0$, $t > 0$, it will be $x_2(t) \equiv 0$, $t > 0$. Note that the initial condition exists in the impulse terms at the initial instant. Thus, the state may be driven to zero by admissible control in any short period. But, the time period generally cannot be zero. This shows that generally no optimal solution exists for time-optimal problems for fast subsystems.

<u>Definition 9-3.1</u>. Consider system (9-3.2) with control constraints (9-3.3). Let $\varepsilon > 0$ be any prescribed scalar. If there exist admissible control $u^*_\varepsilon(t)$ and time point t_1 such that the state $x(t)$ is driven to zero at t_1, i.e., $x(t_1) = 0$, but there exists no admissible control such that the state $x(t)$ is driven to zero at any period shorter than $t-\varepsilon$, the control $u^*_\varepsilon(t)$ is called ε-suboptimal control and the time t_1 is called the ε-suboptimal time.

By definition, the ε-suboptimal control always exists for a fast subsystem. Clearly, the ε-optimal control is practical only for small ε, and it is acceptable if is sufficiently small.

In general we have the following theorem.

Theorem 9-3.1. Let t_1 be the optimal time for subsystem (9-3.5) and $\bar{u}^*(t)$ be the optimal control, $x_1^*(t_1) = 0$. Then for any $\varepsilon > 0$, there exists an ε-suboptimal control $u_\varepsilon^*(t)$ such that the state of system (9-3.2) is driven to zero at time $t_1 + \varepsilon$, $x(t_1+\varepsilon) = 0$.

Proof. Let t_1 be the optimal time for subsystem (9-3.5) and $\bar{u}^*(t)$ be the optimal control, $0 \leqslant t \leqslant t_1$, such that $x_1^*(t_1) = 0$. Noting the slow substate solution we know

$$x_1^*(t_1) = e^{A_1 t_1} x_1(0) + \int_0^{t_1} e^{A_1(t_1-t)} B_1 \bar{u}^*(t)dt = 0.$$

Let the new control be

$$u_\varepsilon^*(t) = \begin{cases} \bar{u}^*(t) & 0 \leqslant t \leqslant t_1 \\ 0 & t_1 < t \leqslant t_1 + \varepsilon, \end{cases}$$

then we have

$$x_1^*(t_1+\varepsilon) = 0, \qquad x_2^*(t_1+\varepsilon) = 0.$$

By definition it is easy to know that $u^*(t)$ is the ε-suboptimal control. Q.E.D.

Therefore, the solvability of time-optimal control for singular system (9-3.2) is governed by that for its slow subsystem.

Example 9-3.1. Consider the following two-order system

$$\begin{bmatrix} 1 & 0 \\ 0 & 0 \end{bmatrix} \dot{x} = \begin{bmatrix} -1 & 2 \\ 0 & 1 \end{bmatrix} x + \begin{bmatrix} 1 \\ -1 \end{bmatrix} u \tag{9-3.8}$$

$$x(0) = \begin{bmatrix} -1 \\ 0 \end{bmatrix}$$

with constraint

$$|u| \leqslant 2. \tag{9-3.9}$$

System (9-3.8) has the EF1:

$$\begin{bmatrix} 1 & 0 \\ 0 & 0 \end{bmatrix} \begin{bmatrix} \dot{x}_1 \\ \dot{x}_2 \end{bmatrix} = \begin{bmatrix} -1 & 0 \\ 0 & 1 \end{bmatrix} \begin{bmatrix} x_1 \\ x_2 \end{bmatrix} + \begin{bmatrix} 3 \\ -1 \end{bmatrix} u$$

$$\begin{bmatrix} x_1(0) \\ x_2(0) \end{bmatrix} = \begin{bmatrix} -1 \\ 0 \end{bmatrix}$$

in which the slow subsystem

$$\dot{x}_1 = -x_1 + 3u, \qquad x_1(0) = -1$$

has the time optimal control

$$\bar{u}^*(t) = 2, \qquad 0 \leqslant t \leqslant \ln\frac{7}{6}$$

with optimal terminal time

$$t_1 = \ln\frac{7}{6}.$$

Combining it with Theorem 9-3.1, we know that for any prescribed scalar $\varepsilon > 0$, the ε -suboptimal control is

$$u_\varepsilon^*(t) = \begin{cases} \bar{u}^*(t) = 2 & 0 \leqslant t \leqslant \ln\frac{7}{6} \\ 0 & \ln\frac{7}{6} < t \leqslant \ln\frac{7}{6} + \varepsilon \end{cases}$$

and under this control the state of its closed-loop system reaches zero at time $\ln\frac{7}{6} + \varepsilon$.

The time-optimal control problem for general nonlinear singular systems is much more complex than in normal system cases.

9-4. Optimal Control for Discrete-Time Singular Systems

In this section we consider the discrete-time singular system

$$Ex(k+1) = Ax(k) + Bu(k) \tag{9-4.1}$$

where the state $x(k) \in \mathbb{R}^n$, the control input $u(k) \in \mathbb{R}^m$ at time k; and E, A, B are constant matrices of appropriate dimensions. System (9-4.1) is regular and rank$E < n$.

For system (9-4.1) the general cost functional has the form

$$J = J(x(0), x(1), \ldots, x(L), u(0), u(1), \ldots, u(L+h-1)), \tag{9-4.2}$$

which is a scalar functional in $u(0)$, $u(1)$, ... , $u(L+h-1)$, where h is the nilpotent index of the coefficient of the backward subsystem for (9-4.1). The controls after step L are included because they are included in $x(L)$. The cost functional often stands for the overall loss or benefit of the control process. The solution for the optimal control problem is assumed to uniquely exist.

Constraints for system (9-4.1) are often considered the bounded control, i.e.,

$$|u(k)| \leqslant a_k, \quad k = 0, 1, 2, \ldots \tag{9-4.3}$$

Further, terminal conditions often imposed on the state of such systems, include $x(L) \subset \mathbf{S}$, $\mathbf{S}$ is any prescribed set.

As previously shown, any regular system (9-4.1) is r.s.e. to(refer to (5-2.12)):

$$\begin{aligned} E_{11}x_1(k+1) + E_{12}x_2(k+1) &= A_{11}x_1(k) + B_1u(k) \\ E_{22}x_2(k+1) &= x_2(k) \end{aligned} \tag{9-4.4}$$

where

$$Q_1EP_1 = \begin{bmatrix} E_{11} & E_{12} \\ 0 & E_{22} \end{bmatrix}, \qquad Q_1AP_1 = \begin{bmatrix} A_{11} & 0 \\ 0 & I_{n_2} \end{bmatrix}$$

$$Q_1B = \begin{bmatrix} B_1 \\ 0 \end{bmatrix}, \qquad P_1^{-1}x(k) = \begin{bmatrix} x_1(k) \\ x_2(k) \end{bmatrix}$$

where $x_1(k) \in \mathbb{R}^{n_1}$, $x_2(k) \in \mathbb{R}^{n_2}$, $n_1 + n_2 = n$, E_{22} is nilpotent, Q_1 and P_1 are nonsingular, and (E_{11}, A_{11}, B_1) is normalizable.

Note that the solution of substate $x_2(k)$ is identically zero:

$$x_2(k) = 0, \qquad k = 0, 1, 2, \dots$$

and consequently the substate $x_1(k)$ has the dynamics

$$E_{11}x_1(k+1) = A_{11}x_1(k) + B_1u(k). \tag{9-4.5}$$

Thus cost functional (9-4.2) becomes

$$\bar{J} = J = \bar{J}(x_1(0), x_1(1), \dots, x_1(L), u(0), u(1), \dots, u(L+h-1)). \tag{9-4.6}$$

By decomposition, subsystem (9-4.5) is normalizable. There exists a matrix $K \in \mathbb{R}^{m\times n}$ so that $\deg(|sE_{11}-(A_{11}+B_1K)|) = \operatorname{rank}E_{11}$. Therefore, under the control

$$u(k) = Kx_1(k) + v(k) \tag{9-4.7}$$

where $v(k)$ is the new control input, when (9-4.7) is applied to system (9-4.1), the closed-loop system

$$E_{11}x_1(k+1) = (A_{11}+B_1K)x\ (k) + B_1v(k)$$

is r.s.e. to

$$\begin{aligned} \bar{x}_1(k+1) &= \bar{A}_1\bar{x}_1(k) + \bar{B}_1v(k) \\ 0 &= \bar{x}_2(k) + \bar{B}_2v(k) \end{aligned} \tag{9-4.8}$$

where

$$\bar{P}^{-1}x_1(k) = \begin{bmatrix} \bar{x}_1(k) \\ \bar{x}_2(k) \end{bmatrix}, \qquad \bar{x}_1(k) \in \mathbb{R}^{\bar{n}_1}, \quad \bar{x}_2(k) \in \mathbb{R}^{\bar{n}_2},$$

and $\bar{n}_1 + \bar{n}_2 = n_1$, $\bar{P}_1$ is nonsingular.

From (9-4.7) and (9-4.8) we know that constraints (9-4.3) and cost functional (9-4.4) become, respectively,

$$\tilde{J} = \bar{J} = J(\bar{x}_1(0), \bar{x}_1(1), \ldots, \bar{x}_1(L), v(0), v(1), \ldots, v(L)) \tag{9-4.9}$$

and

$$|f(\bar{x}_1(k), v(k))| \leqslant a_k, \quad k = 0, 1, \ldots, \tag{9-4.10}$$

$$f(\bar{x}_1(k), v(k)) = K\bar{P}_1[I/0]\bar{x}_1(k) + (I - K\bar{P}_1[0/I]\bar{B}_1)v(k).$$

It has been proven that if $\{v^*(k),\ k = 0, 1, \ldots, L\}$ is the optimal control solution for system

$$\bar{x}_1(k+1) = \bar{A}_1\bar{x}_1(k) + \bar{B}_1 v(k)$$

with constraints (9-4.10) so that cost functional (9-4.9) is minimized. Then

$$u^*(k) = Kx^*(k) + v^*(k) = K\bar{P}_1\begin{bmatrix} I \\ 0 \end{bmatrix}\bar{x}_1^*(k) + (I - K\bar{P}_1\begin{bmatrix} 0 \\ I \end{bmatrix}\bar{B}_2)v^*(k),$$
$$k = 0, 1, 2, \ldots, L$$

is the optimal control for system (9-4.1) with constraints (9-4.6), minimizing (9-4.2), and the following holds:

$$\min J = \min \tilde{J}.$$

Thus, via appropriate transformation, the optimal control problem for singular systems is changed into that for normal systems. This method shows the possibility of solving the problem using linear system theory. The dynamic programming is a powerful method in the optimal control problem for the discrete-time normal system case. This method, fortunately, is also applicable here.

Next we will consider using dynamic programming to solve the optimal quadratic regulation problem for system (9-4.1). Consider system (9-4.1) with cost functional

$$J = \sum_{k=0}^{L} J_k \tag{9-4.11}$$

$$J_k = x^\tau(k)S(k)x(k) + u^\tau(k)R(k)u(k), \quad k = 0, 1, \ldots, L$$

where $S(k)$ and $R(k)$ are symmetric positive definite matrices. The general physical sense of (9-4.11) is the overall loss (or the overall benefit, in which case the cost functional should be maximized).

According to the preceding discussion, by applying transformation (9-4.4)-(9-4.10) we obtain the optimal control for system (9-4.1) under cost functional (9-4.11):

$$u^*(k) = K\bar{P}_1\begin{bmatrix} I \\ 0 \end{bmatrix}\bar{x}_1^*(k) + (I-K\bar{P}_1\begin{bmatrix} 0 \\ I \end{bmatrix}B_2)v^*(k), \quad k = 0, 1, \ldots, L \tag{9-4.12}$$

in which the vector $v^*(k)$ and $\bar{x}_1^*(k)$ are given by the optimal control and trajectory for the system

$$\bar{x}_1(k+1) = \bar{A}_1\bar{x}_1(k) + \bar{B}_1v(k) \tag{9-4.13}$$

with cost functional

$$J = \sum_{k=0}^{L} \bar{J}_k$$

$$\bar{J}_k = \bar{x}_1^\tau(k)\bar{S}(k)\bar{x}_1(k) + 2\bar{x}_1^\tau(k)H(k)v(k) + v^\tau(k)\bar{R}(k)v(k), \quad k = 0, 1, \ldots, L.$$

Here

$$\bar{S}(k) = V_1^\tau(k)V_1(k) + K_1^\tau R(k)K_1$$

$$H(k) = K_1^\tau R(k)(I-K_2\bar{B}_2) - V_1^\tau(k)V_1(k)\bar{B}_2$$

$$\bar{R}(k) = \bar{B}_2^\tau V_2^\tau(k)V_2(k)\bar{B}_2 + (I-K_2\bar{B}_2)^\tau R(K)(I-K_2\bar{B}_2)$$

$$k = 0, 1, \ldots, L$$

and $V(k) = [V_1(k), V_2(k)]$ is nonsingular satisfying $\bar{P}_1^\tau S(k)\bar{P}_1 = V^\tau(k)V(k)$, $K\bar{P}_1 = [K_1, K_2]$, subject to the constraints

$$|K_1\bar{x}_1(k) + (I-K_2\bar{B}_2)v(k)| \leqslant a_k, \quad k = 0, 1, \ldots, L. \tag{9-4.15}$$

Furthermore, noticing that $\bar{J}_k$ may be written as

$$\bar{J}_k = \bar{x}_1^\tau(k)\tilde{S}(k)\bar{x}_1(k) + w^\tau(k)\bar{R}(k)w(k)$$

$$\tilde{S}(k) = \bar{S}(k) - H(k)\bar{R}^{-1}(k)H^\tau(k)$$

$$w(k) = v(k) - \bar{R}^{-1}(k)H^\tau(k)\bar{x}_1(k),$$

$$k = 0, 1, \ldots, L,$$

system (9-4.13) and cost functional (9-4.14) become

$$\bar{x}_1(k+1) = \tilde{A}_1(k)\bar{x}_1(k) + \bar{B}_1w(k) \tag{9-4.16}$$

where $\tilde{A}_1(k) = \bar{A}_1 - \bar{B}_1\bar{R}^{-1}(k)H^\tau(k)$, and

$$J = \sum_{k=0}^{L} \bar{J}_k = \sum_{k=0}^{L} g_k + x_1^\tau(0)\tilde{S}(0)x_1(0) \tag{9-4.17}$$

where

$$g_k = \bar{x}_1^{\tau}(k+1)\tilde{S}(k+1)\bar{x}_1(k+1) + w^{\tau}(k)\bar{R}(k)w(k), \quad k = 0, 1, 2, \dots, L$$

$$\tilde{S}(L+1) = 0.$$

The constraints now are

$$| f_k(\bar{x}_1(k), w(k)) | \leqslant a_k, \quad k = 0, 1, \dots, L \tag{9-4.18}$$

$$f_k(\bar{x}_1(k), w(k)) = (K_1 - (I - K_2B_2)\bar{R}^{-1}(k)H^{\tau}(k))\bar{x}_1(k) + (I - K_2B_2)w(k).$$

Clearly, the optimal control $v^*(k)$ is given by

$$v^*(k) = w^*(k) - \bar{R}^{-1}(k)H^{\tau}(k)\bar{x}_1^*(k), \quad k = 0, 1, \dots, L$$

where $w^*(k)$ and $\bar{x}_1^*(k)$ are the solutions for optimal problem (9-4.16) - (9-4.18). Since (9-4.16)-(9-4.18) is an optimal problem for discrete-time normal system (9-4.16), the problem may be solved by using the dynamic programming principle, which states that if $w^*(k)$ is the solution of min J, it is the solution starting from any step, or equivalently,

$$J^* = \min J = \min_{w(i)} \sum_{k=0}^{L} g_k + \bar{x}_1^{\tau}(0)\tilde{S}(0)\bar{x}_1(0)$$

$$= \min_{w(i)} \{ \sum_{k=0}^{L-1} g_k + \min_{w(L)} g_L \} + \bar{x}_1^{\tau}(0)\tilde{S}(0)\bar{x}_1(0)$$

$$= \dots \dots$$

$$= \min_{w(0)} \{ g_0 + \min_{w(1)} \{ g_1 + \dots + \min_{w(L)} g_L \}\} + \bar{x}_1^{\tau}(0)\tilde{S}(0)\bar{x}_1(0)$$

where $w(i) \in U$, U is the admissible control set determined by $| f_i | \leqslant a_i$, $i = 0, 1, \dots, L$.

Assume that no constraint is imposed on control $u(k)$, $k = 0, 1, \dots, L$. From (9-4.15) and (9-4.18) we know that there is no constraint on the new control $w(k)$. The algorithm of computing J^* and the optimal control $w^*(k)$ follows.

1. Choosing the last step control $w^*(L)$, so that the last step cost is minimized:

$$\min_{w(L)} g_L = \min_{w(L)} \{ \bar{x}_1^{\tau}(L+1)\tilde{S}(L+1)\bar{x}_1(L+1) + w^{\tau}(L)\bar{R}(L)w(L) \}.$$

From (9-4.16) we know that g_L has the form

$$g_L = \bar{x}_1(L)\tilde{A}_1^{\tau}(L)\tilde{S}(L+1)\bar{x}_1(L) + 2\bar{x}_1(L)\tilde{A}_1^{\tau}(L)\tilde{S}(L+1)\bar{B}_1w(L)$$
$$+ w^{\tau}(L)(\bar{R}(L) + \bar{B}_1^{\tau}\tilde{S}(L+1)\bar{B}_1)w(L).$$

Note that $\bar{x}_1(L)$ is independent of $w(L)$. To find $w(L)$ minimizing g_L we differentiate g_L with respect to $w(L)$ to obtain

$$\frac{\partial g_L}{\partial w(L)} = 2\bar{B}_1^{\tau}\tilde{S}(L+1)\tilde{A}_1(L)\bar{x}_1(L) + 2(\bar{R}(L)+\bar{B}_1^{\tau}\tilde{S}(L+1)\bar{B}_1)w(L) = 0,$$

from which we can solve

$$w^*(L) = -\,M(L)\tilde{A}_1(L)\bar{x}_1(L) \tag{9-4.19}$$

where

$$M(L) = (\bar{R}(L) + \bar{B}_1^{\tau}\tilde{S}(L+1)\bar{B}_1)^{-1}B_1^{\tau}\tilde{S}(L+1) \tag{9-4.20}$$

and the minimum value of g_L is given by

$$g_L^* = \min g_L = \bar{x}_1^{\tau}(L)\tilde{A}_1^{\tau}(L)F(L)\tilde{A}_1(L)\bar{x}_1(L)$$

where

$$F(L) \triangleq \tilde{S}(L+1) - \tilde{S}(L+1)\bar{B}_1M(L), \tag{9-4.21}$$

which is symmetric by (9-4.20).

2. Determining the second-to-last-step optimal control $w^*(L-1)$ so as to minimize the second-to-last-step cost:

$$\min_{w(L-1)} (g_{L-1} + \min_{w(L)} g_L) = \min_{w(L-1)} (g_{L-1} + g_L^*).$$

Since

$$g_{L-1} + g_L^* = \bar{x}_1^{\tau}(L)(\tilde{S}(L)+\tilde{A}_1(L)F(L)A_1(L))\bar{x}_1(L) + w^{\tau}(L-1)\bar{R}(L-1)w(L-1),$$

using the same method as in Step 1 we know that the optimal control $w^*(L-1)$ minimizing $g_{L-1} + g_L^*$ is

$$w^*(L-1) = -\,M(L-1)\tilde{A}_1(L-1)\bar{x}_1(L-1) \tag{9-4.22}$$

where

$$\left.\begin{aligned} M(L-1) &= [\bar{R}(L-1) + \bar{B}_1^{\tau}G(L)\bar{B}_1]^{-1}\bar{B}_1^{\tau}G(L) \\ G(L) &= \tilde{S}(L) + \tilde{A}_1^{\tau}(L)F(L)\tilde{A}_1(L) \end{aligned}\right. \tag{9-4.23}$$

and the minimum value of $g_{L-1} + g_L^*$ is

$$\bar{x}_1^{\tau}(L-1)\tilde{A}_1^{\tau}(L-1)F(L-1)A_1(L-1)\bar{x}_1(L-1)$$

in which

$$F(L-1) = G(L) - G(L)\bar{B}_1M(L-1). \tag{9-4.24}$$

3. Continuing this process, we may finally obtain the optimal control that minimi-

zing J is

$$w^*(k) = -M(k)\tilde{A}_1(k)\bar{x}_1(k), \tag{9-4.25}$$

where

$$M(k) = [\bar{R}(k) + \bar{B}_1^\tau G(k+1)\bar{B}_1]^{-1}\bar{B}_1 G(k+1)$$

$$G(k) = \tilde{S}(k) + \tilde{A}_1^\tau(k)F(k)\tilde{A}_1(k)$$

$$F(k) = G(k+1) - G(k+1)\bar{B}_1 M(k),$$

with the terminal condition $G(L+1) = S(L+1) = 0$. The minimum value of J is

$$J^* = \bar{x}_1^\tau(0)\tilde{A}_1^\tau(0)F(0)\tilde{A}_1(0)\bar{x}_1(0) + \bar{x}_1^\tau(0)\tilde{S}(0)\bar{x}_1(0)$$

$$= \bar{x}_1^\tau(0)G(0)\bar{x}_1(0). \tag{9-4.26}$$

To summarize, we have proven the following.

Theorem 9-4.1. Consider system (9-4.1) with cost functional (9-4.11). If no constraint is imposed on control input, the optimal control that minimizes (9-4.11) is

$$u^*(k) = Kx_1^*(k) + v^*(k) = K_1\bar{x}_1^*(k) + (I-K_2B_2)v^*(k)$$

$$= [K_1-(I-K_2B_2)\bar{R}^{-1}(k)H^\tau(k)]\bar{x}_1^*(k) + (I-K_2B_2)w^*(k),$$

$$k = 0, 1, \ldots, L,$$

where $w^*(k)$ is determined by (9-4.25) and $\bar{x}_1^*(k)$ is the optimal trajectory. Moreover, the minimum value of the cost functional is given by (9-4.26).

Combining this statement with (9-4.25) we see that the optimal control law is in the form of state feedback

$$u^*(k) = \bar{K}\bar{x}_1^*(k) = \bar{K}[I \quad 0]\bar{P}_1^{-1}x_1^*(k), \quad k = 0, 1, \ldots, L$$

$$\bar{K} = K_1 - (I-K_2B_2)(\bar{R}^{-1}(k)H^\tau(k) + M(k)\tilde{A}_1(k).$$

In computation of feedback gain matrices, M(k) is independent of the initial condition x(0). Thus it may be calculated off-line and stored for the process.

Furthermore, equation (9-4.25) indicates that the computation of feedback matrices M(k) and G(k) is backward. They must be computed before the process. In fact, their computation forms the design procedure of the optimal control.

Example 9-4.1. Consider the discrete-time singular system

$$\begin{bmatrix} 1 & 0 & 0 \\ 0 & 0 & 1 \\ 0 & 0 & 0 \end{bmatrix} x(k+1) = \begin{bmatrix} -1 & 1 & 2 \\ 0 & 1 & 0 \\ 0 & 0 & 1 \end{bmatrix} x(k) + \begin{bmatrix} 1 \\ 0 \\ 1 \end{bmatrix} u(k) \tag{9-4.27}$$

with cost functional

$$J = x^{\tau}(0)x(0)+u^2(0)+x^{\tau}(1)x(1)+0.5u^2(1)+2x^{\tau}(2)x(2)+\frac{1}{3}u^2(2). \tag{9-4.28}$$

We find the optimal control according to the previously described procedure. In this problem, the control is unconstrained, and system (9-4.27) is impulse controllable.

The matrix $K = [0 \;\; 1 \;\; 0]$ satisfies $\deg(|sE - (A+BK)|) = \mathrm{rank}E$. Let $u(k) = Kx(k) + v(k)$. Then the closed-loop system is

$$\begin{bmatrix} 1 & 0 & 0 \\ 0 & 0 & 1 \\ 0 & 0 & 0 \end{bmatrix} x(k+1) = \begin{bmatrix} -1 & 1 & 2 \\ 0 & 1 & 0 \\ 0 & 1 & 1 \end{bmatrix} x(k) + \begin{bmatrix} 1 \\ 0 \\ 1 \end{bmatrix} v(k). \tag{9-4.29}$$

Under the coordinate transformation

$$Q_1 = \begin{bmatrix} 1 & 0 & -1 \\ 0 & 1 & -1 \\ 0 & 0 & 1 \end{bmatrix}, \qquad P_1 = \begin{bmatrix} 1 & 0 & 0 \\ 0 & -1 & 1 \\ 0 & 1 & 0 \end{bmatrix}$$

$$P_1^{-1}x(k) = \begin{bmatrix} x_1(k) \\ x_2(k) \end{bmatrix}, \qquad x_1(k) \in \mathbb{R}^2, \qquad x_2(k) \in \mathbb{R},$$

system (9-4.29) is r.s.e. to

$$\begin{aligned} x_1(k+1) &= \begin{bmatrix} -1 & 1 \\ 0 & -1 \end{bmatrix} x_1(k) + \begin{bmatrix} 0 \\ -1 \end{bmatrix} v(k) \\ 0 &= x_2(k) + v(k). \end{aligned} \tag{9-4.30}$$

Cost functional (9-4.20) becomes

$$\begin{aligned} J &= 2x^{\tau}(0)x(0)+2x^{\tau}(0)K^{\tau}v(0) +v^2(0)+\frac{3}{2}x^{\tau}(1)x(1)+x^{\tau}(1)K^{\tau}v(1)+\frac{1}{2}v^2(1) \\ &\quad + \frac{7}{3}x^{\tau}(2)x(2) + \frac{2}{3}x^{\tau}(2)x(2)+\frac{1}{3}v^2(2) \\ &= \frac{5}{3}x_1^{\tau}(0)x_1(0)+3(v(0)-\frac{2}{3}x_1^{\tau}(0)[{}^{\;0}_{-1}])^2+ \frac{3}{5}x_1^{\tau}(1)x_1(1)+ \frac{3}{2}(v(1)-\frac{3}{5}x_1^{\tau}(1)[{}^{\;0}_{-1}])^2 \\ &\quad + \frac{14}{3}x_1^{\tau}(2)x_1(2)+ \frac{13}{3}(v(2)- \frac{7}{13}x_1^{\tau}(2)[{}^{\;0}_{-1}])^2 \\ &= \frac{5}{3}x_1^{\tau}(0)x_1(0) + 3w^2(0) + \frac{3}{5}x_1^{\tau}(1)x_1(1) + \frac{3}{2}w^2(1) + \frac{14}{13}x_1^{\tau}(2)x_1(2) + \frac{13}{3}w^2(2) \end{aligned} \tag{9-4.31}$$

where

$$\begin{aligned} w(0) &= v(0) + [0 \;\; \tfrac{2}{3}]x_1(0) \\ w(1) &= v(1) + [0 \;\; \tfrac{3}{5}]x_1(1) \\ w(2) &= v(2) + [0 \;\; \tfrac{7}{13}]x_1(2). \end{aligned} \tag{9-4.32}$$

Substituting the represention of w(0), w(1), and w(2) into (9-4.30), we obtain

$$x_1(k+1) = A(k)x_1(k) + \begin{bmatrix} 0 \\ -1 \end{bmatrix} w(k) \tag{9-4.33}$$

where

$$A(0) = \begin{bmatrix} -1 & 1 \\ 0 & -\frac{5}{3} \end{bmatrix}, \quad A(1) = \begin{bmatrix} -1 & 1 \\ 0 & -\frac{8}{5} \end{bmatrix}, \quad A(2) = \begin{bmatrix} -1 & 1 \\ 0 & -\frac{20}{13} \end{bmatrix}.$$

1. Minimizing $\frac{13}{3}w^2(2)$ yields $w^*(2) = 0$.

2. From

$$\min_{w(1)} \left[\frac{14}{13}x_1^\tau(2)x_1(2) + \frac{3}{2}w^2(1)\right]$$

$$= \min_{w(1)} \left[\frac{4}{13}\left(A(1)x_1(1)+\begin{bmatrix} 0 \\ -1 \end{bmatrix}w(1)\right)^\tau\left(A(1)x_1(1)+\begin{bmatrix} 0 \\ -1 \end{bmatrix}w(1)\right) + \frac{3}{2}w^2(1)\right],$$

we have

$$w^*(1) = [0 \quad -0.67]x_1(1)$$

and

$$\min_{w(1)} \left[\frac{14}{13}x_1^2(2) + \frac{3}{2}w^2(1)\right] = x_1^\tau(1)\begin{bmatrix} 1.08 & -1.08 \\ -1.08 & 2.68 \end{bmatrix}x_1(1) = J_1^*.$$

3.

$$\min_{w(0)} \left[J_1^* + 3w^2(0) + 0.6x_1^\tau(1)x_1(1)\right]$$

$$= \min_{w(0)} \left[3w^2(0) + x_1^\tau(1)\begin{bmatrix} 1.68 & -1.08 \\ -1.08 & 3.28 \end{bmatrix}x_1(1)\right]$$

results in

$$w^*(0) = [-0.17 \quad -1.04]x_1(0)$$

and

$$J^* = \min J = x_1^\tau(0)\begin{bmatrix} 3.68 & -0.87 \\ -0.87 & 6.90 \end{bmatrix}x_1(0) \tag{9-4.34}$$

Therefore, by combining these equations with $u(k) = Kx(k) + v(k)$ and (9-4.32), the optimal control is given by

$$u^*(0) = [-0.17 \quad -0.71 \quad -1.71]x^*(0)$$

$$u^*(1) = [0 \quad -0.27 \quad -1.27]x^*(1)$$

$$u^*(2) = [0 \quad 0.46 \quad -0.54]x^*(2).$$

The overall minimum value of the cost functional is determined by (9-4.34).

9-5. Notes and References

Section 9-2 is in reference to Cheng et al. (1987); Section 9-3 is based on Campbell (1982b); Section 9-4 is in reference to Bender and Laub (1987a), and Dai (1988g).

The optimal control problem has been considered by several researchers, a partial list of other papers not mentioned previously includes: Bender and Laub (1987b); Cobb (1983a); Lewis (1985b); Lovass-Nagy and Schilling (1986); Luenberger (1987); and Pandolfi (1981).

At present, some problems exist in solving the optimal control problem such as the existence and uniqueness of the optimal solution; how to define the cost functional since distributions as solutions for continuous-time singular systems are one kind of generalized function; whether a maximum principle exists for singular systems as for normal systems.

CHAPTER 10

SOME FURTHER TOPICS

10-1. Large-Scale Singular Systems

Large-scale systems, characterized by complexity or high order, are often composed of several subsystems. The classic method usually proves itself incapable to cope with such systems. Thus the large-scale system theory developed in recent years. There are many topics in large-scale system theory. The following are three main problems considered in large-scale system theory.

1. Modeling. The task of modeling for large-scale systems is to reduce the system order so that a high order system may be approximated by a model of lower order, which is convenient for analysis and design.

2. Hierarchical control. Hierarchical, or multilevel, control naturally appears in real systems such as economic and management systems where hierarchical structure exists in the system. The subsystems at a given level, the level below it, can be controlled, or coordinated, by the higher level so that the whole system achieves an overall goal.

3. Decentralized control. For large-scale systems with several local control stations, the classic central control strategy is often not economical or even impossible, especially when information exchange is impossible among these stations. In such cases, the decentralized control demonstrates its advantage.

These problems are also of interest in large-scale singular systems. In this discussion, we shall introduce some basic problems in decentralized control for singular systems.

10-1.1. Decentralized dynamic compensation

Consider the following singular system with d local control stations:

$$E\dot{x} = Ax + \sum_{i=1}^{d} B_i u_i \tag{10-1.1}$$

$$y_i = C_i x$$

$$i = 1, 2, \dots, d$$

where $x \in \mathbb{R}^n$ is its state; $u_i \in \mathbb{R}^{m_i}$, $y_i \in \mathbb{R}^{r_i}$ are its local control input and measure output, respectively, at the ith local station; E, A, B_i, C_i are constant matrices of appropriate dimensions. It is assumed that (E,A) is regular, rankE < n.

In system (10-1.1), information exchanges among local stations are supposed to be impossible. Each station chooses its control input based only on its own measure output. Therefore, the dynamic compensation law at the local station should have the form of

$$\begin{aligned} \dot{x}_{c_i} &= A_{c_i} x_{c_i} + B_{c_i} y_i \\ u_i &= F_{c_i} x_{c_i} + F_i y_i \\ & i = 1, 2, \ldots, d \end{aligned} \tag{10-1.2}$$

where $x_{c_i} \in \mathbb{R}^{n_{c_i}}$ is its state; and A_{c_i}, B_{c_i}, F_{c_i}, F_i, i = 1, 2, ... , d, are constant matrices of appropriate dimensions.

Under the local compensation of (10-1.2), the closed-loop system is

$$\begin{aligned} E\dot{x} &= Ax + \sum_{i=1}^{d} B_i u_i \\ \dot{x}_{c_i} &= A_{c_i} x_{c_i} + B_{c_i} y_i \\ u_i &= F_{c_i} x_{c_i} + F_i y_i \\ y_i &= C_i x \\ & i = 1, 2, \ldots, d \end{aligned}$$

or more concisely,

$$\begin{aligned} E\dot{x} &= Ax + Bu \\ \dot{x}_c &= A_c x_c + B_c y \\ u &= F_c x_c + Fy \\ y &= Cx \end{aligned} \tag{10-1.3}$$

where

$$\begin{aligned} B &= [B_1, B_2, \ldots, B_d], \qquad C = [C_1/C_2/\ldots/C_d], \\ A_c &= \mathrm{diag}(A_{c1}, A_{c2}, \ldots, A_{cd}), \\ B_c &= \mathrm{diag}(B_{c1}, B_{c2}, \ldots, B_{cd}), \\ F_c &= \mathrm{diag}(F_{c1}, F_{c2}, \ldots, F_{cd}), \\ F &= \mathrm{diag}(F_1, F_2, \ldots, F_d), \end{aligned} \tag{10-1.4}$$

and

$$u = [u_1/u_2/\dots/u_d] \in \mathbb{R}^m$$

$$y = [y_1/y_2/\dots/y_d] \in \mathbb{R}^r \qquad (10\text{-}1.5)$$

$$x_c = [x_{c1}/x_{c2}/\dots/x_{cd}] \in \mathbb{R}^{n_c}$$

$$m = \sum_{i=1}^{d} m_i, \qquad r = \sum_{i=1}^{d} r_i, \qquad n_c = \sum_{i=1}^{d} n_{ci}.$$

Our problem is to find local compensation law (10-1.2) such that the overall closed-loop system (10-1.3) is stable, i.e.,

$$\sigma\left(\begin{bmatrix} E & 0 \\ 0 & I \end{bmatrix}, \begin{bmatrix} A+BFC & BF_c \\ B_cC & A_c \end{bmatrix}\right) \subset \mathbb{C}^-. \qquad (10\text{-}1.6)$$

This is the dynamic local stabilization problem for decentralized singular systems.

In decentralized control systems, the notation of fixed mode is important.

Definition 10-1.1. Denote

$$\mathbb{K} = \{ F \mid F = \mathrm{diag}(F_1, F_2, \dots, F_d) \in \mathbb{R}^{m\times r},\ F_i \in \mathbb{R}^{m_i \times r_i}, i = 1, 2, \dots, d \}. \qquad (10\text{-}1.7)$$

Then the greatest common divisor of the characteristic polynomial $|sE-(A+BFC)|$ for all $F \in \mathbb{K}$:

$$\psi(E,A,B,C,\mathbb{K}) = \underset{F\in\mathbb{K}}{\mathrm{gcd}} \left(|sE-(A+BFC)|\right)$$

is defined as the fixed polynomial for singular system (10-1.1), where gcd represents the greatest common divisor; and the set

$$\Lambda(E,A,B,C,\mathbb{K}) = \bigcap_{F\in\mathbb{K}} (E,A+BFC)$$

is termed the set of fixed finite modes for system (10-1.1); and let

$$\nu = \mathrm{rank}E - \max_{F} \deg(|sE-(A+BFC)|).$$

Then we will say that system (10-1.1) has ν fixed infinite modes. According to the definition, we have the following properties.

Property 1. $\psi(E,A,B,C,\mathbb{K}) \mid |sE-A|$;

Property 2. $\Lambda(E,A,B,C,\mathbb{K}) \subset \sigma(E,A)$;

Property 3. $\nu \leqslant \mathrm{rank}E$.

Example 10-1.1. Consider the singular system

$$\begin{bmatrix} 1 & 0 & 0 & 0 \\ 0 & 1 & 0 & 0 \\ 0 & 0 & 0 & 1 \\ 0 & 0 & 0 & 0 \end{bmatrix} \dot{x} = \begin{bmatrix} 1 & -1 & 0 & 0.5 \\ 0 & -1 & -2 & 0 \\ 0 & 0 & 1 & 0 \\ 0 & 0 & 0 & 1 \end{bmatrix} x + \begin{bmatrix} 1 \\ 0 \\ 0 \\ 0 \end{bmatrix} u_1 + \begin{bmatrix} -1 \\ 0 \\ 0 \\ 1 \end{bmatrix} u_2$$

$$y_1 = [1 \quad 0 \quad 0 \quad 0]x \qquad (10\text{-}1.8)$$

$$y_2 = [0 \quad 0 \quad 1 \quad 1]x.$$

In this system, for any $F = \mathrm{diag}(f_1, f_2)$, we have

$$|sE - (A+BFC)| = (s+1)(s-1-f_1)(f_2s+1+f_2). \qquad (10\text{-}1.9)$$

Therefore,

$$\psi(E,A,B,C,\mathbb{K}) = s+1$$

$$\Lambda(E,A,B,C,\mathbb{K}) = -1$$

$$\nu = \mathrm{rank}E - \max \deg(|sE-(A+BFC)|) = 3-3 = 0.$$

This system has one fixed finite mode and no fixed infinite mode.

Concerning the decentralized dynamic stabilization, the following theorems have been proven.

Theorem 10-1.1. For any nonnegative number $n_{ci} \geqslant 0$, $i = 1, 2, \ldots, d$, $n_c = \sum_{i=1}^{d} n_{ci}$, if we denote

$$\bar{E} = \mathrm{diag}(E, I_{n_c}), \qquad \bar{A} = \mathrm{diag}(A, 0),$$

$$\bar{B} = \mathrm{diag}(B, I_{n_c}), \qquad \bar{C} = \mathrm{diag}(C, I_{n_c}), \qquad (10\text{-}1.10)$$

$$\bar{F} = \begin{bmatrix} F & F_c \\ B_c & A_c \end{bmatrix},$$

$$\bar{\mathbb{K}} = \{ F \mid F = \begin{bmatrix} F & F_c \\ B_c & A_c \end{bmatrix} \in \mathbb{R}^{(n+m)x(m+r)} \}$$

where A_c, B_c, F_c, and F are determined by (10-1.4), it will be $\psi(E,A,B,C,\mathbb{K}) = \psi(\bar{E},\bar{A},\bar{B},\bar{C},\bar{\mathbb{K}})$.

Theorem 10-1.2. There exists a local dynamic compensation control (10-1.2) such that closed-loop system (10-1.3) is stable if and only if all the fixed finite modes are stable, i.e., $\Lambda(E,A,B,C,\mathbb{K}) \subset \mathbb{C}^-$. Furthermore, the closed-loop system not only is stable but also has no impulse terms in its state response iff it has only stable fixed finite modes and no fixed infinite modes.

Example 10-1.2. As seen in Example 10-1.1, the only one fixed finite mode for system (10-1.8) is stable. By Theorem 10-1.3 there exists a local dynamic compensation control (10-1.2) such that the closed-loop system is stable. In fact, from Example 10-1.1 we know that if $f_1 = -2$, $f_2 = 1$, and $u_1 = -2y_1$, $u_2 = y_2$, the closed-loop system is stable. Thus, system (10-1.8) may be stabilized via local static output feedback.

10-1.2. Local state feedback stabilization

Consider the following subsystems:

$$\begin{aligned} E_i\dot{x}_i &= A_ix_i + B_iu_i + w_i \\ y_i &= C_ix_i \\ i &= 1, 2, \ldots, d \end{aligned} \tag{10-1.11}$$

where $x_i \in \mathbb{R}^{n_i}$, $u_i \in \mathbb{R}^{m_i}$, and $y_i \in \mathbb{R}^{r_i}$ are the local state, local control input, and local measure output of the ith subsystem; E_i, A_i, B_i, C_i, $i = 1, 2, \ldots, d$, are constant matrices; and w_i represents the interconnection of the ith subsystem with other subsystems.

Suppose that subsystems are interconnected via input and output:

$$w_i = B_iz_i, \qquad z_i = \sum_{i \neq j}^{d} F_{ij}y_j, \qquad i = 1, 2, \ldots, d, \tag{10-1.12}$$

which is called the outer connection.

For the outer connected d subsystems described by (10-1.11), let the local state feedback be

$$u_i = K_i^1x_i + K_i^2\dot{x}_i, \quad i = 1, 2, \ldots, d. \tag{10-1.13}$$

The closed-loop system is

$$(E-BK_2)\dot{x} = (A+BK_1+BFC)x \tag{10-1.14}$$

where

$$E = \mathrm{diag}(E_1, E_2, \ldots, E_d),$$

$$A = \mathrm{diag}(A_1, A_2, \ldots, A_d),$$

$$B = \mathrm{diag}(B_1, B_2, \ldots, B_d),$$

$$C = \mathrm{diag}(C_1, C_2, \ldots, C_d),$$

$$K_i = \mathrm{diag}(K_1^i, K_2^i, \ldots, K_d^i), \quad i = 1, 2,$$

$$F = \begin{bmatrix} 0 & F_{12} & \cdots & F_{1d} \\ F_{21} & 0 & \cdots & F_{2d} \\ & \cdots & \cdots & \\ F_{d1} & F_{d2} & \cdots & 0 \end{bmatrix},$$

$$x = [x_1/x_2/\ldots/x_d].$$

Concerning the stabilization via local state feedback control, we can prove the following theorems (Dai, 1988i).

Theorem 10-1.4. Assume that each subsystem in (10-1.11) is controllable. Then there exist local state feedback controls (10-1.13) such that the closed-loop system (10-1.14) is stable.

Theorem 10-1.5. Assume that each subsystem in (10-1.11) is controllable. Then there exist local P-state feedbacks:

$$u_i = K_i x_i, \qquad i = 1, 2, \ldots, d$$

such that, when applied to system (10-1.11), the closed-loop system

$$E\dot{x} = (A+BK_1+BFC)x$$

is stable, $\sigma(E, A+BK_1+BFC) \subset \mathbb{C}^-$.

Example 10-1.3. Consider the singular system

$$\begin{bmatrix} 1 & 0 & 0 \\ 0 & 1 & 0 \\ 0 & 0 & 0 \end{bmatrix} \dot{x}_1 = \begin{bmatrix} 1 & 1 & 0 \\ 0 & 1 & 0 \\ 0 & 0 & 1 \end{bmatrix} x_1 + \begin{bmatrix} 0 \\ 1 \\ 1 \end{bmatrix} u_1 + w_1$$

$$\begin{bmatrix} 1 & 0 & 0 \\ 0 & 1 & 0 \\ 0 & 0 & 0 \end{bmatrix} \dot{x}_2 = \begin{bmatrix} 0 & 1 & 0 \\ 0 & 0 & 0 \\ 0 & 0 & 1 \end{bmatrix} x_2 + \begin{bmatrix} 0 \\ 1 \\ -1 \end{bmatrix} u_2 + w_2 \qquad (10\text{-}1.15)$$

$$y_1 = x_1$$

$$y_2 = [1 \quad 1 \quad 1]x_2$$

with interconnection

$$w_1 = [0 \quad 1 \quad 1]^\tau z_1, \qquad z_1 = [1 \quad 0 \quad 0]y_1$$

$$w_2 = [0 \quad 1 \quad -1]^\tau z_2, \qquad z_2 = y_2.$$

It is easy to testify that each subsystem in (10-1.15) is controllable. Therefore, it may be stabilized via local feedback control.

In fact, under the local feedback control

$$u_1 = [14 \quad \tfrac{10}{3} \quad \tfrac{8}{3}]x_1$$

$$u_2 = [0 \quad 1 \quad 1]x_2,$$

the closed-loop system has stable pole set $\{-1, -1, -1, -1\}$.

10-2. Adaptive Control

Adaptive control has the capacity to adapt not only the deviations in system structural parameters, but also some structural uncertainties such as the order of a system, so that the closed-loop system has the expected properties.

At present, there are mainly two approaches in the study of adaptive control: self-tuning and model reference. Here we only introduce elementary results in model reference adaptive control for singular systems.

Only single input, single output singular systems are studied.

Definition 10-2.1. The transfer function $G(s)$ is said to be (strictly) positive real if all poles of $G(s)$ are stable and

$$\mathrm{Re}(G(jw)) > 0, \quad \text{for all real finite } w \tag{10-2.1}$$

($\gamma > 0$ exists such that $\mathrm{Re}(G(jw)) \geqslant \gamma > 0$ for all real finite w).

Example 10-2.1. Consider the transfer function

$$G(s) = \frac{bs+c}{s+a}.$$

Then we have

$$\mathrm{Re}(G(jw)) = \frac{ac+bw^2}{a^2+w^2}.$$

By definition, it is easy to testify that $G(s)$ is positive real if $a > 0$, $b \geqslant 0$, and $c > 0$; Furthermore, when $a > 0$, $\mathrm{Re}(G(jw)) > \min(b, c/a) = \gamma$. Thus, $G(s)$ is strictly positive real if $a > 0$, $b > 0$, and $c > 0$.

Meyer, Kalman, and Yacubovitch have proven the following well-known positive lemma.

Lemma 10-2.1. Let $G(s)$ be a strictly proper rational function with a realization (A,b,c):

$$G(s) = c(sI-A)^{-1}b.$$

Then $G(s)$ is positive (strictly positive) iff there exist a vector η and symmetric positive definite matrices P and Q such that

1. $A^{\tau}P + PA = -Q$.

2. $Pb - c^{\tau} = 0$ ($Pb - c^{\tau} = \sqrt{\gamma}\eta$, $\gamma > 0$).

Consider the system described by the input-output relationship:

$$y(s) = G(s)u(s) \tag{10-2.2}$$

in which G(s) may be not proper. Thus G(s) may be the transfer function of a singular system.

In the model reference control problem, our aim is to find the control u(s) so that the output y(s) is the prescribed signal:

$$y(s) = y_r(s), \qquad y_r(s) = G_m(s)r(s)$$

in which $y_r(s)$ is the reference output, or the expected signal; r(s) is the reference input; and $G_m(s)$ is a given transfer function, or the model.

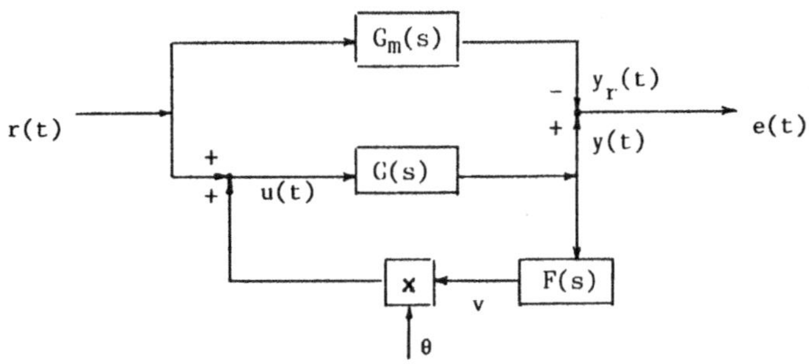

Figure 10-2.1. Model Reference Control

The general form of model reference adaptive control is shown in Figure 10-2.1, in which θ is the adjustable parameter. The task of model reference adaptive control is to choose the parameter θ such that e(t) tends to zero when time tends to infinity. Following the principle in normal systems, it is supposed that there exists a vector transfer function F(s) and a constant vector θ^* such that

$$G_m(s) = \frac{G(s)}{1 - G(s)\theta^{*\tau}F(s)} .$$

Thus from Figure 10-2.1 we see that the error e(t):

$$\begin{aligned} e(s) &= y(s) - y_r(s) \\ &= G(s)(r(s) + \theta^{\tau}v(s)) - G_m(s)r(s) \end{aligned}$$

$$= \frac{G_m(s)}{1 + G_m(s)\theta^{*\tau}F(s)}[r(s) + \theta^{\tau}v(s)] - G_m(s)r(s)$$

$$= G_m(s)\theta^{\tau}v(s) - G_m(s)\theta^{*\tau}F(s)G(s)[r(s)+\theta^{\tau}v(s)]$$

$$= G_m(s)\phi^{\tau}v(s) \tag{10-2.3}$$

where $\phi = \theta - \theta^*$ is the adjustable error vector.

If there exists a rational function $L(s)$ such that the transfer matrix

$$\bar{G}_m(s) = G_m(s)L(s)$$

is strictly positive real, we choose

$$\phi = L(s)\psi \ , \tag{10-2.4}$$

where ψ is a vector to be determined. According to (10-2.3) we have

$$e(s) = G_m(s)L(s)\psi^{\tau}v(s) = \bar{G}_m(s)\psi^{\tau}v(s).$$

On the other hand, let (A,b,c) be the minimal realization of $\bar{G}_m(s)$, $c = [1, 0,\ldots, 0]$. Then

$$\begin{aligned} \dot{x} &= Ax + b(\psi^{\tau}v) \\ e &= cx. \end{aligned} \tag{10-2.5}$$

Therefore, from Lemma 10-2.1, there exist symmetric positive definite matrices P and Q satisfying

1. $A^{\tau}P + PA = -Q.$

2. $Pb = c^{\tau}.$

Let $\lambda > 0$, and T be any symmetric positive definite matrix, and

$$\psi = -\frac{1}{\lambda}Tvc. \tag{10-2.6}$$

Next we verify that in this case there will be $\lim_{t\to\infty} e(t) = 0$. Consider the Lyapunov function:

$$V = x^{\tau}Px + \lambda\psi^{\tau}T^{-1}\psi \ .$$

Since

$$\dot{V} = x^{\tau}(PA + A^{\tau}P)x + 2x^{\tau}Pb(\psi^{\tau}v) + 2\lambda\,\psi^{\tau}T^{-1}\dot{\psi} \ ,$$

we know $\dot{V} = -x^{\tau}Qx$. Thus system (10-2.5) is stable and $\lim_{t\to\infty} e(t) = 0$.

In summary, we know from this discussion that if we choose the control

$$u = r + \phi \tag{10-2.7}$$

where ϕ is determined in (10-2.4) and (10-2.6), such a control u makes the closed-loop system stable, i.e., $\lim_{t\to\infty} e(t) = 0$.

Figure 10-2.2 shows the control strategy.

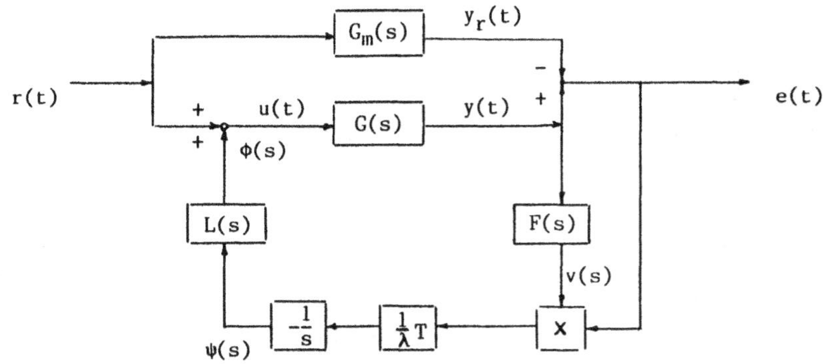

Figure 10-2.2. Adaptive Control Scheme (10-2.6)

As may be seen from Figure 10-2.2, in this adaptive process, the output y(s) is used to compare with the reference signal $y_r(s)$, to evaluate the error e(t) from which we determine the control, and to reduce the error e(t) to zero.

10-3. Stochastic Singular Systems

So far, singular systems studied are the deterministic systems in which all the items are deterministic; even in the presence of external disturbance, the disturbance is supposed to be deterministic. In real problems, control system design based on such system models can meet with most of our demands. However, on the other hand, any real systems are unavoidably influenced by stochastic factors such as environmental influence. Therefore, such systems have not only the deterministic inputs, but also unavoidably stochastic inputs. Meanwhile, subject to the limited precision in measurement methods, stochastic error always exists in measure output. Although these stochastic factors may be neglected in most cases, in some cases where high precision is required, such factors must be considered to obtain a good performance for the control system. Thus the system model naturally becomes the stochastic model that will be studied in this section. We briefly discuss the problem of filtering and LQG control for discrete-time singular systems.

Consider the following discrete-time stochastic singular system:

$$\begin{aligned} Ex(k+1) &= Ax(k) + Bu(k) + Dw(k) \\ y(k) &= Cx(k) + Fw(k) \end{aligned} \tag{10-3.1}$$

$$k = 0,\ 1,\ 2,\ \ldots$$

where the state $x(k) \in \mathbb{R}^n$, the control input $u(k) \in \mathbb{R}^m$, the measure output $y(k) \in \mathbb{R}^r$, the external stochastic disturbance $w(k) \in \mathbb{R}^p$; $E,\ A \in \mathbb{R}^{n\times n}$, $B \in \mathbb{R}^{n\times m}$, $D \in \mathbb{R}^{n\times p}$, $C \in \mathbb{R}^{r\times n}$, $F \in \mathbb{R}^{r\times p}$ are constant matrices. This system is assumed to be regular and rank$E < n$.

The following assumptions are also made on system (10-3.1).

Assumption 1. The disturbance $\{w(k)\}$ satisfies $\mathcal{E}w(k) = 0$, $\mathcal{E}w(k)w^\tau(j) = \Delta_{kj}I$, where

$$\Delta_{kj} = \begin{cases} 1 & \text{when } k = j \\ 0 & \text{when } k \neq j. \end{cases}$$

Assumption 2. $[x(0)/y(0)]$ and $\{w(k)\}$ are independent.

Assumption 3. System (10-3.1) is Y-controllable and Y-observable.

Let $a(0),\ a(1),\ \ldots,\ a(k),\ \ldots$ be a vector series. The following notation will be used:

$$a^k = [a(0)/a(1)/\ldots/a(k)].$$

The filtering and LQG problems may be stated as follows.

10-3.1. State filtering

The so-called state filtering problem here means the linear unbiased least-squares state estimation. We find a vector $\hat{x}(k)$ in the form

$$\hat{x}(k) = h(k) + H(k)y^k \tag{10-3.2}$$

where $h(k)$, $H(k)$ are deterministic vector and matrix, respectively, such that the error covariance is minimized, i.e.,

$$\min\ (x(k) - \hat{x}(k))(x(k) - \hat{x}(k))^\tau. \tag{10-3.3}$$

$u(k)$ is supposed to be admissible, i.e., $u(k)$ is in the form of (10-3.2).

To obtain the estimation, we first rewrite system (10-3.1) in appropriate form.

Subject to the assumption of Y-observability, there exists a matrix G satisfying

$$\deg(|zE - (A-GC)|) = \text{rank}E = q. \tag{10-3.4}$$

Thus two nonsingular matrices Q_1 and P_1 may be selected such that

$$Q_1EP_1 = \mathrm{diag}(I_q, 0), \qquad Q_1(A-KC)P_1 = \mathrm{diag}(A_1, I_{n-q}).$$

Note that by plus and minus the same term $Ky(k)$ on the right side of (10-3.1), system (10-3.1) may be rewritten as

$$\begin{aligned} Ex(k+1) &= (A-KC)x(k) + Bu(k) + Ky(k) + (D-KF)w(k) \\ y(k) &= Cx(k) + Fw(k). \end{aligned} \tag{10-3.5}$$

If we take the state coordinate transformation

$$P_1^{-1}x(k) = \begin{bmatrix} x_1(k) \\ x_2(k) \end{bmatrix}, \qquad x_1(k) \in \mathbb{R}^q, \quad x_2(k) \in \mathbb{R}^{n-q} \tag{10-3.6}$$

system (10-3.5), and thus system (10-3.1), is r.s.e. to

$$\begin{aligned} x_1(k+1) &= A_1x_1(k) + B_1u(k) + K_1y(k) + D_1w(k) \\ 0 &= x_2(k) + B_2u(k) + K_2y(k) + D_2w(k) \\ y(k) &= C_1x_1(k) + C_2x_2(k) + Fw(k) \end{aligned} \tag{10-3.7}$$

where

$$Q_1B = \begin{bmatrix} B_1 \\ B_2 \end{bmatrix}, \quad Q_1K = \begin{bmatrix} K_1 \\ K_2 \end{bmatrix}, \quad Q_1(D-KF) = \begin{bmatrix} D_1 \\ D_2 \end{bmatrix}, \quad CP_1 = [C_1 \quad C_2].$$

From the form (10-3.7) it is easy to prove the following theorem (Dai, 1988k).

<u>Theorem 10-3.1</u>. Assume that system (10-3.1) is Y-observable, $u(k)$ is the admissible control. Then the linear, unbiased least-squares state estimation for system (10-3.1) is given by

$$\hat{x}(k) = P_1 \begin{bmatrix} \hat{x}_1(k) \\ \hat{x}_2(k) \end{bmatrix}$$

and $\hat{x}_1(k)$, $\hat{x}_2(k)$ are determined by the following recursive algorithm:

$$\begin{aligned} \hat{x}_1(k+1) &= A_1\hat{x}_1(k) + B_1u(k) + K_1y(k) + G_1(k+1)[y(k+1)-C_1\hat{x}_1(k)-C_2\hat{x}_2(k)] \\ &\quad + B_1G_2(k)[y(k)-C_1\hat{x}_1(k-1)-C_2\hat{x}_2(k-1)] \end{aligned} \tag{10-3.8}$$

$$\hat{x}_1(0) = \mathcal{E}x_1(0) - R_{x1oyo}R_{yo}^+[y(0)-\mathcal{E}y(0)], \quad \hat{x}_1(-1) = 0$$

where $R_{x1oyo} = \mathcal{E}(x_1(0)-\mathcal{E}x_1(0))(y(0)-\mathcal{E}y(0))^\tau$, $R_{yo} = \mathcal{E}(y(0)-\mathcal{E}y(0))(y(0)-\mathcal{E}y(0))^\tau$,

and

$$\begin{aligned} \hat{x}_2(k) &= -B_2u(k) - K_2y(k) - D_2G_2(k)[y(k)-C_1\hat{x}_1(k-1)-C_2\hat{x}_2(k-1)] \\ \hat{x}_2(-1) &= 0, \end{aligned} \tag{10-3.9}$$

where

$$G_1(k+1) = [A_1\ 0\ B_1]\Phi(k)[C_1\ \ C_2\ 0]^{\tau}([C_1\ C_2\ 0]\Phi(k)[C_1\ C_2\ 0]^{\tau} + FF^{\tau})^{+}$$

$$G_2(k+1) = F^{\tau}([C_1\ C_2\ 0]\Phi(k)[C_1\ \ C_2\ 0]^{\tau} + FF^{\tau})^{+}$$

$$\Phi(k+1) = \begin{bmatrix} A_1 & 0 & D_1 \\ 0 & 0 & 0 \\ 0 & 0 & 0 \end{bmatrix}\Phi(k)\begin{bmatrix} A_1^{\tau} & 0 & 0 \\ 0 & 0 & 0 \\ D_1^{\tau} & 0 & 0 \end{bmatrix} + \begin{bmatrix} 0 \\ -D_2 \\ I \end{bmatrix}[0\ -D_2^{\tau}\ I] \tag{10-3.10}$$

$$- \begin{bmatrix} G_1(k+1) \\ -D_2G_2(k+1) \\ G_2(k+1) \end{bmatrix}\left(\begin{bmatrix} A_1 & 0 & D_1 \\ 0 & 0 & 0 \\ 0 & 0 & 0 \end{bmatrix}\Phi(k)\begin{bmatrix} C_1^{\tau} \\ C_2^{\tau} \\ 0 \end{bmatrix} + \begin{bmatrix} 0 \\ -D_2 \\ I \end{bmatrix}F^{\tau}\right)^{\tau}$$

$$\Phi(0) = R_{xlo} - R_{xloyo}R_{yo}^{+}R_{yoxlo},$$

where $R_{xlo} = \mathcal{E}(x_1(0)-\mathcal{E}x_1(0))(x_1(0)-\mathcal{E}x_1(0))^{\tau}$; $R_{yoxlo} = \mathcal{E}(y(0)-\mathcal{E}y(0))(x_1(0)-\mathcal{E}x_1(0))^{\tau}$.

The estimation error covariance matrix is

$$[P_1\ \ 0]\Phi(k)[P_1^{\tau}/0]. \tag{10-3.11}$$

Remark 1. We see that the algorithm has the conventional form, a dynamic model plus a modification term. The only difference is that two time modification terms are needed here.

Remark 2. (10-3.11) shows that when u(k) is admissible, the estimation error covariance matrix (10-3.11) is independent of u(k).

Remark 3. As seen in the estimation algorithm, R_{xlo}, R_{yo}, R_{xloyo}, R_{yoxlo}, and $\mathcal{E}x_1(0)$ are needed here. However, these items are usually not available. In such case, we can use $\hat{x}_1(0) = 0$ and $\Phi(0) = \alpha I$, where $\alpha > 0$ is a sufficiently large scalar, in lieu of these in the algorithm, under certain conditions such as observability of the system.

10-3.2. LQG problem

Let u(k) be the control law in the form of

$$u(k) = \bar{h}(k) + \bar{H}(k)y^k \tag{10-3.12}$$

where $\bar{h}(k)$, $\bar{H}(k)$ are deterministic vector and matrix, respectively. Obviously, control (10-3.12) is a linear control, and u(k) will be termed admissible if (10-3.12) is satisfied.

The LQG problem is to select an admissible control u^L such that the cost function

$$J(u) = \mathcal{E}\sum_{k=0}^{L}(x^{\tau}(k)R(k)x(k) + u^{\tau}(k)S(k)u(k)) \tag{10-3.13}$$

is minimized. Here R(k) and S(K) are symmetric nonnegative definite matrices for any $k = 0, 1, \ldots, L$.

To solve the LQG problem for singular system (10-3.1), we first impose a preliminary output feedback control, then change the problem into a LQG problem for a normal system and finally solve the problem using linear system theory.

Under the Y-controllability and Y-observability assumptions for system (10-3.1) there exists a matrix K satisfying $\deg(|zE - (A+BKC)|) = \text{rank}E$. For the K chosen in such a way, we choose the control

$$u(k) = Ky(k) + \bar{u}(k), \tag{10-3.14}$$

where $\bar{u}(k)$ is admissible. Then u(k) is also admissible. Substituting (10-3.14) into (10-3.1) we obtain the closed-loop system

$$\begin{aligned} Ex(k+1) &= (A+BKC)x(k) + B\bar{u}(k) + (D+BKF)w(k) \\ y(k) &= Cx(k) + Fw(k) \end{aligned} \tag{10-3.15}$$

for which, as before, there exist nonsingular matrices Q and P such that it is r.s.e. to

$$\begin{aligned} x_1(k+1) &= A_1x_1(k) + B_1\bar{u}(k) + D_1w(k) \\ 0 &= x_2(k) + B_2\bar{u}(k) + D_2w(k) \\ y(k) &= C_1x_1(k) + C_2x_2(k) + Fw(k). \end{aligned} \tag{10-3.16}$$

Here the coefficients are determined in (10-3.7), where K is replaced by -BK.

If we define the new variable

$$v(k) = \begin{bmatrix} x_1(k) \\ w(k) \end{bmatrix} \tag{10-3.17}$$

from (10-3.16) we know that v(k) satisfies:

$$\begin{aligned} v(k+1) &= \begin{bmatrix} A_1 & D_1 \\ 0 & 0 \end{bmatrix} v(k) + \begin{bmatrix} B_1 \\ 0 \end{bmatrix} \bar{u}(k) + \begin{bmatrix} 0 \\ 1 \end{bmatrix} w(k+1) \\ y(k) &= \bar{C}v(k) + \bar{F}w(k) + \bar{G}\bar{u}(k) \end{aligned} \tag{10-3.18}$$

where

$$\bar{C} = [C_1 \quad 0], \quad \bar{G} = -C_2B_2, \quad \bar{F} = F - C_2D_2, \tag{10-3.19}$$

and

$$\begin{bmatrix} x(k) \\ u(k) \end{bmatrix} = \begin{bmatrix} x(k) \\ KCx(k) + KFw(k) + \bar{u}(k) \end{bmatrix}$$

$$= \begin{bmatrix} P & 0 \\ 0 & I \end{bmatrix} \begin{bmatrix} I & 0 & 0 \\ 0 & -D_2 & -B_2 \\ KC_1 & K(F-C_2D_2) & I-KC_2B_2 \end{bmatrix} \begin{bmatrix} v(k) \\ \bar{u}(k) \end{bmatrix}. \qquad (10\text{-}3.20)$$

Under the assumption that $R(k)$ and $S(k)$ are symmetric nonnegative definite and P is nonsingular, so we may choose two matrices $L_1(k)$, $L_2(k)$ such that

$$\begin{aligned} P^{\tau}R(K)P &= L_1^{\tau}(k)L_1(k) \geqslant 0 \\ S(k) &= L_2^{\tau}(k)L_2(k) \geqslant 0. \end{aligned} \qquad (10\text{-}3.21)$$

If we denote

$$V_1(k) = \begin{bmatrix} L_1(k)\mathrm{diag}(I, -D_2) \\ [L_2(k)KC_1, \quad L_2(k)K(F-C_2D_2)] \end{bmatrix}, \qquad V_2(k) = \begin{bmatrix} L_1(k)[0/-B_2] \\ L_2(k)(I-KC_2B_2) \end{bmatrix}$$

$$\bar{R}(k) = V_1^{\tau}(k)V_1(k), \quad N(k) = V_1^{\tau}(k)V_2(k), \quad \bar{S}(k) = V_2^{\tau}(k)V_2(k), \qquad (10\text{-}3.22)$$

the direct computation, also noticing

$$\begin{bmatrix} \bar{R}(k) & N(k) \\ N^{\tau}(k) & \bar{S}(k) \end{bmatrix} \geqslant 0,$$

shows that the cost functional is

$$J(u) = \mathcal{E} \sum_{k=0}^{L} (x^{\tau}(k)R(k)x(k) + u^{\tau}(k)S(k)u(k))$$

$$= \mathcal{E} \sum_{k=0}^{L} [v^{\tau}(k) \quad \bar{u}^{\tau}(k)] \begin{bmatrix} \bar{R}(k) & N(k) \\ N^{\tau}(k) & \bar{S}(k) \end{bmatrix} \begin{bmatrix} v(k) \\ \bar{u}(k) \end{bmatrix}. \qquad (10\text{-}3.23)$$

Hence, the LQG problem for system (10-3.1) is changed into the same problem for the normal system (10-4.18) with the cost functional of (10-3.23). Therefore, we have the following theorem (Dai, 1988k).

Theorem 10-3.2. Assume that system (10-3.1) is Y-controllable and Y-observable. Then the optimal stochastic control that minimizes the cost (10-3.13) is

$$u^*(k) = Ky(k) - U(k)\hat{v}(k) + (I-W^+(k)W(k))\omega(k) = Ky(k) + \bar{u}^*(k)$$

$$k = 0, 1, \ldots, L \qquad (10\text{-}3.24)$$

where $\omega(k)$ is any admissible control and $U(k)$, $W(k)$ are determined by

$$U(k) = W^{+}(k)\left([B_1/0]V(k+1)\begin{bmatrix} A_1 & D_1 \\ 0 & 0 \end{bmatrix} + N^{\tau}(k)\right)$$

$$W(k) = \bar{S}(k) + [B_1^{\tau} \quad 0]V(k+1)[B_1/0] \tag{10-3.25}$$

and $V(k)$ is the unique symmetric nonnegative definite solution of the Riccati equation

$$V(k) = \left(\begin{bmatrix} A_1 & D_1 \\ 0 & 0 \end{bmatrix} - \begin{bmatrix} B_1 \\ 0 \end{bmatrix} U(k)\right)^{\tau} V(k+1)\left(\begin{bmatrix} A_1 & D_1 \\ 0 & 0 \end{bmatrix} - \begin{bmatrix} B_1 \\ 0 \end{bmatrix} U(k)\right) + U^{\tau}(k)\bar{S}(k)U(k) + \bar{R}(k)$$

$$V(L+1) = 0.$$

The vector $\hat{v}(k)$ is given by the following recursive algorithm:

$$\hat{v}(k) = \begin{bmatrix} \hat{v}_1(k) \\ \hat{v}_2(k) \end{bmatrix}$$

$$\hat{v}_1(k+1) = A_1\hat{v}_1(k) + D_1\hat{v}_2(k) + B_1\bar{u}^*(k) + K_1(k+1)[y(k+1) - \bar{C}\bar{u}^*(k) - C_1\hat{v}_1(k))$$

$$\hat{v}_1(0) = \mathcal{E}x_1(0)$$

and

$$\hat{v}_2(k) = K_2(k)[y(k) - \bar{G}\bar{u}^*(k) - C_1\hat{v}_1(k)].$$

The coefficient matrices satisfy

$$\begin{bmatrix} K_1(k+1) \\ K_2(k+1) \end{bmatrix} = \begin{bmatrix} [A_1 \quad D_1]\Psi(k)\bar{C}^{\tau} \\ F^{\tau} \end{bmatrix} (\bar{C}\Psi(k)\bar{C}^{\tau} + \bar{F}^{\tau}\bar{F})^{\tau}$$

$$\Psi(k+1) = \begin{bmatrix} A_1 & D_1 \\ 0 & 0 \end{bmatrix} \Psi(k) \begin{bmatrix} A_1^{\tau} & 0 \\ D_1^{\tau} & 0 \end{bmatrix} - \begin{bmatrix} K_1(k+1) \\ K_2(k+1) \end{bmatrix} \left(\begin{bmatrix} A_1 & D_1 \\ 0 & 0 \end{bmatrix}\Psi(k)\bar{C}^{\tau} + \begin{bmatrix} 0 \\ F \end{bmatrix}\bar{F}^{\tau}\right)^{\tau} + \begin{bmatrix} 0 \\ I \end{bmatrix}[0 \quad I]$$

$$\Psi(0) = \mathrm{diag}(R_{xlo} - R_{xloyo}R_{yo}^{+}R_{yoxlo}, \quad I).$$

Remark 4. To solve the LQG problem, the estimation of $w(k)$ is needed in our algorithm. This is because $x(k)$ and $w(k)$ are corelated instead of independent. For this reason, the past two step estimation error modification terms are needed in the filtering and LQG problem.

10-4. Notes and References

At present, little work has been done on the problems in this chapter for singular systems. In this chapter, we showed some trivial results in decentralized control, adaptive control, and stochastic control for singular systems, stemming from displaying the difference and similarity between singular and normal systems. However, the general probelms of decentralized, adaptive control and stochastic control for singular systems are far from such a simple problem, especially for singular systems described in a state space model.

APPENDIX A

DISTRIBUTION AND DIRAC FUNCTION

Let $\varphi(x)$ be an infinite times differentiable function defined on the real axis $\mathbb{R}$ and $\varphi(x)$ has exactly zero value outside a bounded closed set $\mathbb{A} \subset \mathbb{R}$. Such functions are termed test functions. Let $\mathbb{A}$ be the smallest closed set such that, if $x_0 \in \mathbb{A}$, for any neighborhood $U(x_0)$ of x_0, there exists a point $x \in U(x_0)$ for which $\varphi(x) \neq 0$. Then $\mathbb{A}$ is termed the support of a function $\varphi(x)$. We use $\mathbb{D}$ to denote the set of test functions; obviously, $\mathbb{D}$ is a linear space.

Definition A-1. Let $\{\varphi_i(x)\}$ be a sequence of test functions. If the functions $\varphi_i(x)$ have the same support $\mathbb{A}$ and for any $k = 0, 1, 2, \ldots,$ the sequence $\varphi_i^{(k)}(x)$ converges uniformly to zero, the sequence $\{\varphi_i\}$ is said to converge to the zero function and is denoted by

$$\lim_{i \to \infty} \varphi_i(x) = 0, \quad \text{or} \quad \varphi_i(x) \xrightarrow{\mathbb{D}} 0.$$

Definition A-2. Any linear and continuous functional defined on $\mathbb{D}$ is called a distribution.

If f(x) is a distribution, f(x) has the following properties.

1. $\langle f,(a\varphi_1+b\varphi_2)\rangle = a\langle f,\varphi_1\rangle + b\langle f,\varphi_2\rangle, \quad \forall\ \varphi_1, \varphi_2 \in \mathbb{D};\ a, b \in \mathbb{R}.$
2. If $\varphi_i \xrightarrow{\mathbb{D}} 0, \quad \langle f,\varphi_i\rangle \to 0\ (i \to \infty).$

Here $\langle f,\varphi\rangle$ is a real number defined by f, φ according to a certain law.

Example A-1. Let f(x) be a locally integrable and real function defined on $\mathbb{R}$. Then using the inner product for f, g

$$\langle f,g\rangle = \int_{\mathbb{R}} f(x)g(x)dx = \int_{\mathbb{A}} f(x)g(x)dx, \tag{A.1}$$

we can uniquely define a distribution g.

Assume that f(x) is a distribution. If there exist constant $M > 0$ and $k \in \mathbb{R}$ satisfying

$$|f(x)| \leqslant M|x|^k, \qquad \text{when } x \to \infty,$$

f(x) will be termed the temperate distribution.

For example, the function $f(x) = e^{-a|x|}$, $a > 0$, and the polynomial functions are temperate distributions.

By definition, the function in the normal sense are distributions. Thus, the concept of distribution is the generalization of function in the normal sense.

Example A-2. Let $\langle f,g\rangle$ be defined by (A.1). Then

$$\langle \delta,\varphi\rangle = \varphi(0) \tag{A.2}$$

uniquely defines a function $\delta(x)$.

Assume

$$f_\varepsilon(x) = \begin{cases} \frac{1}{\varepsilon} & 0 \leqslant x \leqslant \varepsilon, \quad \varepsilon > 0, \\ 0 & \text{for other values.} \end{cases}$$

Then for any test function $\varphi(x) \in \mathbb{D}$, we have

$$\lim_{\varepsilon\to 0} \langle f_\varepsilon,\varphi\rangle = \lim_{\varepsilon\to 0} \int_{-\infty}^{\infty} f_\varepsilon(x)\varphi(x)dx = \lim_{\varepsilon\to 0} \int_0^{\varepsilon} \varphi(x)dx$$

$$= \varphi(0) = \langle \delta , \varphi\rangle.$$

Thus

$$\delta = \lim_{\varepsilon\to 0} f_\varepsilon .$$

This is the Dirac function.

Definition A-3. Assume that $f(x)$ is a distribution and denote

$$\mathbb{B} = \bigcup_{\mathbb{B}_i} \{ \mathbb{B}_i \mid f(x) = 0, \ x \in \mathbb{B}_i, \ \mathbb{B}_i \text{ is open} \} .$$

Then the complement of $\mathbb{B}$ is called the support of f, it is a closed set.

Let $\langle f,g\rangle$ be determined by (A.1) and $f(x)$ be differentiable on $\mathbb{R}$ with derivative $f'(x)$. Then we have

$$\langle f',\varphi\rangle = \int_{-\infty}^{\infty} f'(x)\varphi(x)dx = \lim_{T\to\infty} \int_{-T}^{T} \varphi(x)df(x)$$

$$= \lim_{T\to\infty} \left(\varphi(x)f(x)\Big|_{-T}^{T} - \int_{-T}^{T} \varphi'(x)f(x)dx\right)$$

$$= - \int_{-\infty}^{\infty} f(x)\varphi'(x)dx = - \langle f,\varphi'\rangle.$$

This equation holds for any $\varphi \in \mathbb{D}$.

Definition A-4. The derivative f' of f is a distribution that is uniquely determined by

$$\langle f', \varphi \rangle = - \langle f, \varphi' \rangle.$$

Example A-3. Consider the unit jump function

$$I(x) = \begin{cases} 1 & \text{when } x > 0 \\ 0 & \text{when } x \leqslant 0. \end{cases}$$

We have

$$\langle I'(x), \varphi \rangle = - \langle I(x), \varphi' \rangle = - \int_{-\infty}^{\infty} I(x)\varphi'(x)dx = - \int_{0}^{\infty} \varphi'(x)dx$$

$$= \varphi(0) = \langle \delta, \varphi \rangle$$

indicating $\delta(x) = \frac{d}{dt}I(x)$. Thus the derivative of the unit jump function is the Dirac function.

By Definition A-4, the ith order derivative $f^{(i)}(x)$ of f(x) is uniquely defined by

$$\langle f^{(i)}, \varphi \rangle = (-1)^i \langle f, \varphi^{(i)} \rangle, \quad i = 1, 2, \ldots \tag{A.3}$$

The arbitrary order derivative of a normal function may not exist. However, from (A.3) we see that a distribution has any order derivative. In fact, it is for the differentiation of discontinuous function that we intrduce the notion of distribution.

Assume that f is a distribution. If there exists a piecewise continuous function g(x), points $\tau_i \in \mathbb{R}$, $i = 0, \pm 1, \pm 2, \ldots$, and the number of τ_i is finite on any bounded set, such that

$$f = g, \quad x \in (\tau_i, \tau_{i+1}), \quad i = 0, \pm 1, \pm 2, \ldots,$$

the distribution is called piecewise continuous. We use $\mathbb{D}'$ to denote the set of distributions on $\mathbb{R}$ and $\mathbb{R}'_p$ to denote the piecewise continuous distribution on $\mathbb{R}$.

Let f(x) be a real function on $\mathbb{R}$, which has derivative except the point τ_0 and

$$f(\tau_0+0) = \lim_{\varepsilon \to 0+} f(\tau_0+\varepsilon), \quad f(\tau_0-0) = \lim_{\varepsilon \to 0-} f(\tau_0+\varepsilon)$$

exist. Thus

$$\Delta_{\tau_0} f = f(\tau_0+0) - f(\tau_0-0)$$

represents the jump at τ_0 of function f.

Furthermore, let f'(x) be the distribution derivative of function f(x) and $\tilde{f}'(x)$ be its derivative in the normal sense, which is assumed to exist at the point $x \in \mathbb{R}$. Then we have

$$\langle f',\varphi\rangle = \langle f,-\varphi'\rangle = \lim_{\varepsilon\to 0}\left\{-\int_{-\infty}^{\tau_0-\varepsilon} f(x)\varphi'(x)dx - \int_{\tau_0+\varepsilon}^{\infty} f(x)\varphi'(x)dx\right\},\ \varepsilon > 0.$$

Integrating by parts and taking limitations, we obtain

$$\langle f',\varphi\rangle = \lim\left\{-f\varphi\Big|_{-\infty}^{\tau_0-\varepsilon} - f\varphi\Big|_{\tau_0+\varepsilon}^{\infty} + \int_{-\infty}^{\tau_0-\varepsilon}\tilde{f}'\varphi dx + \int_{\tau_0+\varepsilon}^{\infty}\tilde{f}'\varphi dx\right\}$$

$$= \Delta_{\tau_0}f\varphi(\tau_0) + \langle\tilde{f}',\varphi\rangle = \langle\tilde{f}' + \Delta_{\tau_0}f\delta(x-\tau_0),\ \varphi\rangle.$$

Therefore

$$f'(x) = \tilde{f}'(x) + \Delta_{\tau_0}f\delta(x-\tau_0). \tag{A.4}$$

Generally, if f has derivatives on $\mathbb{R}$, except the points τ_i, i = 1, 2, ..., and all $\Delta_{\tau_i}f = f(\tau_i+0) - f(\tau_i-0)$ are finite, it will be

$$f'(x) = \tilde{f}'(x) + \sum_{i=1}^{n}\Delta_{\tau_i}f\,\delta(x-\tau_i).$$

Theorem A-1. Suppose that f(x) is continuous, b_j^i, i = 0, 1, ..., n, j = 1, 2, ..., m, are real numbers. Then if

$$f(x) + \sum_{i=0}^{n}\sum_{j=1}^{m} b_j^i\delta^{(i)}(x-\tau_j) = 0 \tag{A.5}$$

we have

$$f(x) = 0 \tag{A.6}$$

$$b_j^i = 0,\ i = 0, 1, \ldots, n,\ j = 1, 2, \ldots, m. \tag{A.7}$$

Definition A-5. Let the support of distribution f(x) belong to $[0, +\infty)$; $F(x)e^{-sx}$ is a temperate distribution. We define the Laplace transformation of f(x) by

$$L[f] = \langle f, e^{-sx}\rangle. \tag{A.8}$$

Clearly, Definition A-5 is the generalization of the normal Laplace transformation. Equation (A.8) is the Laplace transformation in the normal sense when f(x) is the normal function.

It is easy to verify that the following properties are possessed by Laplace transformation in the normal sense:

$$L[\delta^{(i)}] = s^i$$

$$L[f'(x)] = sL[f] - f(0).$$

APPENDIX B

SOME RESULTS IN MATRIX THEORY

B-1. Eigenvalues, eigenvectors, and characteristic polynomials

<u>Definition B-1</u>. Let the matrix $A \in \mathbb{C}^{n \times n}$. If there exist a complex scalar λ and a nonzero vector $\alpha \in \mathbb{C}^n$ such that

$$A\alpha = \lambda\alpha,$$

the scalar λ is called the eigenvalue of matrix A, and α is called the eigenvector of matrix A belonging to the eigenvalue λ.

The set of eigenvalues of matrix A is called the spectrum of A and is denoted by $\sigma(A)$.

The polynomial $f(s) = |sI-A|$ is the characteristic polynomial of A.

If matrix A has the form

$$A = \begin{bmatrix} 0 & 1 & 0 & \dots & 0 \\ 0 & 0 & 1 & \dots & 0 \\ & \dots & & \dots & \dots \\ 0 & 0 & 0 & \dots & 1 \\ a_o & a_1 & a_2 & & a_{n-1} \end{bmatrix}$$

then $|sI-A| = s^n + a_{n-1}s^{n-1} + \dots + a_1 s + a_o$.

B-2. Cayley-Hamilton Theorem

Let $f(s) = s^n + a_{n-1}s^{n-1} + \dots + a_1 s + a_o$ be the characteristic polynomial of matrix A. Then the Cayley-Hamilton Theorem states that

$$f(A) = A^n + a_{n-1}A^{n-1} + \dots + a_1 A + a_o I = 0.$$

B-3. Jordan form and matrix decomposition

The two matrices A and $\tilde{A}$ are called similar, or algebraically equivalent, if there exists a nonsingular matrix P such that $\tilde{A} = PAP^{-1}$.

Let $s_1, s_2, \dots, s_k$ be the k distinct eigenvalues of matrix A. Then there exists a nonsingular matrix P such that

$$PAP^{-1} = \text{diag}(J_{s_1}, J_{s_1}, \dots, J_{s_k}) \triangleq J \quad (B.1)$$

where

$$J_{s_i} = \text{diag}(J^1_{s_i}, J^2_{s_i}, \dots, J^{h_i}_{s_i}) \in \mathbb{C}^{n(s_i)xn(s_i)}, \quad 1 \leqslant i \leqslant k \quad (B.2)$$

$$J^j_{s_i} = \begin{bmatrix} s_i & 1 & & & \\ & s_i & 1 & & \\ & & \ddots & & \\ & & & & 1 \\ & & & & s_i \end{bmatrix} \in \mathbb{C}^{n_j(s_i)xn_j(s_i)}, \quad 1 \leqslant j \leqslant h_i \quad (B.3)$$

$$\sum_{j=1}^{h_i} n_j(s_i) = n(s_i), \quad 1 \leqslant i \leqslant k.$$

Every matrix $J^j_{s_i}$ ($1 \leqslant i \leqslant k$, $1 \leqslant j \leqslant h_i$) in (B.3) is called a Jordan block and matrix J is the Jordan canonical form of matrix A.

If matrix A has the form

$$A = FG, \quad (B.4)$$

and the matrices F and G are of full column and row rank, respectively, (B.4) is called the full rank decomposition of matrix A.

If matrix A has the form

$$A = P\text{diag}(\sigma_1, \sigma_2, \dots, \sigma_r, 0)Q, \quad (B.5)$$

where P and Q are orthogonal, $\sigma_1 \geqslant \sigma_2 \geqslant \dots \geqslant \sigma_r > 0$, (B.5) is called the singular value decomposition.

If matrix A has the form

$$A = P\text{diag}(I_r, 0)Q, \qquad r = \text{rank}A, \quad (B.6)$$

where P and Q are nonsingular, (B.6) is called the rank decomposition.

B-4. Some special matrices

We use I_n to denote the nxn unit matrix.

The matrix N is called nilpotent if there exists a number n such that $N^n = 0$. The smallest n satisfying $N^n = 0$ is called the nilpotent index of N.

Matrix A is said to be symmetric if $A = A^\tau$; it is symmetric positive definite (positive semidefinite) if for any vector x, $x \neq 0$, $x^\tau Ax > 0$ ($x^\tau Ax \geqslant 0$).

B-5. The minimal polynomial

A polynomial g(s) is said to be the minimal polynomial of matrix A if g(s) is a monic polynomial (or the coefficient of the highest order is unit) with the lowest order satisfying $g(A) = 0$; $g(s)$ exists since the characteristic polynomial of A satisfying $f(A) = 0$.

B-6. Matrix inverse formula

The following matrix equation is called the matrix inverse gormula:

$$(A+BCD)^{-1} = A^{-1} - A^{-1}B(DA^{-1}B + C^{-1})^{-1}DA^{-1} \tag{B.7}$$

where A^{-1}, C^{-1}, and either $(A+BCD)^{-1}$ or $(C^{-1}+DA^{-1}B)^{-1}$ are assumed to exist. A special and useful case of (B.7) is

$$(I+BCD)^{-1} = I - B(DB + C^{-1})^{-1}D.$$

B-7. Exponential matrix

For a square matrix A, the exponential matrix of A is defined as

$$e^{At} = I + \frac{1}{1!} At + \frac{1}{2!} (At)^2 + \dots + \frac{1}{k!} (At)^k + \dots$$

For any nxn matrix A there exist continuous functions $f_o(t)$, $f_1(t)$, ... , $f_{n-1}(t)$ such that

$$e^{At} = f_o(t)I + f_1(t)A + \dots + f_{n-1}(t)A^{n-1}.$$

B-8. Sylvester Equation

For any matrix C the Sylvester equation

$$AX - XB = C$$

has a unique solution if and only if $\sigma(A) \cap \sigma(B) = \emptyset$.

B-9. Pseudo-inverse

A^+ will be used to denote the pseudo-inverse of matrix A. It is the solution of the following equations

$$AXA = A, \quad XAX = X, \quad (AX)^{\tau} = AX, \quad (XA)^{\tau} = XA.$$

B-10. Matrix norm

The norm $\|A\|$ of matrix $A \in \mathbb{C}^{nxm}$ is a scalar satisfying

1. For any matrix $A \neq 0$, $\|A\| > 0$.

2. $\|\alpha A\| = \alpha \|A\|$, $\forall \ \alpha \in \mathbb{C}$.

3. $\|A+B\| \leqslant \|A\| + \|B\|$, for any $A, B \in \mathbb{C}^{nxm}$.

Norm $\|A\|$ is called consistent if for any two matrices A and B of appropriate dimensions, $\|AB\| \leqslant \|A\|\|B\|$.

The norm $\|x\|$ of vector x is the special case of matrix normal when $x \in \mathbb{C}^{nx1}$.

APPENDIX C

SOME RESULTS IN CONTROL THEORY

The following system

$$\dot{x}(t) = Ax(t) + Bu(t), \quad x(0) = x_0$$
$$y(t) = Cx(t) \tag{C.1}$$

where $x(t) \in \mathbb{R}^n$, $u(t) \in \mathbb{R}^m$, and $y(t) \in \mathbb{R}^r$, is the classic linear system in linear system theory which is termed the normal system here for the sake of distinction with singular systems; x_0 is the initial condition.

C-1. State response and stability

The solution of the differential equation in (C.1) is called the state response of system (C.1). It is given by

$$x(t) = e^{At}x_0 + \int_0^t e^{A(t-\tau)}Bu(\tau)d\tau .$$

System (C.1) is called (asymptotically) stable if $\lim_{t\to\infty} x(t) = 0$, $\forall\ x_0$. System (C.1) is stable if and only if $\sigma(A) \subset \mathbb{C}^-$.

C-2. Pole and pole structure

The poles of system (C.1) are eigenvalues of the coefficient matrix A. The pole structure for system (C.1) indicates the eigenstructure in the Jordan form of matrix A.

C-3. Controllability and observability

System (C.1) is controllable iff

$$\text{rank}[sI-A,\ B] = n, \quad \forall\ s \in \mathbb{C},$$

or equivalently,

$$\text{rank}[B, AB, \dots, A^{n-1}B] = n.$$

System (C.1) is observable iff

$$\text{rank}[sI-A/C] = n, \quad \forall\ s \in \mathbb{C},$$

or equivalently,

$$\text{rank}[C/CA/\ldots/CA^{n-1}] = n, \quad \forall\ s \in \mathbb{C}.$$

C-4. Pole placement

Let the state feedback be

$$u(t) = Kx(t) + v(t). \tag{C.2}$$

When applied to system (C.1), its closed-loop system is

$$\dot{x}(t) = (A+BK)x(t) + Bv(t). \tag{C.3}$$

If system (C.1) is controllable, the closed-loop poles of (C.3) may be arbitrarily assigned via feedback (C.2).

C-5. Controllability indices and observability indices

Let system (C.1) be controllable. Its controllability indices are a set of real numbers defined in the following way. Assume that B is of full column rank, otherwise, delete the corelated columns in B. Let $\nu_1 \leqslant \nu_2 \leqslant \ldots \leqslant \nu_m$ be the controllability indices of (A,B). Then ν_1 is the smallest number of i in the following series

$$B, \quad AB, \quad A^2B, \quad \ldots, \quad A^iB,$$

when corelated column appears in it. Delete this column and the corresponding columns that create this column in B and all corresponding columns in this series. Then ν_2 is the second smallest number in this series when corelated columns occur. Delete all the corresponding columns as before, and so on, until the controllability indices of (A,B) are defined. ν_m is called the index of (A,B).

The observability indices of (A,C) are the controllability indices of (A^τ, C^τ).

C-6. Controllability and observability decomposition

For any system (C.1) there exists a nonsingular matrix T such that

$$TAT^{-1} = \begin{bmatrix} A_{11} & 0 & A_{13} & 0 \\ A_{21} & A_{22} & A_{23} & A_{24} \\ 0 & 0 & A_{33} & 0 \\ 0 & 0 & A_{43} & A_{44} \end{bmatrix}, \qquad TB = \begin{bmatrix} B_1 \\ B_2 \\ 0 \\ 0 \end{bmatrix},$$

$$CT^{-1} = [C_1 \quad 0 \quad C_3 \quad 0],$$

where (A_{11}, B_1, C_1) is controllable and observable;

$$\left(\begin{bmatrix} A_{11} & 0 \\ A_{21} & A_{22} \end{bmatrix}, \begin{bmatrix} B_1 \\ B_2 \end{bmatrix}, [C_1 \quad 0] \right)$$

is controllable but unobservable; and

$$\left(\begin{bmatrix} A_{11} & A_{13} \\ 0 & A_{33} \end{bmatrix}, \begin{bmatrix} B_1 \\ 0 \end{bmatrix}, [C_1 \;\; C_3]\right)$$

is observable but uncontrollable.

C-7. Optimal regulation with quadratic critierion

For system (C.1) and quadratic cost functional

$$J = \int_0^\infty (x^\tau Sx + u^\tau Ru)dt,$$

where S and R are symmetric positive definite, system (C.1) is stabilizable. Then the optimal control that minimizes the cost functional J is

$$u = -R^{-1}B^\tau Mx,$$

where the symmetric matrix M is the unique positive definite solution of the following Riccati equation

$$MA + A^\tau M - MBR^{-1}B^\tau M + S = 0.$$

The minimal value of J is

$$\min J = x^\tau(0)Mx(0).$$

C-8. Transfer matrix

The transfer matrix of system (C.1) is

$$G(s) = C(sI-A)^{-1}B.$$

It is a strictly proper rational matrix, i.e., $\lim_{s\to\infty} G(s) = 0$.

APPENDIX D

LIST OF SYMBOLS AND ABBREVIATIONS

D-1. Symbols

The following symbols are adopted in this book:

$\|A\|$	=	the determination of matrix A;
A^τ	=	the transpose of matrix A;
A^{-1}	=	the inverse of matrix A when it exists, $A^{-1}A = AA^{-1} = I$;
$A > (\geqslant)\ 0$	=	Matrix A is symmetric positive (nonnegative) definite;
$[A/B]$	=	is used to denote the matrix $\begin{bmatrix} A \\ B \end{bmatrix}$;
$\langle A\|B\rangle$	=	$Im[B, AB, \ldots, A^{n-1}B]$, n = the dimension of A;
adj(A)	=	the adjoint matrix of a matrix A, $adj(A)A = \|A\|I$;
$\mathbb{C}$	=	the complex plane
$\mathbf{C}^k$	=	set of k times continuously differentiable functions;
$\mathbf{C}_p^k$	=	set of k times piecewise continuously differentiable functions;
$\mathbb{C}^{mxn}$	=	set of mxn complex matrices;
$\mathbb{C}^n$	=	set of complex vectors;
$\bar{\mathbb{C}}^+$	=	the closed right half complex plane, $\bar{\mathbb{C}}^+ = \{s \mid s \in \mathbb{C}, Re(s) \geqslant 0\}$;
$\mathbb{C}^-$	=	the open left half complex plane, $\mathbb{C}^- = \mathbb{C} - \bar{\mathbb{C}}^+$;
deg(.)	=	the degree of a polynomial;
(i)	=	a superscript of a function, is the ith order derivative of a function;
Im(s)	=	the complex part of a complex scalar s;
Im(A)	=	the range of a matrix A, $Im(A) = \{y \mid y = Ax\}$;
I_n	=	the nxn identity (or unit) matrix;
Ker(A)	=	the kernal (null) of matrix A, $Ker(A) = \{x \mid Ax = 0\}$;
L[.]	=	the Laplace transformation of a function;
$\mathbb{R}$	=	set of real numbers (the real axis);
$\mathbb{R}^{mxn}$	=	set of mxn real matrices;

$\mathbb{R}^n$ = set of real vectors;

rankA = rank of matrix A;

Re(s) = the real part of a complex scalar s;

U^+ = the unit circle on the complex plane, $U^+ = \{ z \mid z \in \mathbb{C}, |z| < 1 \}$;

x_τ = the impulse part of a function x at time τ;

$\Delta_\tau x$ = the jump part of the function x at time τ;

$\delta(t-\tau)$ = the Dirac (or Delta) function;

= coverdot = the derivative of a function

$\mathcal{E}$ = the expectation of a random variable;

$\sigma(A)$ = the spectrum of matrix A, i.e., the eigenvalue set of A;

$\sigma(E,A)$ = the set of finite eigenvalues of regular matrix pencil (E,A);

$\in$ = belonging to;

$\subset$ = be included in;

$\cup$ = the union of sets;

$\cap$ = the intersection of sets;

$\oplus$ = the direct sum of two vectors or linear spaces;

$\mid$ = be divided with no remainder, divided exactly;

$\forall$ = for any, for arbitrary;

$\equiv$ = identical, equal everywhere;

$\not\equiv$ = not identically equal to;

$\triangleq$ = by definition, defined as;

$\{\cdot\}$ = notation of a set.

D-2. Some abbreviations

DDSO = disturbance decoupling state observer;

EF1 (2,3) = the first (second, third) equivalent form;

gcd = greatest common divisor;

HOT = highest order term;

IFO = input function observer;

NOR = normal output regulator;

P- = pure proportional;

P-D	= proportional and derivative;
PPS	= pure prediction system;
r.s.e.	= restricted system equivalent;
SOR	= singular output regulator;
SSNC	= structurally stable normal compensator;
SSNOR	= structurally stable normal output regulator.

APPENDIX E

SUBJECT INDEX

APPENDIX F

BIBLIOGRAPHY

Al-Nasr, N., V. Lovass-Nagy, and G. Rabson (1983a), "General eigenvalue placement in linear control systems by output feedback," Int. J. Systems Sci., Vol.14, No.5, pp.519-528

Al-Nasr, N., V. Lovass-Nagy, D. O'Connor, and A.L. Powers (1983b), "Output function control in general state space systems containing the first derivative of the input vector," Int. J. Systems Sci., Vol.14, No.9, pp.1029-1042.

Anderson, B.D.O., and S.Vongpanitlerd (1973), Network Analysis and Synthesis, A Modern Systems Theory Approach, Prentice-Hall, Englewood Cliffs, N.J.

Aoki, T., S. Hosoe, and Y. Hayakawa (1983), "Structural controllability for linear systems in descriptor form," Trans. Soc. Instrum. & Control Eng. (Japan), Vol.19, No.8, pp.628-635.

Armentano, V.A. (1984), "Eigenvalue placement for generalized linear systems," Systems & Control Letters, Vol.4, No.4, pp.199-202.

Azcaldiran, K., and F.L. Lewis, "A result on the placement of infinite eigenvalues in descriptor systems," Elect. Eng., Georgia Inst. of Tech., Atlanta, Ga.

Bajic, V.B., and M.M. Milic (1986), "Theorems on the bounds of solutions of semi-state models," Int. J. Control, Vol.43, No.3, pp.859-867.

Bajic, V.B. (1986), "Partial exponential stability of semi-state systems," Int. J. Control, Vol.44, No.5, pp.1383-1394.

Bender, D.J. (1985), "Descriptor systems and geometric control theory," Ph.D. dissertation, Dept. of Elect. Comp. Eng., Univ. of California, Santa Barbara, Calif., Sept.

Bender, D.J., and A.J. Laub (1985), "Controllability and observability at infinite of multivariable linear second-order models," IEEE Trans. Aut. Control, Vol.AC-30, No.12, pp.1234-1237.

Bender, D.J. (1987), "Lyapunov-like equations and reachability/observability Gramians for descriptor systems," IEEE Trans. Aut. Control, Vol.AC-32, No.4, pp.343-348.

Bender, D.J., and J. Laub (1987a), "The linear-quadratic optimal regulation for descriptor systems: Discrete-time case," Automatica, Vol.23, No.1, pp.71-86.

Bender, D.J., and A.J. Laub (1987b), "The linear-quadratic optimal regulator for descriptor systems," IEEE Trans. Aut. Control, Vol.AC-32, No.8, pp.672-688.

Bernhard, P. (1982), "On singular implicit linear dynamic systems," SIAM J. Control and Opt., Vol.20, No.5, pp.612-633.

Blomberg, H., and R. Ylinen (1983), Algebraic Theory for Multivariable Linear Systems, Academic Press, London.

Brayton, B.K., F.G. Gustavson, and G.D. Hachtel (1972), "A new efficient algorithm for solving differential-algebraic systems using implicit backward differentiation formulas," Proc. IEEE 60, pp.98-108.

Brenan, K.E. (1983), "Stability and convergence of difference approximations for higher index differential-algebraic systems with applications in trajectory control," Ph.D. dissertation, Dept. of Math., UCLA.

Brierley, S.D., and E.B. Lee (1985), "Coefficient assignment for generalized linear systems," Proc. 4th Int. Conf. on Syst. Eng., Conventry, England, pp.91-100.

Burl, J.B. (1985), "Singular systems --- A tutorial," Conf. Record, 19th Asilomor Conf. on Circuits, Systems and Computers, Pacific Grove, Calif., pp.529-533.

Campbell, S.L. (1980), Singular Systems of Differential Equations, Pitman Publishing Co., New York.

Campbell, S.L. (1981), "A procedure for analyzing a class of nonlinear semi-state equations that arise in circuit and control problems," IEEE Trans. Circuits and Systems, Vol.CAS-28, pp.256-261.

Campbell, S.L. (1982a), "Consistent initial conditions for singular nonlinear systems," Circuits, Systems and Signal Processing, No.2, p.45.

Campbell, S.L. (1982b), Singular Systems of Differential Equations II, Pitman, New York.

Campbell, S.L. (1983a), "One canonical form for higher-index linear time-varying singular systems," Circuits, Systems and Signal Process(USA), Vol.2, No.3, pp.311-326.

Campbell, S.L. (1983b), "Index two linear time-varying singular systems of differential equations," SIAM J. Algebraic and Discrete Meth., Vol.4, No.2, p.237.

Campbell, S.L. (1985), "The numerical solution of high index linear time varying singular systems of differential equations," SIAM J. Sci. & Stat. Comput., Vol.6, No.2, pp. 334-348.

Campbell, S.L., and K. Clark (1981), "Order and the index of singular time invariant linear systems," Systems and Control Letters, Vol.1, pp.119-122.

Campbell, S.L., and L.R. Petzold (1983), "Canonical forms and solvable singular systems of differential equations," SIAM J Algebraic & Discrete Meth., Vol.4, p.517.

Campbell, S.L., and J. Rodriguez (1985), "Bounded solutions of discrete singular Systems on infinite time intervals," IEEE Trans. Aut. Control, Vol.AC-30, No.2, pp.165-168.

Campbell, S.L., and N.J. Rose (1982), "A second-order singular linear system arising in electric power systems analysis," Int. J. Systems Sci., Vol.13, No.1, pp.101-108.

Campbell, S.L, C.D. Meyer, Jr., and N.J. Rose (1976), "Applications of the Drazin inverse to linear systems of differential equations with singular constant coefficients," SIAM J. Appl. Math., Vol.31, No.3.

Chen, Y.C., J.D. Aplevich, and W.J. Wilson (1986), "Simultaneous estimation of state variable and parameters for multivariable linear systems with singular pencil models," IEE Proc. D, Vol.133, No.2, pp.65-72.

Cheng, Z., H. Hong, and J. Zhang (1987), "The optimal state regulation of generalized dynamical systems with quadratic cost functional," Preprints of the 10th World Congress, Munich, West Germany, Jul. 27-31, Vol.9, pp.127-131.

Choi, C.H., and A.J. Laub (1987), "Improving the efficiency of matrix operations in the numerical solution of large implicit systems of linear differential equations," Int. J. Control, Vol.46, No.3, pp.991-1008.

Christodoulou, M.A. (1986), "Decoupling in the design and synthesis of singular systems," Automatica, Vol.22, No.2, pp.245-250.

Christodoulou, M.A. (1987), "Realization of singular systems via Markov parameters and moments," Int. J. Control, Vol.45, No.1, pp.237-246.

Christodoulou, M.A., and B.G. Mertios (1985), "Realization of singular systems via Markov parameters," Int. J. Control, Vol.42, No.6, pp.1433-1442.

Christodoulou, M.A., and P.N. Paraskevopoulos (1984), "Formula for feedback gains in eigenvalue control of singular systems," Electron Letters (GB), Vol.20, No.1, pp.18-19.

Christodoulou, M.A., and P.N. Paraskevopoulos (1985), "Solvability, controllability and observability of singular systems," J. Opt. Theory and Appl., Vol.45, No.1, pp. 53-72.

Chung, H., and Y. Sun (1986), "Analysis of singular systems using Taylor series approach," Electron Lett. (GB), Vol.22, No.21, pp.1099-1100.

Cline, T.B., R.E. Larson, D.G. Luenberger, D.N. Stengel, and K.D. Wall (1976), "Descriptor variable representation of large-scale deterministic systems," Tech. Memo. 5168-1, Systems Control, Inc., Palo Alto, Calif.

Cobb, J.D. (1980), "Descriptor variable and generalized singularly perturbed systems: A geometric approach," Ph.D. thesis, Dept. of Elect. Eng., Univ. of Illinois, Urbana, IL.

Cobb, J.D. (1981), "Feedback and pole placement in descriptor variable systems," Int. J. Control, Vol.33, No.6, pp.1135-1146.

Cobb, J.D. (1982), "on the solutions of linear differential equations with singular coefficients," J. Diff. Eq., Vol.46, No.3, pp.310-323.

Cobb, J.D. (1983a), "Descriptor variable systems and optimal state regulation," IEEE Trans. Aut. Control, Vol.AC-28, No.5, pp.601-611.

Cobb, J.D. (1983b), "A further interpretation of inconsistent initial conditions in descriptor variable systems," IEEE Trans. Aut. Control, Vol.AC-28, No.9, pp.920-922.

Cobb, J.D. (1984), "Controllability, observability and duality in singular systems," IEEE Trans. Aut. Control, Vol.AC-29, No.12, pp.1076-1082.

Cullen, D.J. (1984), "The equivalence of linear systems," Ph.D. thesis, Australian Natl. Univ., Sydney, Australia.

Cullen, D.J. (1986a), "An algebraic condition for controllability at infinite," Systems & Control Letters, No.6, pp.321-324.

Cullen, D.J. (1986b), "State-space realizations at infinity," Int. J. Control, Vol.43, No.4, pp.1075-1088.

Dai, L. (1986a), "Normal compensators for linear singular systems," M.Sc. dissertation, Inst. of Systems Sci., Academia Sinica, Beijing, P.R.C., April.

Dai, L. (1986b), "State feedback and model matching in singular systems," presented at the 2nd Natl. Youth Conf. on Systems and Control Sci., Fujian, P.R.C., July, accepted by Int. J. Control, 1988.

Dai, L. (1987a), "State estimation schemes for singular systems," Prep. 10th IFAC World Congress, Munich, West Germany, July 27-31, Vol.9, pp.211-215.

Dai, L. (1987b), "System equivalence, controllability and observability in singular systems," J. Grad. School, USTC, Academia Sinica, No.1, pp.42-50.

Dai, L. (1988a), "Observers for discrete singular systems," IEEE Trans. Aut. Control, Vol.AC-33, No.2, pp.187-191.

Dai, L. (1988b), "Robustness for descriptor variable systems: A quantitative analysis method," SIAM J. Control and Opt., Vol.26, No.3.

Dai, L. (1988c), "Input function observers for linear singular systems," Acta Math. Scientia, accepted.

Dai, L. (1988d), "Solutions, controllability and observability in discrete singular systems," Acta Math. Scientia, accepted.

Dai, L. (1988e), "Feedback control in singular systems," Control Theory and Appl., accepted.

Dai, L. (1988f),"Decoupling in singular systems," unpublished.

Dai, L. (1988g), "On the optimal control for discrete-time singular systems," unpublished.

Dai, L. (1988h), "Strong decoupling in singular systems."

Dai, L. (1988i), "An H^{∞} method for decentralized stabilization in singular systems," Proc. IEEE/SMC, Shenyang, P.R.C., Aug. 8-13, Vol. 1, pp.722-725.

Dai, L. (1988j), "Impulsive modes and causulity in singular systems."

Dai, L. (1988k), "Filtering and LQG problems for discrete-time singular systems."

Dai, L., and C. Wang (1987a), "Dynamic compensation for discrete singular systems," Systems and Control Letters, Vol.9, pp.329-333.

Dai, L., and C. Wang (1987b), "Normal dynamical compensators for singular systems," Acta Math. Acientia, Vol.7, No.3, pp.273-282.

Dai, L., and C. Wang (1987c), "Structurally stable normal compensators for singular systems," J. Syst. Sci. & Math. Sci., Vol.7, No.1, pp.89-93.

Dai, L., and C. Wang (1987d), "Stable and structurally stable regulators for singular systems," Acta Mathematicae Applicatae Sinica, Vol.3, No.2, pp.122-135.

D'Alessandro, P., M. Dalla Mora, and E.De Santis (1985), "On consistency of linear linearly constrained discrete time systems," J. Franklin Inst., Vol.319, No.4, pp.423-430.

Dolezal, V. (1986), "Generalized solutions of semistate equations and stability," Circuits, Systems and Signal Process, Vol.5, No.4, pp.391-403.

Dziarla, B., and R. Newcomb (1979), "The Drazin inverse and semi-state equations," Proc. 4th Int. Symp. Math. Theory and Network Syst., Delft, The Netherlands, pp.283-289.

El-Tohami, M., V. Lovass-Nagy, and R. Mukundan (1983),"On the design of observers for generalized state space systems using singular value decomposition," Int. J. Control, Vol.38, No.3, pp.673-685.

El-Tohami, M., V. Lovass-Nagy, and D.L. Powers (1984), "On input function observers for generalized state-space systems," Int. J. Control, Vol.40, No.5, pp.903-922.

El-Tohami, M., V. Lovass-Nagy, and D.L. Powers (1985), "On minimal-order inverse of discrete-time descriptor systems," Int. J. Control, Vol.41, No.4, pp.991-1004.

El-Tohami, M., V. Lovass-Nagy, and R. Mukundan (1987), "Design of observers for time-varying discrete-time descriptor systems," Int. J. Control, Vol.46, No.3, pp.841-848.

Eremenko, V.A. (1980), "Reduction of a linear system of differential equations with a singular matrix multiplying the derivatives," Ukranian Math. J., Vol.32, pp.89-94.

Favini, A. (1979), "Laplace transformation method for a class of degenerate evolution problems," Rend. Math., Vol.12, pp.511-536.

Favini, A. (1980), "Controllability conditions of linear degenerate evolution systems," Appl. Math. & Opt., 6, pp.153-167.

Favini, A. (1981), "Abstract potential operators and spectral methods for a class of degenerate evolution problems," J. Diff. Eq., Vol.39, pp.212-225.

Fletcher, L.R. (1986), "Regularizability of descriptor systems," Int. J. Syst. Sci., Vol.47, No.6, pp.843-847.

Fletcher, L.R., J. Kautsly, and N.K. Nichols (1986), "Eigenstructure assignment in descriptor systems," IEEE Trans. Aut. Control, Vol.AC-31, No.12, pp.1138-1141.

Fortman, F.E., and K.L. Hitz (1977), An Introduction to Linear Control Systems, Dekker, New York.

Fuhymann, P.A. (1977), "On strict system equivalence and similarity," Int. J. Control, Vol. 25, pp.5-10.

Gantmacher, F.R. (1974), The Theory of Matrices, Chelsea, New York.

Gear, C.W. (1971), "Simultaneous numerical solution of differential-algebraic equations," IEEE Trans. Circuit Theory, Vol.CT-18, pp.89-95.

Gear, C.W., and L.R. Petzold (1982), "Differential/Algebraic systems and matrix pencils," Lecture Notes in Mathematics, 973, Springer-Verlag, Berlin, pp.75-89.

Gontian Y., and T.J. Tarn (1982), "Strong controllability and strong observability of generalized dynamic systems," Proc. 20th AACC Control and Comp., Dec.

Haggman, B.C., and P.R. Bryant (1984), "Solution of singular constrained differential equations: A generalization of circuits containing capacitor ---- Only loops and inductor ---- Only cutsets," IEEE Trans. Circuits and Systems, Vol.CAS-31, No.12, pp.1015-1028.

Hall, L.M. (1977), "Regular singular differential equations whose conjugate equation has polynomial solution," SIAM J. Math. Anal., 8, pp.778-784.

Harris, C.J., and Billings, S.A. (1981), Self-Tuning and Adaptive Control: Theory and Applications, Peter Peregrinus Ltd., Stevenage, UK.

Hayton, G.E., P. Fretwell, and A.C. Pugh (1986), "Fundamental equivalence of generalized state space systems," IEEE Trans. Aut. Control, Vol.AC-31, No.5, pp.431-439.

Kailath, T. (1980), Linear Systems, Prentice-Hall, Englewood Cliffs, N.J.

Kalyuthnaya, T.S. (1978), "Controllability of general differential systems," Diff. Eq., Vol.14, pp.314-320.

Karcanias, N., and G.E. Hayton (1981a), "Generalized autonomous dynamical systems, algebraic duality and geometric theory," 8th World IFAC Congress, Kyoto, Japan.

Karcanias, N., and G.E. Hayton (1981b), "State space and transfer function invariant infinite zero: A unified approach," Joint Aut. Control Conf. (JACC), Charlottesville, Va.

Kautsky, J., and N.K. Nichols (1986), "Algorithms for robust pole assignment in singular systems," Proc. 1986 American Control Conf., Seattle, Wash., Vol.1, pp.433-436.

Kecs, W., and P.P. Teodorescu (1974), Application of the Theory of Distribution in Mechanics, Editura Academiei and Abacus Press, Romania.

Kiruthi, G., H. Yadznai, and R.W. Newcomb (1980), "A hysteresis circuit seen through semi-state equations," Proc. Midwest Circuit Theory Symp., Toledo, Ohio.

Kucera, V. (1984), "Design of internally proper and stable control systems," Preprints 9th IFAC World Congress, Budapest, Hungary, Vol.8, p.94.

Kucera, V. (1986), "Stationary LQG control of singular systems," IEEE Trans. Aut. Control, Vol.AC-31, No.1, pp.31-39.

Jamshidi, M. (1983), Large-Scale Systems, Modeling and Control, North-Holland, New York.

Larson, R.E., D.G. Luenberger, and D.N. Stengel (1978), "Descriptor variable theory and spatial dynamics programming," System Control, Inc., Topic. Rep.

Leis, J.R., and M.A. Kramer (1985), "Sensitivity analysis of systems of differential and algebraic equations," Comput. & Chem. Eng., Vol.9, No.1, pp.93-96.

Lewis, F. (1983a), "Descriptor systems: Expanded descriptor equation and Markov parameters," IEEE Trans. Aut. Control, Vol.AC-28, No.5, pp.623-627.

Lewis, F.L. (1983b), "Inversion of descriptor systems," Proc. 1983 American Control Conf., San Fransico, Calif., June 22-24, Vol.8, pp.1153-1158.

Lewis, F.L. (1984), "Descriptor systems: Decomposition into forward and backward subsystems," IEEE Trans. Aut. Control, Vol.AC-29, No.2, p.167.

Lewis, F.L. (1985a), "Fundamental, reachability and observability matrices for discrete descriptor systems," IEEE Trans. Aut. Control, Vol.AC-30, No.5, pp.502-505.

Lewis, F.L. (1985b), "Preliminary notes on optimal control for descriptor systems," Proc. 24th IEEE Conf. Decision Control, Fort Lauderdale, Fla., 1985, pp.502-505.

Lewis, F.L. (1986), "A survey of linear singular systems," Circuits Syst. & Signal Process, Vol.5, No.1, pp.3-36.

Lewis, F.L., and B.G. Mertzios (1987), "Analysis of singular systems using orthogonal function," IEEE Trans. Aut. Control, Vol.AC-32, No.6, pp.527-530.

Lewis, F.L., and K. Ozcaldiran (1984), "Reachability and controllability for descriptor systems," Proc. 27th Midwest Symp. Circuits Syst., Morgantown, W.V., Vol.2.

Lewis, F.L., and K. Ozcaldiran (1985), "On the eigenstructure assignment of singular systems," Proc. 24th IEEE Conf. Decision and Control, Fort Lauderdale, Fla., Dec.

Liniger, W. (1979), "Multistep and one-leg method for implicit mixed differential algebraic systems," IEEE Trans. Aut. Control, Vol.AC-26, pp.755-762.

Locke, G. (1981), "General dynamic systems ---- Problems, state of development, applications," Nachrichtentech Electron. (Germany), Vol.31, No.11, pp.461-465.

Lovass-Nagy, V., D.L. Powers, and H.-C. Yan (1986), "On controlling generalized state-space (descriptor) systems," Int. J. Control, Vol.43, No.4, pp.1271-1282.

Lovass-Nagy, V., R. Schilling, and H.-C. Yan (1986), "A note on optimal control of generalized state space (descriptor) systems," Int. J. Control, Vol.44, No.3, pp.613-624.

Luenberger, D.G. (1977), "Dynamic equation in descriptor form," IEEE Trans. Aut. Control, Vol.AC-22, pp.312-321.

Luenberger, D.G. (1978), "Time invariant descriptor systems'" Automatica, Vol.14, pp.473-480.

Luenberger, D.G. (1979), "Nonlinear descriptor systems," J. Economic Dynamics and Control, 1, pp.219-242.

Luenberger, D.G. (1987), "Dynamic equilibria for linear systems and quadratic costs," Automatica, Vol.23, No.1, pp.117-122.

Luenberger, D.G., and A. Arbel (1977), "Singular dynamic Leontief systems," Economitrica, pp.991-995.

Martens, J.P., G.De Mey, and M.C. Vanwormhoudt (1984), "CADLEM: A simulation program for nonlinear electronic networks," Int. J. Electron., Vol.57, No.1, pp.97-110.

Mertzios, B.G. (1984), "On the sensitivity analysis of linear time-invariant singular systems," IEEE Trans. Circuit and Systems, Vol.CAS-31, No.11, pp.978-982.

Mertzios, B.G., and M.A. Christodoulou (1986), "Decoupling and pole-zero assignment of singular systems with dynamic state feedback," Circuits Syst. & Signal Process (USA), Vol.5, No.1, pp.49-68.

Mertzios, B.G., and B.Z. Syrmos (1987), "Transfer function matrix of singular systems," IEEE Trans. Aut. Control, Vol.AC-32, No.9, pp.829-831.

Mesquita, A., and R.J. Dias (1985), "Solution of singular linear systems," Proc. 28th Midwest Symp. on Circuits and Systems, Louiville, Ky., pp.577-580.

Millic, M.M., and V.B. Bajic (1986), "Some properties of solutions of the semi-state model for nonlinear nonstationary systems," Circuits Syst. & Signal Process, Vol.5, No.1, pp.109-123.

Mukundan, R., and W.D. Ayawansa (1983), "Feedback control of singular systems ---- Proportional and derivative feedback of the state," Int. J. Systems Sci., Vol.14, pp.615-632.

Murota, K. (1983), "Structural controllability of a system with some fixed coefficients," Trans. Soc. Instrum. & Control Eng. (Japan), Vol.19, No.9, pp.683-690.

Newcomb, R. (1981), "The semi-state description of nonlinear time variable circuits," IEEE Trans. Circuits & Systems, Vol.CAS-28, No.1, pp.62-71.

Newcomb, R.W. (1982), "Semi-state design theory: binary and swept hysteresis," Circuits, Systems and Signal Processing, 1, p.203.

O'Malley, R.E., Jr. (1979), "A singular singularly-perturbed linear boundary value probelm, " SIAM J. Math. Anal., Vol.10, pp.695-708.

O'Malley, R.E. Jr., and J.E. Flaherty (1980), "Analytical and numerical methods for nonlinear singular singularly perturbed initial value problems," SIAM J. Appl. Math., Vol.38, pp. 225-248.

O'Reilly, J. (1983), Observers for Linear Systems, Academic Press, New York.

Owens, D.H., and R.P. JOnes (1978), "Iterative solution of constrained differential algebraic systems," Int. J. Control, Vol.27, No.6, pp.957-974.

Ozcaldiran, K. (1985), "Control of descriptor systems," Ph.D. thesis, School of Elect. Eng., Georgia Inst. of Tech., Atlanta, Ga, June.

Ozcaldiran, K. (1986), "A geometric characterization of the reachable and the controllable subspace of descriptor systems," Circuits Syst. & Signal Process (USA), Vol.5, No.1, pp.17-48.

Ozcaldiran, K., and F. Lewis (1987), "A geometric approach to eigenstructure assignment for singular systems," IEEE Trans. Aut. Control, Vol.AC-32, No.7, pp.629-631.

Pandolfi, L. (1980), "Controllability and stabilizability for linear systems of algebraic and differential equations," J. Opt. Theory and Appl., Vol.30, No.4.

Pandolfi, L. (1981), "On the regulator problem for linear degenerate control systems," J. Opt. Theory and Appl., Vol.33, No.2.

Paraskevopoulos, P.N. (1984), "Analysis of singular systems using orthogonal functions," IEE Proc. D, Vol.131, No.1, pp.37-38.

Paraskevopoulos, P.N., and M.A. Christodoulou (1984), "On the computation of the transfer function matrix of singular systems," J. Franklin Inst., Vol.317, No.6, pp.403-411.

Paraskevopoulos, P.N., M.A. Christodoulou, and A.K. Boglu (1984), "An algorithm for the transfer function matrix for singular systems," Automatica (GB), Vol.20, pp.259-260.

Paraskevopoulos, P.N., M.A. Christodoulou, and M.A. Tsakiris (1984), "Eigenvalue-eigenvector sensitivity analysis of linear time-invariant singular systems," IEEE Trans. Aut. Control, Vol.AC-29, No.4, pp.344-346.

Petzold, L.R. (1982), "Differential/Algebraic equations are not ODE's," SIAM J. Sci. Stat. Comp., 3, 367.

Pugh, A.C. (1976), "The McMillan degree of a polynomial system matrix," Int. J. Control, Vol.24, pp.129-135.

Pugh, A.C., and A.K. Shelton (1978), "On a new definition of strict system equivalence," Int. J. Control, Vol.27, No.5, pp.657-672.

Pugh, A.C., G.E. Hayton, and P. Fretwell (1987), "Transformation of matrix pencils and implications in linear system theory," Int. J. Control, Vol.45, No.2, pp.529-548.

Rodriguez, J., and D. Sweet (1986), "A characterization of semistate systems," Circuits Syst. & Signal Process, Vol.5, No.1, pp.125-137.

Rosenbrock, H.H.(1970), State-Space and Multivariable Theory, Wiley, New York.

Rosenbrock, H.H. (1974a), "Order, degree and complexty," Int. J. Control, Vol.19, No.2, pp.323-331.

Rosenbrock, H.H. (1974b), "Structural properties of linear dynamic systems," Int. J. Control, Vol.20, pp.191-202.

Rosenbrock, H.H., and A.C. Pugh (1974), "Contributions to a hierarchical theory of systems," Int. J. Control, Vol.19, No.5, pp.845-867.

Saidahmed, M.T.F. (1985), "Generalized reduced order observer for singular systems," Proc. 28th Midwest Symp. on Circuits & Systems, Louisville, Ky., pp.581-585.

Schwartz, L. (1968), Application of Distribution to the Theory of Elementary Particles in Quantum Mechanics, Gordon and Breach, Sci. Publishers Inc., New York.

Shafai, B., and R.L. Carroll (1987), "Design of a minimal-order observer for singular systems," Int. J. Control, Vol.45, No.3, pp.1075-1082.

Shayman, M.A., and Z. Zhou (1987), "Feedback control and classification of generalized linear systems," IEEE Trans. Aut. Control, Vol.AC-32, No.6, pp.483-494.

Sincovec, R.F., A.M. Erisman, E.L. Yip, and A. Epton (1981), "Analysis of descriptor systems using numerical algorithms," IEEE Trans. Aut. Control, Vol.AC-26, No.1, pp.139-147.

Singh, S.P., and R.-W. Liu (1973), "Existence of state equation representation of linear large-scale dynamic systems," IEEE Trans. Circuit Theory, Vol.CT-20, No.5, pp.239-246.

Smith, R.J. (1966), Circuits, Devices, and Systems, Wiley and Sons, New York.

Stengel, D.N., R.E. Larson, D.G. Luenberger, and T.B. Cline (1979), "A descriptor variable approach to modeling and optimization of large-scale systems," Proc. Eng. Foundation Conf. on Systems Eng. for power: Organization Forms for Large-Scale Systems, Vol.7, Davos, Switzland.

Stewart, G.W. (1972), "On the sensitivity of the eigenvalue problem $Ax = \lambda Bx$," SIAM J Num. Anal., 9, pp.669-686.

Tan, S.H., and J. Vandewalle (1986), "A canonical form for regular pencils of matrices," Systems and Control Letters, Vol.7, No.5, pp.373-376.

Tan, S., and J. Vandewalle (1987), "Realization algorithm for determining generalized state-space representations," Int. J. Control, Vol.45, No.4, pp.1137-1146.

Trzaska, Z. (1986), "Computation of the block-pulse solution of singular systems," IEE Proc. D, Vol.133, No.4, pp.191-192.

Van Der Weiden, A.J.J., and O.H. Bosgra (1980), "The determination of structural properties of a linear multivariable system by operations of system similarity ---- 2: Non-proper systems in generalized state-space form," Int. J. Control, Vol.32, No.3, pp.489-537.

Van Dooren, P. (1981), "The generalized eigenstructure problem in linear system theory," IEEE Trans. Aut. Control, Vol.AC-26, No.1, pp.111-129.

Vardulakis, A.I.G., D.N.J. Limebeer, and N. Karcanias (1982), "Structure and Smith-MacMillan form of a rational matrix at infinity," Int. J. Control, Vol.35, No.4, pp.701-726.

Verghese, G.C. (1978), "Infinite-frequency behavior in generalized dynamical systems," Ph.D. thesis, Dept. of Elect. Eng., Stanford Univ., Stanford, Calif.

Verghese, G.C. (1981), "Further notes on singular descriptions," Proc. 1981 JAC Conf., June 17-19, Vol.1.

Verghese, G.C., and T. Kailath (1981), "Rational matrix structure," IEEE Trans. Aut. Control, Vol.AC-26, No.2, pp.434-439.

Verghese, G.C., B.C. Levy, and T. Kailath (1981), "A generalized state space for singular systems," IEEE Trans. Aut. Control, Vol.AC-26, No.4, pp.811-831.

Verghese, G.C., P. Van Dooren, and T. Kailath (1979), "Properties of the system matrix of a generalized state-space systems," Int. J. Control, Vol.30, pp.235-243.

Verhaegen, M., and P. Van Dooren (1986), "A reduced observer for descriptor systems," Systems and Control Letters, Vol.8, No.1, pp.29-38.

Wang, C., and L. Dai (1986a), "Singular dynamic systems -- A survey," Control Theory and Appl., No.1, pp.2-12.

Wang, C., and L. Dai (1986b), "The normal state observers in singular systems," J. Syst. Sci. & Math. Sci., Vol.6, No.4, pp.307-313.

Wang, C., and L. Dai (1987a), "State observer structure in singular systems," Acta Math. Appl. Sinica, Vol.10, No.1, pp.121-124.

Wang, C., and L. Dai (1987b), "State observation decoupling disturbance for singular systems," Control Theory and Appl., No.3, pp.23-30.

Wang, E., and C. Wang (1986), "Optimal recurrent filtering for linear discrete stochastic singular systems (I)," Preprints Ann. Natl. Conf. on Control Theory and Its Appl., Heilongjiang, P.R.C., Vol.1, pp.118-121.

Wang, Y. (1984), "Disturbance resistence and output regulation for linear singular systems," M.Sc. dissertation, Inst. of Systems Science, Academia Sinica, Beijing, P.R.C., June.

Wang, Y., S. Shi, and Z. Zang (1987), "Pole placement and compensator design of generalized systems," Syst. & Control Letters, Vol.8, No.3, pp.205-210.

Wilkinson, J.H. (1979), "Linear differential equations and Kronecker's Canonical form," Recent Advances in Numerical Analysis, C.de Boor and G. Golub, eds., Academic Press, New York.

Wolovich, W.A. (1974), Linear Multivariable Systems, Springer-Verlag, New York.

Wong, K.T. (1974), "The eigenvalue problem λTx + Sx," J. Diff. Eq., Vol.16, pp.270-280.

Wonham, W.M. (1974), "Linear Multivariable Control ---- A Geometric Approach," Lecture Notes in Economics and Mathematical Systems, Vol.101, Springer-Verlag, New York.

Xu, K. (1984), "Realization of disturbance resistance of a generalized state space system by state feedback," Preprints 9th IFAC World Congress, Vol.8, pp.105-109.

Yamada, T. (1983), "Structural controllability and observability of linear time-invariant descriptor systems," Ph.D. dissertation, Stanford Univ., Stanford, Calif.

Yamada, T., and D.G. Luenberger (1985a), "Generic controllability theorems for descriptor systems," IEEE Trans. Aut. Control, Vol.AC-30, No.2, pp144-152.

Yamada, T., and D.G. Luenberger (1985b), "Algorithms to verify generic causality and controllability of descriptor systems," IEEE Trans. Aut. Control, Vol.AC-30, No. 9, pp.874-880.

Yang, C., and H. Tan (1988), "The minimal realization problem of singular systems," Control Theory and Appl., Vol.5, No.1, pp.72-77.

Yip, E.L., and R.F. Sincovec (1981), "Solvability, controllability and observability of continuous descriptor systems," IEEE Trans. Aut. Control, Vol.AC-26, No.3, pp.702-706.

Zaghloul, M.E., and M.T. Saidahmed (1984), "Analysis and design of singular systems," Proc. 27th Midwest Symp. on Circuits and Syst., Morgantown, W.V., Vol.2, pp.676-678.

Zeigler, B.P. (1981), "Impact of general systems orientation: Present and future," 1981 Winter Simulation Conf. Proc., Atlanta, Ga., Vol.1, pp.222-225.

Zhao, K. (1988), "A realization method of singular systems," Control Theory and Appl., Vol.5, No.1, pp.105-108.

Zhou, Z., M.A. Shaymann, and T.J. Tarn (1987), "Singular systems: A new approach in the time domain," IEEE Trans. Aut. Control, Vol.AC-32, No.1, pp.42-50.

Lecture Notes in Control and Information Sciences

Edited by M. Thoma and A. Wyner

Vol. 81: Stochastic Optimization
Proceedings of the International Conference, Kiew, 1984
Edited by I. Arkin, A. Shiraev, R. Wets
X, 754 pages, 1986.

Vol. 82: Analysis and Algorithms of Optimization Problems
Edited by K. Malanowski, K. Mizukami
VIII, 240 pages, 1986.

Vol. 83: Analysis and Optimization of Systems
Proceedings of the Seventh International Conference of Analysis and Optimization of Systems
Antiba, June 26-27, 1986
Edited by A. Bensoussan, J. L. Lions
XVI, 901 pages, 1986.

Vol. 84: System Modelling and Optimization
Proceedings of the 12th IFIP Conference
Budapest, Hungary, September 2–6, 1985
Edited by A. Prékopa, J. Szelezsán, B. Strazicky
XII, 1046 pages, 1986.

Vol. 85: Stochastic Processes in Underwater Acoustics
Edited by Charles R. Baker
V, 205 pages, 1986.

Vol. 86: Time Series and Linear Systems
Edited by Sergio Bittanti
XVII, 243 pages, 1986.

Vol. 87: Recent Advances in System Modelling and Optimization
Proceedings of the IFIP-WG 7/1 Working Conference
Santiago, Chile, August 27-31, 1984
Edited by L. Contesse, R. Correa, A. Weintraub
IV, 199 pages, 1987.

Vol. 88: Bruce A. Francis
A Course in H_∞ Control Theory
XI, 156 pages, 1987.

Vol. 88: Bruce A. Francis
A Course in H_∞ Control Theory
X, 150 pages, 1987.
Corrected - 1st printing 1987

Vol. 89: G. K. H. Pang/A. G. J. McFarlane
An Expert System Approach to Computer-Aided Design of Multivariable Systems
XII, 223 pages, 1987.

Vol. 90: Singular Perturbations and Asymptotic Analysis in Control Systems
Edited by P. Kokotovic, A. Bensoussan, G. Blankenship
VI, 419 pages, 1987.

Vol. 91 Stochastic Modelling and Filtering
Proceedings of the IFIP-WG 7/1 Working Conference
Rome, Italy, Decembre 10-14, 1984
Edited by A. Germani
IV, 209 pages, 1987.

Vol. 92: L. T. Grujić, A. A. Martynyuk, M. Ribbens-Pavella
Large-Scale Systems Stability Under Structural and Singular Perturbations
XV, 366 pages, 1987.

Vol. 93: K. Malanowski
Stability of Solutions to Convex Problems of Optimization
IX, 137 pages, 1987.

Vol. 94: H. Krishna
Computational Complexity of Bilinear Forms
Algebraic Coding Theory and Applications to Digital Communication Systems
XVIII, 166 pages, 1987.

Vol. 95: Optimal Control
Proceedings of the Conference on Optimal Control and Variational Calculus
Oberwolfach, West-Germany, June 15-21, 1986
Edited by R. Bulirsch, A. Miele, J. Stoer and K. H. Well
XII, 321 pages, 1987.

Vol. 96: H. J. Engelbert/W. Schmidt
Stochastic Differential Systems
Proceedings of the IFIP-WG 7/1 Working Conference
Eisenach, GDR, April 6-13, 1986
XII, 381 pages, 1987.

Lecture Notes in Control and Information Sciences

Edited by M. Thoma and A. Wyner

Vol. 97: I. Lasiecka/R. Triggiani (Eds.)
Control Problems for Systems
Described by Partial Differential Equations
and Applications
Proceedings of the IFIP-WG 7.2
Working Conference
Gainesville, Florida, February 3-6, 1986
VIII, 400 pages, 1987.

Vol. 98: A. Aloneftis
Stochastic Adaptive Control
Results and Simulation
XII, 120 pages, 1987.

Vol. 99: S. P. Bhattacharyya
Robust Stabilization Against
Structured Perturbations
IX, 172 pages, 1987.

Vol. 100: J. P. Zolésio (Editor)
Boundary Control and Boundary Variations
Proceedings of the IFIP WG 7.2 Conference
Nice, France, June 10-13, 1987
IV, 398 pages, 1988.

Vol. 101: P. E. Crouch,
A. J. van der Schaft
Variational and Hamiltonian
Control Systems
IV, 121 pages, 1987.

Vol. 102: F. Kappel, K. Kunisch,
W. Schappacher (Eds.)
Distributed Parameter Systems
Proceedings of the 3rd International Conference
Vorau, Styria, July 6–12, 1986
VII, 343 pages, 1987.

Vol. 103: P. Varaiya, A. B. Kurzhanski (Eds.)
Discrete Event Systems:
Models and Applications
IIASA Conference
Sopron, Hungary, August 3-7, 1987
IX, 282 pages, 1988.

Vol. 104: J. S. Freudenberg/D. P. Looze
Frequency Domain Properties of Scalar
and Multivariable Feedback Systems
VIII, 281 pages, 1988.

Vol. 105: Ch. I. Byrnes/A. Kurzhanski (Eds.)
Modelling and Adaptive Control
Proceedings of the IIASA Conference
Sopron, Hungary, July 1986
V, 379 pages, 1988.

Vol. 106: R. R. Mohler (Editor)
Nonlinear Time Series and
Signal Processing
V, 143 pages. 1988.

Vol. 107: Y. T. Tsay, L.-S. Shieh, St. Barnett
Structural Analysis and Design
of Multivariable Systems
An Algebraic Approach
VIII, 208 pages, 1988.

Vol. 108: K. J. Reinschke
Multivariable Control
A Graph-theoretic Approach
274 pages, 1988.

Vol. 109: M. Vukobratović/R. Stojić
Modern Aircraft Flight Control
VI, 288 pages, 1988.

Vol. 110: In preparation

Vol. 111: A. Bensoussan, J. L. Lions (Eds.)
Analysis and Optimization
of Systems
XIV, 1175 pages, 1988.

Vol. 112: Vojislav Kecman
State-Space Models of Lumped
and Distributed Systems
IX, 280 pages, 1988

Vol. 113: M. Iri, K. Yajima (Eds.)
System Modelling and Optimization
Proceedings of the 13th IFIP Conference
Tokyo, Japan, Aug. 31 – Sept. 4, 1987
IX, 787 pages, 1988.

Vol 114: A Bermúdez (Editor)
Control of Partial Differential Equations
Proceedings of the IFIP WG 7.2
Working Conference
Santiago de Compostela, Spain, July 6–9, 1987
IX, 318 pages, 1989

Vol. 115: H.J. Zwart
Geometric Theory for Infinite
Dimensional Systems
VIII, 156 pages, 1989.

Lecture Notes in Control and Information Sciences

Edited by M. Thoma and A. Wyner

Vol. 116: M.D. Mesarovic, Y. Takahara
Abstract Systems Theory
VIII, 439 pages, 1989

Vol. 117: K.J. Hunt
Stochastic Optimal Control Theory
with Application in Self-Tuning Control
X, 308 pages, 1989.

Vol. 118: L. Dai
Singular Control Systems
IX, 332 pages, 1989

GPSR Compliance

The European Union's (EU) General Product Safety Regulation (GPSR) is a set of rules that requires consumer products to be safe and our obligations to ensure this.

If you have any concerns about our products, you can contact us on ProductSafety@springernature.com

In case Publisher is established outside the EU, the EU authorized representative is:

Springer Nature Customer Service Center GmbH
Europaplatz 3
69115 Heidelberg, Germany

Batch number: 10704763

Printed by Printforce, the Netherlands